Rational Approximations and Orthogonality

Translations of
MATHEMATICAL
MONOGRAPHS

Volume 92

Rational Approximations and Orthogonality

E. M. Nikishin
V. N. Sorokin

American Mathematical Society
Providence, Rhode Island

Е. М. Никишин и В. Н. Сорокин

Рациональные аппроксимации и ортогональность

«НАУКА», МОСКВА, 1988

Translated from the Russian by Ralph P. Boas
Translation edited by Sergei Gelfand

1980 *Mathematics Subject Classification* (1985 *Revision*). Primary 41A21, 42C05, 11J70, 31A05, 31A15, 11J13, 11J88, 30E25, 30E05; Secondary 44A60.

NOTATION. This book presents a wide range of problems connected with rational approximations of numbers and analytic functions. These problems touch on many topics in contemporary mathematical analysis: analytic functions, orthogonal polynomials, spectral theory of operators, potential theory, etc. The development of the theory of Padé approximants, and the current application of this theory in related disciplines, motivate the book, which is an introduction to this circle of problems.

The book is intended for students and future research workers who are interested in function theory and number theory.

Library of Congress Cataloging-in-Publication Data

Nikishin, E. M. (Evgeniĭ Mikhaĭlovich)
[Ratsional'nye approksimatsii i ortogonal'nost'. English]
Rational approximation and orthogonality/E. M. Nikishin and V. N. Sorokin.
p. cm.—(Translations of mathematical monographs; v. 92)
Translation of: Ratsional'nye approksimatsii i ortogonal'nost'.
Includes bibliographical references and index.
ISBN 0-8218-4545-4
1. Approximation theory. 2. Analytic functions. I. Sorokin, V. N. (Vladimir Nikolaevich) II. Title. III. Series.
QA221.N5513 1991 91-18793
511'.4—dc20 CIP

The paper used in this book is acid-free and falls within the guidelines
established to ensure permanence and durability. ♾
This publication was typeset using $\mathcal{A}\mathcal{M}\mathcal{S}$-TeX,
the American Mathematical Society's TeX macro system.

10 9 8 7 6 5 4 3 2 1 96 95 94 93 92 91

Contents

Preface

This is a book about rational approximation, principally of analytic functions and, to some extent, of real numbers. In both theories (numerical and functional) we encounter many general features that originate from the concept of the Euclidean algorithm. In the functional domain, these lead to the Padé approximants; in the numerical domain, to the theory of continued fractions and Diophantine approximation. In turn, the theory of Padé approximants for classes of functions of Markov type (integrals of Cauchy type for measures on the line) coincides with the theory of orthogonal polynomials and the theory of discrete Sturm-Liouville operators. These, and more general objects (simultaneous approximations of several functions) and their study, require the introduction of potential theory, and of boundary problems for analytic functions. Consequently, the study of rational approximations involves a variety of mathematical concepts and objects. The book aims to introduce the reader to these concepts and objects.

We thank A. I. Aptekarev and V. I. Parusnikov for their careful reading of the manuscript and for many remarks and suggestions. We also thank our colleagues in the Mechanical and Mathematical Department of the M. V. Lomonosov Moscow State University and the V. A. Steklov Mathematical Institute, who have assisted us in the preparation of the book.

NOTATION.

GCD(p, q)	greatest common divisor of the integers p and q
LCM(p, q)	least common multiple of the integers p and q
$[\alpha]$	integral part of the real number α
$\deg \alpha$	degree of the algebraic number α
$H(\alpha)$	height of the algebraic number α
L^2_μ	Lebesgue space
l_1	the Banach space of sequences such that $\sum \lvert x_n \rvert < +\infty$
l_2	the Banach space of sequences such that $\sum \lvert x_n \rvert^2 < +\infty$
$\lim_{r \to 1-}$	limit from the left
$\lim_{\varepsilon \to 0+}$	limit from the right
$\stackrel{\text{def}}{=}$	formula of definition
$\triangleright$	end of proof
$\in$	"is a member of"

CHAPTER 1

Rational Approximation of Numbers

§1. Metrics on a field of rational numbers

1. Let X be a metric space with metric d. We recall that a sequence x_n, $n \in \mathbf{N}$, of elements of X is said to be *fundamental* if for every $\varepsilon > 0$ there exists a positive integer N such that $d(x_n, x_m) < \varepsilon$ for all n and $m \geq N$. A metric space X is said to be *complete* if every fundamental sequence of elements of X has a limit in X.

Let us outline the well-known process of completing an arbitrary metric space X. We denote by $\mathscr{X}$ the set of fundamental sequences in X. We say that two sequences $x = \{x_n\} \in \mathscr{X}$ and $y = \{y_n\} \in \mathscr{X}$ are *equivalent* if $d(x_n, y_n) \to 0$ as $n \to \infty$. In this way $\mathscr{X}$ is partitioned into equivalence classes of sequences. We denote the set of these classes by $\overline{X}$. Let x and y be any representatives of the classes $\overline{x}$ and $\overline{y}$. We set

$$\overline{d}(\overline{x}, \overline{y}) = \lim_{n\to\infty} d(x_n, y_n). \tag{1.1}$$

The limit (1.1) exists and is independent of the choice of x and y; the function $\overline{d}$ is a metric.

DEFINITION. We say that $\overline{X}$ is the *completion* of X.

The space $\overline{X}$ is a complete metric space; X is embedded in $\overline{X}$ in a natural way. Let X_0 be the subspace of $\overline{X}$ consisting of the classes of sequences that are equivalent to the constant sequence

$$\{x, x, \dots, x, \dots\}, \qquad x \in X, \tag{1.2}$$

and let d_0 be the restriction of $\overline{d}$ to $X_0 \times X_0$. Then the space X_0, with the metric d_0, is isometric to X with the metric d: the one-to-one mapping that carries x to the class containing (1.2) preserves distances. We shall frequently just write $X \subset \overline{X}$. If X is a complete metric space, we have $X = \overline{X}$ in the sense just explained. In particular, we always have $\overline{\overline{X}} = \overline{X}$.

One of the methods of constructing the field $\mathbf{R}$ of real numbers is to complete the field $\mathbf{Q}$ of rational numbers in the metric

$$d(x, y) = |x - y|, \qquad x, y \in \mathbf{Q}, \tag{1.3}$$

where $|\cdot|$ denotes the absolute value. Later we shall consider metrics for $\mathbf{Q}$ that are different from (1.3), and the corresponding completions of $\mathbf{Q}$.

2. Let us recall that a field is a nonempty set F on which there are defined two binary operations, addition and multiplication, such that F is an Abelian group under addition, $F\backslash\{0\}$ is an Abelian group under multiplication (0 is the zero element), and the distributive law $x\cdot(y+z) = x\cdot y + x\cdot z$ is satisfied for all $x, y, z \in F$.

DEFINITION. A *norm* $\|\cdot\|$ on the field F is a function with nonnegative values, such that the following properties hold for $x, y \in F$:

1) $\|x\| = 0 \Leftrightarrow x = 0$;
2) $\|x\cdot y\| = \|x\|\cdot\|y\|$;
3) $\|x+y\| \leq \|x\| + \|y\|$.

Property 2) states that the norm is compatible with the multiplicative structure, and 3) is the triangle inequality. (The norm concept for a field is, in general, not the same as the norm concept for a linear space.)

We mention some properties of the norm that follow immediately from the definition: $\|1\| = 1$; $\|-x\| = \|x\|$, where 1 is the unit element of the field; $(-x)$ is the additive inverse of x: $(-x)+x = 0$.

In defining a metric d on a field F, we shall always suppose that the metric is generated by a norm:

$$d(x, y) = \|x-y\|, \qquad x, y \in F. \tag{1.4}$$

The field F is an Abelian group under addition, and the requirement (1.4) corresponds to the property that d is invariant under translations on this group:

$$d(x+z, y+z) = d(x, y), \qquad x, y, z \in F.$$

The completion $\overline{F}$ of a field F in the metric (1.4) has the structure of a field. If $x = \{x_n\}$, $y = \{y_n\}$ are any representatives of two equivalence classes in $\overline{F}$, the sum and product of these classes are the equivalence classes equivalent respectively to $x+y = \{x_n+y_n\}$ and $x\cdot y = \{x_n\cdot y_n\}$. This is a satisfactory definition (independent of the choice of x and y) and actually provides the structure of a field on $\overline{F}$. If $F\subset\overline{F}$, we consider F to be a subfield of $\overline{F}$.

DEFINITION. Two norms on a field F are said to be *equivalent* if a fundamental sequence with respect to one norm is fundamental with respect to the other; and if, for every fundamental sequence, the set of sequences equivalent to it with respect to one norm is the same as the set of sequences equivalent to it with respect to the other.

If $\|\cdot\|$ is a norm on a field F, then $\|\cdot\|^\alpha$ is also a norm on F for $0<\alpha\leq 1$. If $\|\cdot\|_1$ and $\|\cdot\|_2 = \|\cdot\|_1^\alpha$ $(\alpha > 0)$ are norms on F, they are equivalent.

It follows from the definition that the completions of a field F under metrics generated by equivalent norms are equivalent. Consequently, the problem of finding completions of $\mathbf{Q}$ in the metric (1.4) leads to the problem of finding all pairwise nonequivalent norms on that field. We already know one norm, namely $|\cdot|_\infty = |\cdot|$, the absolute value of a number. We now present an example of a norm on $\mathbf{Q}$ that is not equivalent to $|\cdot|_\infty$.

Let p be a prime; then a rational number $\alpha \neq 0$ can be written in the form $\alpha = \beta p^\nu$, $\nu \in \mathbf{Z}$, where $\beta = u/v$, u and $v > 0$ are relatively prime integers (their GCD is 1), and neither one is divisible by p. We set $|\alpha|_p = p^{-\nu}$ and, by definition, $|0|_p = 0$. The function $|\cdot|$ so defined is a norm. In fact, properties 1) and 2) of the definition are evident. Let us verify 3). Let $\alpha_1 = \beta_1 p^{\nu_1}$ and $\alpha_2 = \beta_2 p^{\nu_2}$ be representations of the form described above, with $\nu_1 \leq \nu_2$. Then $\alpha_1 + \alpha_2 = p^{\nu_1} \cdot (\beta_1 + \beta_2 p^{\nu_2 - \nu_1})$, where the denominator of $\beta_1 + \beta_2 p^{\nu_2-\nu_1}$ is not divisible by p. Hence

$$|\alpha_1 + \alpha_2|_p \leq p^{-\nu_1} = \max(|\alpha_1|_p, |\alpha_2|_p) \leq |\alpha_1|_p + |\alpha_2|_p.$$

If $\|\cdot\| = |\cdot|_p$, there is a stronger inequality than 3), namely

$$\|x + y\| \leq \max(\|x\|, \|y\|). \tag{1.5}$$

DEFINITION. A norm $\|\cdot\|$ on a field F is said to be *non-Archimedean* if (1.5) is satisfied for $x, y \in F$; and *Archimedean* otherwise.

The norm $|\cdot|_p$ is not equivalent to $|\cdot|_\infty$, since $|p^k|_p = p^{-k} \to 0$ as $k \to \infty$. In addition, $|\cdot|_p$ and $|\cdot|_q$ are not equivalent when $p \neq q$.

DEFINITION. If p is a prime, $|\cdot|_p$ is called the *p-adic* norm on $\mathbf{Q}$.

3. Consequently, $\mathbf{Q}$ has both the Archimedean norm $|\cdot|_\infty$ and a countable number of non-Archimedean norms $|\cdot|_p$ ($p = 2, 3, 5, 7, \ldots$, the primes). There is also a trivial norm (non-Archimedean): $|0|_0 = 0$, $|x|_0 = 1$ for $x \neq 0$.

THEOREM 1.1 (Ostrowski). *Every nontrivial norm $\|\cdot\|$ on $\mathbf{Q}$ is equivalent to $|\cdot|_p$, either for some prime p or for $p = \infty$.*

PROOF. Let us suppose that there is an $n \in \mathbf{Z}_+$ for which $\|n\| > 1$. Let n_0 be the smallest such integer n; then $n_0 \geq 2$ (since $\|0\| = 0$, $\|1\| = 1$). We determine $\alpha > 0$ from the equation $\|n_0\| = n_0^\alpha$.

We can write an integer $n > n_0$ in the form

$$n = a_0 + a_1 n_0 + a_2 n_0^2 + \cdots + a_k n_0^k, \tag{1.6}$$

where $a_i \in \mathbf{Z}$, $0 \leq a_i < n_0$, $i = 0, \ldots, k$, $a_k \neq 0$. The formula (1.6) is the representation of n in the n_0-adic system. Since $\|a_i\| \leq 1$, we have

$$\|n\| \leq \sum_{i=0}^{k} \|a_i n_0^i\| \leq \sum_{i=0}^{k} n_0^{\alpha i} \leq n_0^{\alpha k} \sum_{i=0}^{\infty} n_0^{-\alpha i} = C n_0^{\alpha k} \leq C n^\alpha. \tag{1.7}$$

Inequality (1.7) holds for all $n \in \mathbf{N}$. Therefore if n and N are in $\mathbf{N}$ we have $\|n^N\| = \|n\|^N \leq C n^{N\alpha}$ or equivalently $\|n\| \leq C^{1/N} n^\alpha$. Letting $N \to \infty$, we obtain

$$\|n\| \leq n^\alpha, \qquad n \in \mathbf{N}. \tag{1.8}$$

We now prove the opposite inequality for an arbitrary positive integer n. We determine a number $k \in \mathbf{Z}_+$ from the inequalities

$$n_0^k \le n < n_0^{k+1}.$$

By using (1.8), we then obtain

$$\begin{aligned}\|n\| &\ge \|n_0^{k+1}\| - \|n_0^{k+1} - n\| \ge n_0^{(k+1)\alpha} - (n_0^{k+1} - n)^\alpha \\ &\ge n_0^{(k+1)\alpha}(1 - (1 - 1/n_0)^\alpha) = C_1 n_0^{(k+1)\alpha} > C_1 n^\alpha,\end{aligned}$$

where $C_1 > 0$. Therefore, for $n, N \in \mathbf{N}$ we have $\|n^N\| \ge C_1 n^{N\alpha}$ or equivalently $\|n\| \ge C_1^{1/N} n^\alpha$. Again letting $N \to \infty$, we obtain

$$\|n\| \ge n^\alpha, \qquad n \in \mathbf{N}. \tag{1.9}$$

It follows from (1.8) and (1.9) that $\|n\| = n^\alpha$, $n \in \mathbf{N}$, and since the norm is multiplicative it follows that this equality holds for rational x: $\|x\| = |x|^\alpha$.

Thus the norm $\|\cdot\|$ is equivalent to the Archimedean norm $|\cdot|_\infty$.

Let us now suppose that $\|n\| \le 1$ for every positive integer n. We select the smallest $n_0 \in \mathbf{N}$ for which $\|n_0\| < 1$. This is possible, since the norm is nontrivial. We have $n_0 \ge 2$. This n_0 must be a prime, since if $n_0 = ab$, where a, $b \in \mathbf{Z}$, $1 < a, b < n_0$, we would have

$$\|n_0\| = \|a\| \cdot \|b\| = 1,$$

since $\|a\| = \|b\| = 1$ by the assumption that n_0 is minimal.

Let q be a prime and $q \ne n_0$. Let us show that $\|q\| = 1$. Suppose, on the contrary, that $\|q\| < 1$. We find a positive integer N such that $\|n_0^N\| < 1/2$, $\|q^N\| < 1/2$. The numbers n_0^N and q^N are relatively prime. Therefore, there exist integers a and b for which $an_0^N + bq^N = 1$. Hence

$$1 = \|1\| \le \|a\|\,\|n_0^N\| + \|b\|\,\|q^N\| < 1/2 + 1/2 = 1.$$

The resulting contradiction shows that $\|q\| = 1$.

Let $n = n_0^\nu p_1^{\alpha_1} \cdots p_k^{\alpha_k}$ be the expression of the positive integer n as a product of primes. Then

$$\|n\| = \|n_0\|^\nu \|p_1\|^{\alpha_1} \cdots \|p_k\|^{\alpha_k} = \|n_0\|^\nu.$$

If we choose $\alpha > 0$ so that $\|n_0\| = n_0^{-\alpha}$, we obtain

$$\|n\| = |n|_{n_0}^\alpha, \qquad n \in \mathbf{N}. \tag{1.10}$$

It follows from (1.10) that for every $x \in \mathbf{Q}$ we have $\|x\| = |x|_{n_0}^\alpha$, i.e., $\|\cdot\|$ is equivalent to the non-Archimedean norm $|\cdot|_{n_0}$. ▷

4. Definition. The completion of $\mathbf{Q}$ under the norm $|\cdot|_p$, where p is a prime, is called the *field* $\mathbf{Q}_p$ *of p-adic numbers.*

Every $x \in \mathbf{Q}_p$ $(x \neq 0)$ has a unique representation of the form

$$x = \sum_{j=k}^{\infty} a_j p^j , \tag{1.11}$$

where $k \in \mathbf{Z}$, $a_j \in \mathbf{Z}$, $0 \leq a_j < p$, $a_k \neq 0$. Here we are to understand that we have a fundamental sequence of partial sums

$$x_n = \sum_{j=k}^{k+n-1} a_j p^j , \qquad n \in \mathbf{N}, \tag{1.12}$$

defining a p-adic number x. We may think of the right-hand side of (1.11) as an ordinary series of elements of $\mathbf{Q}_p$ that converges in the metric of the field; the sum of the series means, as usual, the limit of the partial sums (1.12).

Elements $x \in \mathbf{Q}_p$ for which $|x|_p \leq 1$ are called p-adic integers. We denote the set of these integers by $\mathbf{Z}_p$. A number x is in $\mathbf{Z}_p$ if $x = 0$, or if $k \geq 0$ in (1.11). The set $\mathbf{Z}_p$ is a ring under addition and multiplication. The set of invertible elements or units of the ring, i.e., those $x \in \mathbf{Z}_p$ for which $1/x \in \mathbf{Z}_p$, is denoted by $\mathbf{Z}^*$. It is clear that $x \in \mathbf{Z}_p^*$ if $k = 0$ in (1.11).

REMARK

For more details on p-adic numbers, see [16, Chapter 1, §§3,4,5]; also [50, Chapter 1]. A large collection of problems can be found there.

PROBLEMS

1. Prove that the completion $\overline{X}$ of a metric space X is a complete metric space.

2. Let n and m be integers with greatest common divisor d. Prove that there are integers a and b such that $an + bm = d$.

SOLUTION. $G = \{an+bm: a, b \in \mathbf{Z}\}$ is a cyclic subgroup of $\mathbf{Z}$; consequently $G = \{kd: k \in \mathbf{Z}\}$, where d is the smallest positive element of G. Evidently d is the GCD of n and m.

3. Prove that every p-adic number $x \neq 0$ has a unique representation (1.11). (See [50].)

§2. The continued fraction of a real number

1. Let p and $q = q_0$ be integers, $q > 0$. Forming quotients and remainders, we obtain

$$\begin{aligned}
p &= a_0 q_0 + q_1 , && 0 < q_1 < q_0 , \\
q_0 &= a_1 q_1 + q_2 , && 0 < q_2 < q_1 , \\
q_{i-1} &= a_i q_i + q_{i+1} , && 0 < q_{i+1} < q_i , \qquad i = 2, 3, \ldots .
\end{aligned} \tag{2.1}$$

These are steps in the *Euclidean algorithm*. Since the sequence $q_1, q_2, \ldots$ of remainders consists of nonnegative integers and is strictly decreasing, the algorithm eventually terminates:

$$q_{N-1} = a_N q_N , \qquad q_{N+1} = 0.$$

Then q_N is the GCD of p and q.

It follows from (2.1) that

$$\frac{p}{q} = a_0 + \frac{q_1}{q} = a_0 + \cfrac{1}{a_1 + \cfrac{q_2}{q_1}} = \cdots = a_0 + \cfrac{1}{a_1 + \cfrac{1}{a_2 + \ddots \cfrac{1}{a_N}}} \tag{2.2}$$

where a_0 is an integer and $a_1, \ldots, a_N$ are nonnegative integers; $a_N > 1$ if $N > 0$. The expression (2.2) is called an *N-term continued fraction*.

2. Let α be a real number. Let us recall that the largest integer a_0 that does not exceed α is called the integral part of α and is denoted by $a_0 = [\alpha]$. The fractional part of α is $\{\alpha\} = \alpha - [\alpha]$.

Let us describe the algorithm that we shall apply to $\alpha = \alpha_0$. If α is an integer, then $\alpha = a_0$ and the algorithm terminates. Otherwise we write $\alpha = a_0 + 1/\alpha_1$, where $\alpha_1 = 1/\{\alpha_0\} > 1$. Let $a_1 = [\alpha_1]$. If α_1 is an integer, we have $\alpha_1 = a_1$, i.e., $\alpha = a_0 + 1/a_1$, and the algorithm terminates. Otherwise,

$$\alpha = a_0 + \cfrac{1}{a_1 + \cfrac{1}{\alpha_2}},$$

where $\alpha_2 = 1/\{\alpha_1\} > 1$. Continuing this process, at the kth step we obtain

$$\alpha = a_0 + \cfrac{1}{a_1 + \cfrac{1}{a_2 + \ddots + \cfrac{1}{\alpha_k}}} \tag{2.3}$$

where a_0 is an integer, $a_1, \ldots, a_{k-1}$ are positive integers, and $\alpha_k > 1$, $k = 1, 2, \ldots$. There are two alternatives: 1) there exists an $N \in \mathbf{Z}_+$ for which α_N is an integer; or 2) α_N is nonintegral for all $N \in \mathbf{Z}_+$. In case 1), α is equal to the fraction (2.2), and consequently is rational. Conversely, for a rational α the algorithm is the Euclidean algorithm and therefore leads to the expression of α as a finite continued fraction (2.2). In that case, $k = 1, \ldots, N$ in (2.3) and the α_k are rational. Therefore case 2) occurs when α is irrational, the index k in (2.3) can take any positive integral value, and every α_k is irrational.

The numbers α_k are called the *remainders* of α.

Thus, with an irrational α we can formally associate an infinite continued fraction

$$\alpha \sim a_0 + \cfrac{1}{a_1 + \cfrac{1}{a_2 + \ddots + \cfrac{1}{a_k} + \ddots}} \tag{2.4}$$

where a_0 is an integer and $a_1, a_2, \dots$ are positive integers.

3. We obtained the continued fractions (2.2) and (2.4) by applying the Euclidean algorithm to the real number α. However, we can study continued fractions independently of the algorithm. Then we call the expressions (2.2) and (2.4) continued fractions, where $a_0, a_1, \dots$ are formal variables (*incomplete quotients* or *elements* of the fraction), which, depending on the class of problems that are to be solved, can take various numerical values. A continued fraction (2.2) or (2.4) will be written in the form $[a_0; a_1, \dots, a_N]$ or $[a_0; a_1, a_2, \dots]$.

A finite continued fraction

$$\pi_k = [a_0; a_1, \dots, a_k], \qquad k = 0, 1, \dots, \tag{2.5}$$

is called a *section of the continued fraction* (2.2) or (2.4); k in (2.5) ranges from N for an N-term fraction to infinity for an infinite fraction. This convention will be used later without special notice. A fraction π_k is a rational function of its elements and can be written as a quotient

$$\pi_k = p_k/q_k \tag{2.6}$$

of two polynomials in the variables $a_0, \dots, a_k$, with integral coefficients (the polynomials are linear in each variable). This representation is not unique. We shall specify a form of (2.6) which is called a *convergent* of the continued fraction (2.2) or (2.4).

PROPOSITION 2.1. *For each* $k = 0, 1, \dots,$ *there exists a representation* (2.6) *for which* p_k *and* q_k *satisfy the recursive relations*

$$\begin{aligned} p_{k+1} &= a_{k+1}p_k + p_{k-1}, \\ q_{k+1} &= a_{k+1}q_k + q_{k-1}, \end{aligned} \tag{2.7}$$

where $k = 1, 2, \dots,$ *with the initial conditions*

$$\begin{aligned} p_0 &= a_0, & p_1 &= a_0a_1 + 1, \\ q_0 &= 1, & q_1 &= a_1. \end{aligned} \tag{2.8}$$

PROOF. In fact, (2.8) says that the zeroth and first convergents of the fraction are equal to a_0 and $a_0 + 1/a_1$. Relation (2.7) is immediately verifiable for $k = 1$. We proceed by induction on k. Suppose that the possibility of the representation (2.6) has been established for $k = n-1$. Then for $k = n$ we have

$$\pi_{n+1} = [a_0; a_1, \dots, a_{n-1}, a_n + 1/a_{n+1}] = p_n'/q_n',$$

a presentation of π_{n+1} as an N-term fraction, where p_n' and q_n' are polynomials p_n and q_n in the variables $a_0, a_1, \dots, a_{n-1}, a_n + 1/a_{n+1}$. Using the induction hypothesis, we obtain

$$\begin{aligned} \frac{p_n'}{q_n'} &= \frac{(a_n + 1/a_{n+1})\,p_{n-1} + p_{n-2}}{(a_n + 1/a_{n+1})\,q_{n-1} + q_{n-2}} \\ &= \frac{a_{n+1}(a_n p_{n-1} + p_{n-2}) + p_{n-1}}{a_{n+1}(a_n q_{n-1} + q_{n-2}) + q_{n-1}} = \frac{a_{n+1}p_n + p_{n-1}}{a_{n+1}q_n + q_{n-1}}. \end{aligned} \tag{2.9}$$

Consequently, we can actually take p_{n+1} and q_{n+1} to be the numerator and denominator of the fraction (2.9). ▷

Relations (2.7) remain valid for $k = -1$ and 0 if we set $p_{-2} = 0$, $q_{-2} = 1$; $p_{-1} = 1$, $q_{-1} = 0$ (and only in this case).

We omit the proofs of the following propositions, where we suppose that $a_0 \in \mathbf{Z}$, $a_1, a_2, \dots \in \mathbf{N}$.

PROPOSITION 2.2. $q_n p_{n-1} - p_n q_{n-1} = (-1)^n$, $n = -1, 0, \dots$.

COROLLARY. *The numbers p_n, q_n are relatively prime, $n = -2, -1, \dots$.*

PROPOSITION 2.3. $\pi_{n-1} - \pi_n = (-1)^n/q_n q_{n-1}$, $n \geq 1$.

PROPOSITION 2.4. $\pi_0 < \pi_2 < \pi_4 < \cdots$, $\pi_1 > \pi_3 > \pi_5 > \cdots$.

PROPOSITION 2.5. *Let α_k be the kth remainder for α. Then*

$$\alpha = \frac{p_{k-1}\alpha_k + p_{k-2}}{q_{k-1}\alpha_k + q_{k-2}}, \qquad k \geq 0.$$

PROPOSITION 2.6.

$$\frac{1}{q_{n+1} + q_n} < |q_n\alpha - p_n| \leq \frac{1}{q_{n+1}}, \qquad n \geq 0,$$

and $q_n\alpha - p_n$ is positive for even n and negative for odd n.

COROLLARY. *If α is irrational, the sequence of convergents converges to α.*

COROLLARY. $|\alpha - p_n/q_n| < q_n^{-2}$, $n = 0, 1, \dots$.

4. DEFINITION. The *best approximation* to a real number α is the fraction p/q (p and q integers, $q > 1$) such that the inequality $|q'\alpha - p'| > |q\alpha - p|$ is satisfied for all integers p' and q', $1 \leq q' < q$.

By a fraction, we understand a formal quotient of numbers. It follows from the definition that the best approximation p/q is an irreducible fraction, i.e., p and q are relatively prime.

We denote by $\|\alpha\|$ the distance from α to the nearest integer (this is not to be confused with the norm). The following definition is equivalent to that given above.

DEFINITION. The *best approximation* to a real number α is the fraction p/q $(p, q$ integers, $q > 1)$ such that

1) $\|q\alpha\| = |q\alpha - p|$;
2) $\|q'\alpha\| > \|q\alpha\|$ for positive integers $q' < q$.

An integer α does not have best approximations. The following theorem gives a description of the set of best approximations to a nonintegral number in terms of its continued fraction.

THEOREM 2.1. *Every best approximation to a real number* $\alpha \notin \mathbf{Z}$ *is a convergent of* α. *Every convergent* p_k/q_k *with* $q_k > 1$ *is a best approximation to* α.

PROOF. Let p/q be a best approximation to α. Let us recall the relative positions of the points $\{\pi_n\}$ and α on the real line: the numbers $\pi_0, \pi_2, \ldots$ increase strictly and converge to α; the numbers $\pi_1, \pi_3, \ldots$ decrease strictly and converge to α. The points $\{\pi_n\}$ and α partition the real line into a finite or countable set of intervals. Let us suppose that p/q lies in one of these intervals.

Let $p/q < \pi_0$; then

$$|1 \cdot \alpha - a_0| < |\alpha - p/q| < |q\alpha - p|,$$

which contradicts the definition of best approximation.

Let $p/q > \pi_1$; then

$$|1 \cdot \alpha - a_0| \le \frac{1}{a_1} = \frac{1}{q_1} \le q\left|\frac{p_1}{q_1} - \frac{p}{q}\right| \le q\left|\alpha - \frac{p}{q}\right| = |q\alpha - p|,$$

which again contradicts the definition.

For some positive integer n, let p/q belong to the interval with endpoints π_{n+1} and π_{n-1}. Then, by using Propositions 2.1–2.6, we obtain

$$\frac{1}{qq_{n-1}} \le \left|\frac{p}{q} - \frac{p_{n-1}}{q_{n-1}}\right| < \left|\frac{p_n}{q_n} - \frac{p_{n-1}}{q_{n-1}}\right| = \frac{1}{q_n q_{n-1}}.$$

Therefore $q_n < q$. On the other hand,

$$|q_n\alpha - p_n| \le \frac{1}{q_{n+1}} \le q \cdot \left|\frac{p_{n+1}}{q_{n+1}} - \frac{p}{q}\right| \le q\left|\alpha - \frac{p}{q}\right| = |q\alpha - p|.$$

This is again a contradiction. Consequently $p/q = \pi_n$ for some n.

According to Proposition 2.6,

$$|q_{n+1}\alpha - p_{n+1}| \le 1/q_{n+2} < |q_n\alpha - p_n|, \qquad n = 0, 1, \ldots. \tag{2.10}$$

Let p_n/q_n be a convergent of α, $q_n > 1$. Let us suppose that this fraction is not a best approximation:

$$|q\alpha - p| \le |q_n\alpha - p_n|, \qquad 1 \le q < q_n. \tag{2.11}$$

Let (p, q) be the pair that satisfies (2.11) with the smallest q. If $q > 1$, the fraction p/q is a best approximation to α and, by what was proved above, $p/q = \pi_m$, $m < n$. Then (2.10) and (2.11) contradict each other. If $q = 1$, the proof is similar. ▷

5. Let $GL_2(\mathbf{Z})$ be the set of 2×2 matrices with integral elements and determinant either $+1$ or -1. This set forms a group under matrix multiplication, with

$$I = \begin{pmatrix} 1 & 0 \\ 0 & 1 \end{pmatrix}$$

as the identity element. With each matrix

$$A = \begin{pmatrix} a & b \\ c & d \end{pmatrix}$$

in the group we associate the linear fractional transformation

$$\widetilde{A}\alpha = \frac{a\alpha + b}{c\alpha + d}, \tag{2.12}$$

which is a one-to-one mapping of the extended real line $\widehat{\mathbf{R}} = \mathbf{R} \cup \{\infty\}$ on itself. To the two matrices A and $(-A)$, and only such pairs, there corresponds a single linear fractional transformation. If $\beta = \widetilde{A}\alpha$ is defined by (2.12), the inverse transformation $\alpha = \widetilde{A}^{-1}\beta$ is given by the formula

$$\alpha = \frac{-d\beta + b}{c\beta - a},$$

i.e., $\widetilde{A}^{-1}$ corresponds to the matrices $\pm A^{-1}$. Now let the transformation

$$\widetilde{A'}\alpha = \frac{a'\alpha + b'}{c'\alpha + d'}$$

correspond to the matrix A'; then to the product AA' there corresponds the composition

$$(\widetilde{A'}\widetilde{A})\alpha = \frac{(a'a + b'c)\alpha + (a'b + b'd)}{(c'a + d'c)\alpha + (c'b + d'd)}$$

of the transformations $\widetilde{A'}$ and $\widetilde{A}$. Thus we have shown that the set of transformations $\widetilde{A}$ for $A \in GL_2(\mathbf{Z})$ forms a group under the operation of composition of mappings. This group is isomorphic to the quotient group $GL_2(\mathbf{Z})/\{I, -I\}$. We denote it by Σ.

DEFINITION. Let α and β be real numbers. We say that α is *equivalent* to β $(\alpha \sim \beta)$ if there is an $\widetilde{A} \in \Sigma$ such that $\alpha = \widetilde{A}\beta$.

Since Σ is a group, it follows immediately that $\alpha \sim \beta$ is an equivalence relation on the set of real numbers, which is thus divided into equivalence classes. The set of rational numbers is one of these classes.

In Σ we introduce the especially simple transformations $\widetilde{J}\alpha = -\alpha$, corresponding to the matrix

$$J = \begin{pmatrix} -1 & 0 \\ 0 & 1 \end{pmatrix},$$

and

$$\widetilde{T}\alpha = \alpha + 1, \qquad \widetilde{S}\alpha = 1/\alpha, \tag{2.13}$$

corresponding to the matrices

$$T = \begin{pmatrix} 1 & 1 \\ 0 & 1 \end{pmatrix}, \qquad S = \begin{pmatrix} 0 & 1 \\ 1 & 0 \end{pmatrix}.$$

The following relations are satisfied: $J^2 = I$, $S^2 = I$,

$$T^k = \begin{pmatrix} 1 & k \\ 0 & 1 \end{pmatrix}, \qquad k \in \mathbf{Z},$$

and also

$$J = T^{-1}STST^{-1}S. \tag{2.14}$$

LEMMA 2.1. *The transformations $\widetilde{T}$ and $\widetilde{S}$ are generators of Σ, i.e., for every $\widetilde{A} \in \Sigma$ there exist integers $n_1, n_2, \dots, n_k$ such that*

$$\widetilde{A} = \widetilde{T}^{n_1}\widetilde{S}\widetilde{T}^{n_2}\widetilde{S}\cdots\widetilde{T}^{n_k}\widetilde{S}.$$

PROOF. Let $\widetilde{A}$ have the form (2.12). If $a \neq 0$, there is an integer l_1 such that

$$|b - l_1 a| < |a|. \tag{2.15}$$

In (2.12), we replace α first by $\alpha - l_1$ and then by $1/\alpha$; we obtain

$$\widetilde{A}\widetilde{T}^{-l_1}\widetilde{S}\alpha = \frac{(b - l_1 a)\alpha + a}{(d - l_1 c)\alpha + c} = \frac{a'\alpha + b'}{c'\alpha + d'},$$

where $|a'| < |a|$ by (2.15). If $a' \neq 0$, by a repetition of the procedure we obtain

$$\widetilde{A}\widetilde{T}^{-l_1}\widetilde{S}\widetilde{T}^{-l_2}\widetilde{S}\alpha = \frac{a''\alpha + b''}{c''\alpha + d''},$$

where $|a''| < |a'|$. If $a'' \neq 0$, we repeat the procedure again. Since the sequence of nonnegative integers $|a|$, $|a'|$, $|a''|, \dots$ is strictly decreasing, it is finite, and one of its elements is zero. After a finite number of steps we arrive at a transformation $\widetilde{A}\widetilde{B}_1$ corresponding to a matrix of the form

$$\pm\begin{pmatrix} 0 & b_1 \\ c_1 & d_1 \end{pmatrix}, \tag{2.16}$$

where $\widetilde{B}_1$ is a composition of transformations (2.13).

The determinant of the matrix (2.16) is $-c_1 b_1 = \pm 1$; consequently $|c_1| = |b_1| = 1$. We choose the matrix (2.16) with $b_1 = 1$, and consider two cases: $c_1 = 1$ and $c_1 = -1$. In the first case,

$$\widetilde{A}\widetilde{B}_1\alpha = \frac{1}{\alpha + d_1}. \tag{2.17}$$

In (2.17) we replace α by $\alpha - d_1$, and then α by $1/\alpha$. We obtain

$$\widetilde{A}\widetilde{B}_1\widetilde{T}^{-d_1}\widetilde{S}\alpha = \alpha.$$

In the second case, replacing α by $\widetilde{J}\alpha = -\alpha$ leads back to the first case. If we recall that by (2.14) $\widetilde{J}$ is the composition of the transformations (2.13), we see that we always have $\widetilde{A}\widetilde{B} = \widetilde{I}$, where $\widetilde{B}$ is a composition of the transformations (2.13). ▹

It follows from Propositions 2.2 and 2.5 that an irrational number is equivalent to any of its remainders. This example is typical. More precisely, we have the following theorem.

THEOREM 2.2. *Two irrational numbers α and β are equivalent if and only if their continued fraction expansions are the same from some point on:*

$$\begin{aligned} \alpha &= [a_0; a_1, \dots, a_n, c_1, c_2, \dots], \\ & \qquad\qquad\qquad\qquad\qquad\qquad k, n \in \mathbf{Z}_+. \\ \beta &= [b_0; b_1, \dots, b_k, c_1, c_2, \dots], \end{aligned} \tag{2.18}$$

The proof depends on the following lemma.

LEMMA 2.2. *Let α and β be irrational and*

$$\alpha = \frac{a\beta + b}{c\beta + d},$$

where a, b, c, d are integers and $ad - bc = \pm 1$. If

$$\beta > 1, \qquad c > d > 0, \tag{2.19}$$

the number β is a remainder in the continued fraction expansion of α: $\alpha = [a_0; a_1, \dots, a_k, \beta]$. The integer a_0 and positive integers $a_1, \dots, a_k$ are determined from the equation

$$\frac{a}{c} = [a_0; a_1, \dots, a_k]. \tag{2.20}$$

PROOF. The finite continued fraction (2.2) will be uniquely defined if we suppose that its last element $a_N > 1$ for $N > 0$. If this condition were not satisfied, every rational number could be expanded in a continued fraction in two ways: the expansion (2.2) and one obtained from (2.2) by replacing a_N by $(a_N - 1) + \frac{1}{1}$. We may select an expansion in (2.20) such that

$$q_k p_{k-1} - p_k q_{k-1} = bc - ad. \tag{2.21}$$

Since $q_k = c$ and $p_k = a$, it follows from (2.21) that

$$q_k(p_{k-1} - b) = p_k(q_{k-1} - d). \tag{2.22}$$

The equality (2.22) implies that q_k divides its right-hand side. Since $\mathrm{GCD}(q_k, p_k) = 1$, it follows that $q_k | (q_{k-1} - d)$ ($u|v \Leftrightarrow u$ is an exact divisor of v). By hypothesis, $0 < d < c = q_k$; consequently, $|q_{k-1} - d| < q_k$.

Hence $q_{k-1} - d = 0$, i.e., $q_{k-1} = d$ and $p_{k-1} = b$,

$$\alpha = \frac{p_k\beta + p_{k-1}}{q_k\beta + q_{k-1}} = [a_0; a_1, \dots, a_k, \beta]. \quad \triangleright$$

PROOF OF THE THEOREM. If (2.18) is satisfied, we have $\alpha \sim \gamma$ and $\beta \sim \gamma$, where $\gamma = [c_1; c_2, \dots]$; consequently, $\alpha \sim \beta$.

Let $\alpha \sim \beta$, i.e.,

$$\beta = \frac{a\alpha + b}{c\alpha + d}, \qquad a, b, c, d \in \mathbf{Z}, \qquad ad - bc = \pm 1.$$

Let us suppose for definiteness that $c\alpha + d > 0$. Since, for each positive integer n,

$$\alpha = \frac{p_{n-1}\alpha_n + p_{n-2}}{q_{n-1}\alpha_n + q_{n-2}}, \qquad \alpha_n > 1$$

(α_n is the nth remainder of α), we have

$$\beta = \frac{(ap_{n-1} + bq_{n-1})\alpha_n + (ap_{n-2} + bq_{n-2})}{(cp_{n-1} + dq_{n-1})\alpha_n + (cp_{n-2} + dq_{n-2})}.$$

Since

$$cp_j + dq_j = q_j\left(c\frac{p_j}{q_j} + d\right)$$

and $p_j/q_j \to \alpha$ $(j \to \infty)$, $c\alpha + d > 0$; for sufficiently large n we have

$$cp_{n-1} + dq_{n-1} > cp_{n-2} + dq_{n-2} > 0.$$

Then, by Lemma 2.2, α_n is the remainder of β. $\triangleright$

REMARKS

A detailed exposition of the theory of continued fractions is given in [112]. Continued fractions and their connection with best approximations are discussed in [48, Chapter 1, §§2, 3, 4], [60, Chapter 1], and [117, Chapter 1]. These references contain numerous properties of continued fractions that do not appear in the present book. Continued fractions originated, as A. Ya. Khinchin has remarked ([112, p. 40]), precisely as a method for obtaining best approximations to real numbers.

In [21] one can find applications of continued fractions to the solution of some problems of elementary number theory (finding GCD's and LCM's, solution of congruences).

In [157] there is an extensive bibliography of older work on continued fractions.

PROBLEMS

1. Prove that q_N in (2.1) is equal to the GCD of p and q.

2. Prove Propositions 2.2–2.6. Note that each one follows from its predecessor, and 2.2 from 2.1.

3. The convergence of the chain of fractions $[0; a_1, a_2, \dots]$, where a_n is a positive number (i.e., the convergence of the sequence of convergents) is equivalent to the divergence of the series $a_1 + a_2 + \cdots$. (Compare [112, Theorem 10].)

4. Let α be an irrational number. The inequality $|\alpha - p/q| < 1/(q^2\sqrt{5})$ is satisfied by an infinite number of pairs (p, q), where p is an integer and q is a positive integer. This inequality cannot be improved for the whole class of irrational numbers α (the number $\sqrt{5}$

cannot be replaced by a larger number). *Hint*: Every set of three consecutive convergents in the expansion of α contains at least one that satisfies the inequality (see [48]).

§3. Rational approximations to algebraic numbers

1. Definition. A complex number α that satisfies an algebraic equation

$$a_n\alpha^n + a_{n-1}\alpha^{n-1} + \cdots + a_0 = 0 \tag{3.1}$$

with integral $a_n, a_{n-1}, \dots, a_0$, with $a_n \neq 0$, $n \in \mathbf{N}$, is called an *algebraic number.*

The equation (3.1), of lowest degree, that is satisfied by α is unique if we require that $a_n, a_{n-1}, \dots, a_0$ are relatively prime (do not have any common positive integral divisor larger than 1), and $a_n > 0$. In fact, if n is the smallest index in (3.1), and

$$A = a_n\alpha^n + \cdots + a_0 = 0, \qquad B = b_n\alpha^n + \cdots + b_0 = 0,$$

we have

$$\sum_{j=0}^{n-1}(b_n a_j - a_n b_j)\alpha^j = 0.$$

Consequently, by the minimality of n,

$$b_n a_j = a_n b_j, \qquad j = 0, \dots, n, \tag{3.2}$$

and a_n divides the left-hand side of equality (3.2) for every j. Now $\mathrm{GCD}(a_n, \dots, a_0) = 1$; therefore $a_n | b_n$ and the quotient b_n/a_n divides b_j for every j. It now follows, since $b_n, \dots, b_0$ are relatively prime, that $b_n/a_n = \pm 1$. Since $a_n > 0$ and $b_n > 0$, we have $a_n = b_n$; consequently $a_j = b_j$, $j = 0, \dots, n$, i.e., the polynomials A and B are identical.

Definition. The polynomial (3.1) of smallest degree (whose uniqueness we have established) will be called the *fundamental polynomial* of the algebraic number α. Its degree is $\deg\alpha$, the degree of α. The *height* of the polynomial

$$P(x) = a_n x^n + \cdots + a_0 \tag{3.3}$$

is the number $H(P) = \max|a_j|$ $(0 \leq j \leq n)$. If (3.3) is the fundamental polynomial for the algebraic number α, the *height of* α is defined to be the height of the polynomial: $H(\alpha) = H(P)$.

Every rational number is algebraic of degree 1. The fundamental polynomial is irreducible over the rational field, i.e., is not the product of two polynomials of degree ≥ 1 with rational coefficients.

Definition. A number ξ is called an *algebraic integer* if it satisfies an equation $\xi^n + a_{n-1}\xi^{n-1} + \cdots + a_0 = 0$ with integral coefficients $a_{n-1}, \dots, a_0$.

The leading coefficient of the fundamental polynomial of an algebraic integer is also equal to unity. If α is any algebraic number that satisfies (3.1), the number $\xi = a_n\alpha$ is an algebraic integer.

2. THEOREM 3.1 (Liouville). *Let α be an algebraic number of degree $n \geq 2$. Then there exists a constant $c(\alpha)$ such that*

$$|\alpha - p/q| \geq c(\alpha)q^{-n} \tag{3.4}$$

for all integers p and q $(q > 0)$.

PROOF. Let $P(x)$ be the fundamental polynomial of α; p and q integers, $q > 0$. Then

$$P(\alpha) - P(p/q) = P'(\zeta)(\alpha - p/q),$$

where ζ is in the interval with endpoints α and p/q. Hence

$$\frac{|q^n P(p/q)|}{|P'(\zeta)|} = q^n \left|\alpha - \frac{p}{q}\right|.$$

The irreducibility of $P(x)$ over $\mathbf{Q}$ implies that $P(p/q) \neq 0$. Therefore $q^n P(p/q)$ is a nonzero integer whose absolute value is not less than unity. Therefore

$$1/|P'(\zeta)| \leq q^n |\alpha - p/q|.$$

If $|\alpha - p/q| \geq 1$, the inequality (3.4) is evident. If, however, $|\alpha - p/q| < 1$, we have

$$|P'(\zeta)| \leq nH(\alpha)(|\zeta|^{n-1} + \cdots + 1) \leq n^2 H(\alpha)(|\alpha| + 1)^{n-1} = 1/c(\alpha)$$

($H(\alpha)$ is the height of α) and (3.4) is also satisfied. ▷

In various proofs of Liouville's theorem, we may encounter various expressions for the constant $c(\alpha)$. Here $c(\alpha)$ was calculated by arithmetic operations with absolute values, powers, and the height of α. A constant of this kind is said to be *constructive*.

Theorem 3.1 is also valid for a rational α $(n = 1)$ if we add the hypothesis that $\alpha \neq p/q$.

3. An irrational algebraic number α of the second degree is called a *quadratic irrational*. Let

$$P(x) = ax^2 + bx + c \tag{3.5}$$

be the fundamental polynomial of α. Its roots are

$$\frac{-b \pm \sqrt{b^2 - 4ac}}{2a}, \tag{3.6}$$

one of these is α; and the other, α', is called the algebraic number conjugate to α. The integer $D(\alpha) = b^2 - 4ac > 0$ is called the *discriminant* of α. If $\alpha > 1$ and $-1 < \alpha' < 0$, the number α is said to be *reduced*.

If α is a quadratic irrational, every number equivalent to α is also a quadratic irrational.

LEMMA 3.1. *If a quadratic irrational* α *has discriminant* D, *any number* β *equivalent to* α *also has discriminant* D.

It is sufficient to carry out the proof for $\beta = 1/\alpha$ and $\beta = \alpha + 1$ (see Lemma 2.1). In the first case, the conclusion is evident. In the second case, we have $a(\beta-1)^2 + b(\beta-1) + c = 0$, or $a\beta^2 + (b-2a)\beta + (c-b+a) = 0$; here a, $b-2a$, and $c-b+a$ are relatively prime. Then

$$D(\beta) = (b-2a)^2 - 4a(c-b+a) = b^2 - 4ac = D(\alpha). \quad \triangleright$$

LEMMA 3.2. *For every* $D > 0$ *there exist only finitely many reduced numbers* α *for which* $D(\alpha) = D$.

PROOF. Since $a^2(\alpha-\alpha')^2 = D$ and $\alpha - \alpha' > 1$ (see (3.5) and (3.6)), we have $0 < a < \sqrt{D}$. Since $c^2(1/\alpha - 1/\alpha')^2 = D$, $1/\alpha - 1/\alpha' > 1$, and $c = a\alpha\alpha' < 0$, it follows that $0 < -c < \sqrt{D}$. The coefficient b is uniquely determined by the numbers a, c, and D.

We see that the set of reduced numbers of discriminant D does not exceed $\sqrt{D}\cdot\sqrt{D} = D$. $\triangleright$

THEOREM 3.2 (Euler-Lagrange). *Let* α *be an irrational number. The continued fraction of* α *is periodic if and only if* α *is a quadratic irrational.*

PROOF. Let α_k be the remainders of α; then the α_k are irrational and $\alpha_k > 1$. Periodicity of the fraction for α means that $\alpha_n = \alpha_m$ for some $n < m$.

If $\alpha_n = \alpha_m$ for $n < m$, we have

$$\alpha_n = \frac{p\alpha_m + r}{q\alpha_m + s} = \frac{p\alpha_n + r}{q\alpha_n + s},$$

where p, q, r, and s are integers. It follows from this that α_n is a root of a quadratic equation with integral coefficients. The number α is equivalent to α_n, and therefore α is a quadratic irrational.

Now let α be a quadratic irrational with discriminant D. Then by Lemma 3.1 all the α_n are quadratic irrationals with discriminant D. Let us show that, for sufficiently large n, the numbers α_n are reduced. We have $\alpha_n > 1$ and

$$\alpha = \frac{p_{n-1}\alpha_n + p_{n-2}}{q_{n-1}\alpha_n + q_{n-2}}$$

(see Proposition 2.5), where each p_k/q_k is a convergent for α. Hence

$$\alpha' = \frac{p_{n-1}\alpha'_n + p_{n-2}}{q_{n-1}\alpha'_n + q_{n-2}},$$

where α' is conjugate to α and α'_n is conjugate to α_n. Since $p_n/q_n \to \alpha$ as $n \to \infty$ and $\alpha - \alpha' \neq 0$, we have

$$\alpha'_n = \frac{p_{n-2} - q_{n-2}\alpha'}{q_{n-1}\alpha' - p_{n-1}} = \frac{q_{n-2}}{q_{n-1}} \cdot \frac{p_{n-2}/q_{n-2} - \alpha'}{\alpha' - p_{n-1}/q_{n-1}} < 0$$

for sufficiently large n. Similarly, it follows from

$$1+\alpha_n' = \frac{q_{n-1}-q_{n-2}}{q_{n-1}} + \frac{(-1)^n}{q_{n-1}^2\left(\alpha' - p_{n-1}/q_{n-1}\right)}$$

and from $q_{n-1} > q_n$ that $\alpha_n' + 1 > 0$ for large n. Therefore, when n is sufficiently large, α_n is a reduced quadratic irrational with discriminant D. By Lemma 3.2, there are only finitely many such numbers. Consequently $\alpha_n = \alpha_m$ for some $n > m$. ▷

4. Let us now consider algebraic numbers of order $n \geq 3$. In contrast to the case $n = 2$, at present we know nothing about the structure of their continued fractions. Here are some problems.

A) Is it true that the incomplete quotients of the continued fraction of $\sqrt[3]{2}$ are unbounded?

B) Is there an algebraic number of degree $n \geq 3$ whose incomplete quotients are unbounded?

C) Is there an algebraic number of degree $n \geq 3$ whose incomplete quotients are bounded?

The condition of boundedness of the incomplete quotients is closely connected with the approximation of numbers by rational numbers. Let us introduce a measure of the degree of approximation of a number α:

$$\nu(\alpha) = \underline{\lim}_{q\to\infty}\, q\|q\alpha\| \qquad (q \in \mathbf{N}).$$

If, for example, α is rational, we have $\nu(\alpha) = 0$.

DEFINITION. An irrational number α is said to be *poorly approximable* if $\nu(\alpha) > 0$.

The following definition is equivalent to that just given: An irrational number α is poorly approximable if there is a constant $c(\alpha) > 0$ such that

$$|\alpha - p/q| > c(\alpha)q^{-2} \tag{3.7}$$

for all integers p and q $(q > 0)$.

LEMMA 3.3. *Let* $\alpha = [a_0; a_1, a_2, \ldots]$ *be an irrational number. Then*

$$\overline{\lim_{k\to\infty}}\, a_k = \infty \tag{3.8}$$

if and only if $\nu(\alpha) = 0$. *In other words, an irrational number is poorly approximable only when its incomplete quotients are bounded.*

PROOF. Let $\nu(\alpha) = 0$; then for every $\varepsilon > 0$ there are an integer p and a positive integer q such that $|\alpha - p/q| < \varepsilon q^{-2}$. Let $q_n \leq q < q_{n+1}$, where the q_n are the denominators of the convergents of α. By Theorem 2.1 (on best approximation) we have $|q_n\alpha - p_n| \leq |q\alpha - p|$. Hence

$$\left|\alpha - \frac{p_n}{q_n}\right| < \left|\alpha - \frac{p}{q}\right| \cdot \frac{q}{q_n} < \frac{\varepsilon}{q_n q} \leq \frac{\varepsilon}{q_n^2}. \tag{3.9}$$

On the other hand,

$$\left|\alpha - \frac{p_n}{q_n}\right| > \left|\frac{p_{n+2}}{q_{n+2}} - \frac{p_n}{q_n}\right| = \frac{a_{n+2}}{q_{n+2}q_n}. \tag{3.10}$$

It follows from (3.9) and (3.10) that

$$\frac{a_{n+2}}{q_{n+2}q_n} < \frac{\varepsilon}{q_n^2}.$$

Therefore $1/\varepsilon < q_{n+2}/q_n < a_{n+2} + 1$. Hence we obtain (3.8).
Conversely, if (3.8) is satisfied we have

$$\left|\alpha - \frac{p_n}{q_n}\right| \le \frac{1}{q_n q_{n+1}} = \frac{1}{q_n(a_{n+1}q_n + q_{n-1})} < \frac{1}{a_{n+1}q_n^2},$$

and $q_n\|q_n\alpha\| \to 0$ when the limit is taken over the indices n for which $a_{n+1} \to \infty$. ▷

It follows from Liouville's theorem or the Euler-Lagrange theorem and Lemma 3.3 that (3.7) holds for quadratic irrationals with a constructive constant $c(\alpha)$. If we now turn to an algebraic number α of degree $n \ge 3$, for example, $\alpha = \sqrt[n]{2}$, Liouville's theorem guarantees only that $|\alpha - p/q| > c(\alpha)q^{-n}$ for all p, $q \in \mathbf{Z}$, $q > 0$, whereas the convergents satisfy $|\alpha - p_k/q_k| < q_k^{-2}$. Thus there is a gap between q^2 and q^n. Let $\theta \in [2, n]$ be a number such that the inequality

$$\left|\alpha - \frac{p}{q}\right| > \frac{c}{q^\theta} \tag{3.11}$$

is satisfied, with some constant $c = c(\alpha, \theta) > 0$, for all p, $q \in \mathbf{Z}$, $q > 0$. According to Lemma 3.3, the question of whether we can take $\theta = 2$ in (3.11) is equivalent to the problem of the boundedness of the incomplete quotients of α. Computer calculations [144] have shown that apparently some sequence of the incomplete quotients of $\sqrt[3]{2}$ increases unboundedly:

$$\begin{aligned}\sqrt[3]{2} = [1; 3, 1, 5, 1, 1, 4, 1, 1, 8, 14, 1, 10, 2, 1, 4, 12, 2, 3, 2,\\ 1, 3, 4, 1, 1, 2, 14, 3, 12, 1, 15, 3, 1, 4, 534, 1, 1, 5, 1,\\ 1, 121, 1, 2, 2, 4, 10, 3, 2, 2, 41, 1, 1, 1, 3, 7, 2, 2, 9,\\ 4, 1, 3, 7, 6, 1, 1, 2, 9, 2, 3, 3, 1, 1, 69, 1, 12, \ldots].\end{aligned}$$

For large indices n, the numbers of large incomplete quotients are given in the following table:

n	36	42	92	115	376	421
a_n	534	121	186	372	186	220

n	511	572	587	618	620	726
a_n	255	7451	113	151	4941	108

Liouville's theorem was published in 1844, but the first improvement of his result was obtained by the Norwegian mathematician A. Thue in 1908–1909, first for numbers of the form $\sqrt[n]{a/b}$ $(a, b \in \mathbf{N})$, and later for all algebraic numbers.

THEOREM (Thue). *Let α be an algebraic number of degree $n \geq 2$. Then for each $\varepsilon > 0$ there is a constant $c(\varepsilon) > 0$ such that*

$$\left|\alpha - \frac{p}{q}\right| \geq \frac{c(\varepsilon)}{q^{n/2+1+\varepsilon}}$$

for all $p \in \mathbf{Z}$, $q \in \mathbf{N}$.

It is important to notice that the proof of Thue's theorem does not provide a method for constructing the constant $c(\varepsilon)$; in other words, it is *nonconstructive.*

In 1921, C. L. Siegel strengthened Thue's result by proving that for every algebraic number of degree $n \geq 2$ there is a constant $c(\alpha) > 0$ such that

$$\left|\alpha - \frac{p}{q}\right| \geq \frac{c(\alpha)}{q^{2\sqrt{n}}}$$

for all $p \in \mathbf{Z}$ and $q \in \mathbf{N}$. (In fact, the exponent of the power in Siegel's theorem is even less than $2\sqrt{n}$, but is equivalent to $2\sqrt{n}$ as $n \to \infty$.)

In 1947–1948, F. J. Dyson and A. O. Gel′fond independently improved the exponent $2\sqrt{n}$ by replacing it with $\sqrt{2n}$. Finally, in 1955, K. F. Roth proved the following fundamental theorem.

THEOREM (Roth). *Let α be an algebraic number of degree $n \geq 2$. Then for every $\varepsilon > 0$, there is a constant $c(\varepsilon) > 0$ such that*

$$\left|\alpha - \frac{p}{q}\right| \geq \frac{c(\varepsilon)}{q^{2+\varepsilon}}$$

for all $p \in \mathbf{Z}$ and $q \in \mathbf{N}$.

Roth was awarded a Fields Medal in 1958 for this result.

The constants in the theorems of Siegel, Dyson, Gel′fond, and Roth are, like that in Thue's theorem, nonconstructive.

§5 will be devoted to a proof of Roth's theorem. In §4 we present some preliminary results that are of independent interest.

REMARKS

1. An exposition of the theory of algebraic numbers is given in [30]. Introductory material can be found in [56].

2. For proofs of Liouville's theorem with various expressions for $c(\alpha)$ see [108, p. 28], [117, p. 89], and [112, §9].

3. On quadratic irrationals, see [60, Chapter IV] and [112, §10].

4. The following theorems are valid ([60, Chapter IV]).

THEOREM (Lambert).

$$\frac{e^x - 1}{e^x + 1} = \left[0; \frac{2}{x}, \frac{6}{x}, \frac{10}{x}, \frac{14}{x}, \dots\right].$$

THEOREM (Euler).

$$e = [2; 1, 2, 1, 1, 4, 1, 1, 6, 1, 1, 8, 1, 1, 10, \dots].$$

5. For the history of Roth's theorem see [108].

6. The following problem [143] is open:

CONJECTURE (Lang). *For every algebraic number* α *of degree* ≥ 3 *there is a constant* κ_0 *such that for* $\kappa > \kappa_0$ *and some* $c > 0$

$$\left|\alpha - \frac{p}{q}\right| > \frac{c}{q^2 (\log q)^\kappa}$$

for all $p \in \mathbf{Z}$ *and* $q \in \mathbf{N}$.

7. The first constructive sharpening of Liouville's theorem was obtained by A. Baker. One of his first results [122] was the following theorem.

THEOREM (Baker, 1964). *For all* $p \in \mathbf{Z}$ *and* $q \in \mathbf{N}$

$$\left|\sqrt[3]{2} - \frac{p}{q}\right| > \frac{10^{-6}}{q^{2.955}}.$$

Baker later developed his method much further. In 1968 he was awarded a Fields Medal for these results.

Using Baker's method, N. I. Fel′dman [106] obtained the following constructive strengthening of Liouville's theorem.

THEOREM (Fel′dman, 1971). *For every algebraic number of degree* $n \geq 3$ *there are constructive constants* $\beta(\alpha) > 0$ *and* $\varepsilon(\alpha) > 0$ *such that*

$$\left|\alpha - \frac{p}{q}\right| > \frac{\varepsilon(\alpha)}{q^{n-\beta(\alpha)}} \qquad (p \in \mathbf{Z},\, q \in \mathbf{N}).$$

Effective theorems are discussed in [108, Chapters 8, 9, 10]. Similar problems are contained in [109].

PROBLEMS

1. Let α be an algebraic number; P its fundamental polynomial; and Q a polynomial with integral coefficients ($Q \in \mathbf{Z}[x]$) such that $Q(\alpha) = 0$. Prove that Q is divisible by P in the ring $\mathbf{Z}[x]$, i.e., $Q = HP$, where $H \in \mathbf{Z}[x]$. From this it follows, in particular, that the leading coefficient of the fundamental polynomial of an algebraic integer is equal to unity. *Hint*: Divide Q by P with a remainder and use the minimality of the degree of P; then show that the coefficients of the quotient H are integral.

2. The continued fraction of a number α is exactly periodic (i.e., $\alpha = \alpha_n$ for some positive integer n) if and only if α is a reduced number.

3. Find the number φ (known as the golden section) such that if a unit square is removed from a rectangle with sides 1 and φ, the remaining rectangle is similar to the original rectangle. Expand φ in a continued fraction.

4. Using Liouville's theorem, show that the numbers

$$\zeta = \sum_{k=0}^{\infty} (-1)^k a^{-k!} \qquad (a = 2, 3, \dots)$$

are transcendental (Liouville's example). (See [108, p. 301].)

5. Prove that the set A of algebraic numbers is an algebraically closed field, i.e., the sum, difference, product, and quotient (when the divisor is not zero) of two algebraic numbers is an algebraic number; and finally, every algebraic equation with algebraic coefficients has only algebraic roots. The set $\mathbf{Z}A$ of algebraic integers is a ring. (See [305], [56, §58].)

Hint. Consider, for example, a sum $\alpha + \beta$ of algebraic integers. Let $P(x) = x^n + \cdots$ be the fundamental polynomial of α; $\{\alpha_i\}$, $i = 1, \dots, n$, its roots; $Q(x) = x^m + \cdots$, the fundamental polynomial of β; $\{\beta_j\}$, $j = 1, \dots, m$, its roots. The polynomial

$$R(x) = \prod_{i=1}^{n}\prod_{j=1}^{m}(x - (\alpha_i + \beta_j))$$

is symmetric in the parameters α_i and β_j; consequently, by a well-known theorem, $R(x)$ has integral coefficients.

6. Let $\nu(\cdot)$ be the degree of approximation of a real number (see subsection 4). If θ and θ' are equivalent, we have $\nu(\theta) = \nu(\theta')$.

§4. Polynomials in several variables

1. We use vector and multi-index notation. We denote by $\mathbf{Z}_+^m$ the set of *m-fold multi-indices*, i.e., ordered families of m nonnegative numbers. The order of a multi-index $l = (l_1, \dots, l_m)$ is the sum $|l| = l_1 + \cdots + l_m$. The notation $l \le r$ means that $l_j \le r_j$ for all components of the vector. Let $x = (x_1, \dots, x_m)$ be a set of m variables; then

$$x^l = x_1^{l_1} \cdots x_m^{l_m}.$$

A polynomial in m variables x over a ring $\mathscr{R}$ is a sum

$$F(x) = \sum_{l \le r} c_l x^l, \tag{4.1}$$

where r is a given multi-index, and $c_l \in \mathscr{R}$ are coefficients. Here it is supposed that the degree of F in the variable x_j does not exceed r_j: $\deg_j F \le r_j$. The sum, difference, and product of such polynomials are defined by the usual rules. We introduce an operation of differentiation,

$$D^s = \frac{1}{s_1! \cdots s_m!}\frac{\partial^{|s|}}{\partial x_1^{s_1} \cdots \partial x_m^{s_m}},$$

where $s \in \mathbf{Z}_+^m$. The partial derivative $D^s F$ will also be denoted by F_s.

The operation of differentiation has the fundamental property that it carries a polynomial with integral coefficients to a polynomial with integral coefficients. In fact,

$$D^s F(x) = \sum_{s \le l \le r} \binom{l_1}{s_1} \cdots \binom{l_m}{s_m} c_l x^{l-s}, \tag{4.2}$$

where the binomial coefficients $\binom{l_j}{s_j}$ are integers. Let $H(F) = \max |c_l|$, $l \le r$, denote the *height of the polynomial* F (for $\mathscr{R} = \mathbf{C}$). It follows from (4.2) and the inequalities

$$0 \le \binom{l_j}{s_j} \le 2^{l_j} \le 2^{r_j}$$

that

$$H(D^s F) \le 2^{|r|} H(F). \tag{4.3}$$

2. DEFINITION. A subset A of a real linear space $\mathscr{L}$ is said to be *convex* if, for every two points $a_0, a_1 \in A$, all the points of the line segment with endpoints a_0 and a_1, namely $a_\lambda = \lambda a_1 + (1-\lambda)a_0$, $0 \le \lambda \le 1$, belong to A.

The closed convex hull of $\mathscr{A} \subset \mathscr{L}$, $A = \operatorname{conv}\mathscr{A}$, is the smallest closed convex set that contains all the points of $\mathscr{A}$, i.e., A is the intersection of all closed convex sets that contain $\mathscr{A}$.

For convex sets A and B, we may introduce the sum $A+B = \{a+b : a \in A, b \in B\}$, which is again a convex set.

By an *extreme point* of a closed convex set A, we mean an element $x \in A$ that cannot be written in the form $x = \lambda a + (1-\lambda)b$ with any $a, b \in A$, $a \ne b$, $0 < \lambda < 1$. The set of extreme points is denoted by $\mathscr{K}(A)$. A well-known theorem of the theory of convex sets states that *every bounded closed convex subset of* $\mathbf{R}^m$ *is the closed convex hull of the set of its extreme points.* For example, for a convex polygon the extreme points are the vertices and the conclusion of the theorem is evident.

For a polynomial (4.1) we introduce the set $\mathscr{D}(F) = \{l : c_l \ne 0\}$, and denote by $\mathscr{N}(F)$ its convex hull in $\mathbf{R}^m$. The set $\mathscr{N}(F)$ is called the *Newton polygon* of F. Evidently

$$\mathscr{K}(\mathscr{N}(F)) \subset \mathscr{D}(F).$$

LEMMA 4.1. *The formula*

$$\mathscr{N}(FG) = \mathscr{N}(F) + \mathscr{N}(G) \tag{4.4}$$

is valid for arbitrary polynomials F and G over a ring $\mathscr{R}$ without divisors of zero.

PROOF. We denote the coefficients of x^l in the polynomials $F(x)$ and $G(x)$ by $c_l(F)$ and $c_l(G)$.

Let $c_l(FG) \ne 0$. Since

$$c_l(FG) = \sum_{\alpha+\beta=l} c_\alpha(F)c_\beta(G),$$

we must have, for some α^0 and $\beta^0 = l - \alpha^0$, either $c_{\alpha^0}(F) \ne 0$ and $c_{\beta^0}(G) \ne 0$ or $\alpha^0 \in \mathscr{N}(F)$ and $\beta^0 \in \mathscr{N}(G)$. Therefore

$$\mathscr{D}(FG) \subset \mathscr{N}(F) + \mathscr{N}(G). \tag{4.5}$$

The set on the right of (4.5) is convex; therefore

$$\mathscr{N}(FG) \subset \mathscr{N}(F) + \mathscr{N}(G). \tag{4.6}$$

Let us establish the reverse inclusion. Let l be an extreme point of the set $\mathscr{N}(F) + \mathscr{N}(G)$:

$$l = \alpha^0 + \beta^0, \qquad \alpha^0 \in \mathscr{N}(F),\ \beta^0 \in \mathscr{N}(G). \tag{4.7}$$

Let us show that the representation (4.7) is unique and that the points α^0, β^0 are extreme points of $\mathscr{N}(F)$, $\mathscr{N}(G)$, respectively. In fact, if $l = \alpha^1+\beta^1$ with $\alpha^1 \in \mathscr{N}(F)$, and $\beta^1 \in \mathscr{N}(G)$ is another representation, we have $\alpha^0 \neq \alpha^1$ and $\beta^0 \neq \beta^1$. Set $l^1 = \alpha^1+\beta^0$, $l^0 = \alpha^0+\beta^1$; then $l^1 \neq l$ and $l^0 \neq l$. At the same time, $l = (l^1 + l^0)/2$, which is a contradiction because l is an extreme point. Thus the representation (4.7) is unique. If α^0 is not an extreme point of $\mathscr{N}(F)$ or β^0 is not an extreme point of $\mathscr{N}(G)$, it is evident that l is not an extreme point of $\mathscr{N}(F) + \mathscr{N}(G)$, contrary to the choice of l. Therefore $c_{\alpha^0}(F) \neq 0$ and $c_{\beta^0}(G) \neq 0$, so $c_l(FG) = c_{\alpha^0}(F)c_{\beta^0}(G) \neq 0$, i.e., $l \in \mathscr{N}(FG)$. Therefore we have proved that $\mathscr{K}(\mathscr{N}(F)+\mathscr{N}(G)) \subset \mathscr{N}(FG)$; consequently

$$\mathscr{N}(F)+\mathscr{N}(G) \subset \mathscr{N}(FG). \tag{4.8}$$

Comparing (4.6) and (4.8), we obtain (4.4). ▷

3. We now prove a proposition (known as the Gauss lemma), which is extensively used in number theory.

LEMMA 4.2 (Gauss). *Let F and G be polynomials in several variables, with integral coefficients. Let us suppose that the coefficients of F are relatively prime, and that the coefficients of G have the same property. Then the coefficients of FG are also relatively prime.*

PROOF. Let p be a prime. With each polynomial H with integral coefficients, we associate a polynomial H_p with coefficients in the field of residues mod p : the coefficients of H_p are obtained by replacing the coefficients of H by the corresponding residue classes. By hypothesis, $\mathscr{N}(F_p) \neq \varnothing$, $\mathscr{N}(G_p) \neq \varnothing$, and by Lemma 4.1

$$\mathscr{N}((FG)_p) = \mathscr{N}(F_p) + \mathscr{N}(G_p) \neq \varnothing,$$

since $(FG)_p = F_p \cdot G_p$. Hence $\mathscr{D}((FG)_p) \neq \varnothing$, i.e., there are coefficients of FG that are not divisible by p. ▷

COROLLARY. *Let F and G be polynomials in several variables, with integral coefficients; let A, B, and C be the greatest common divisors of the coefficients of F, G, and FG. Then $C = AB$.*

COROLLARY. *Let F and G be polynomials in several variables; let the coefficients of F be relatively prime integers; and let those of G be rational. If the coefficients of FG are integers, the coefficients of G are also integers.*

4. Let $F(x)$ be a polynomial (4.1), and let f be a nonzero linear functional on $\mathbf{R}^m$. Following Roth, we introduce the notion of the *index of F with respect to f*:

$$\operatorname{Ind}_f F = \min f(l), \qquad l \in \mathscr{D}(F),$$

if $F \neq 0$, and $\mathrm{Ind}_f 0 = +\infty$. The property $\mathrm{Ind}_f F = \beta$ means that the set $\mathscr{D}(F)$ lies in the halfspace $f(l) \geq \beta$. Since the halfspace is a convex set, $\mathscr{N}(F)$ is also in it. Therefore when $F \neq 0$, we have $\mathrm{Ind}_f F = \min f(l)$, $l \in \mathscr{N}(F)$.

LEMMA 4.3. *Let F and G be polynomials in several variables over a ring $\mathscr{R}$ without divisors of zero. Then the following relations are satisfied*:

1. $\mathrm{Ind}_f F_s \geq \mathrm{Ind}_f F - f(s)$, $s \in \mathbf{Z}_+^m$.
2. $\mathrm{Ind}_f (F+G) \geq \min(\mathrm{Ind}_f F, \mathrm{Ind}_f G)$.
3. $\mathrm{Ind}_f FG = \mathrm{Ind}_f F + \mathrm{Ind}_f G$.

PROOF. If at least one of the indices in relations 1, 2, and 3 is equal to $+\infty$, the relations are trivial. Consequently we shall disregard this case.

1. It is clear that $\mathscr{N}(F_s) \subset \mathscr{N}(F) - s$; therefore

$$\mathrm{Ind}_f F_s = \min_{l \in \mathscr{N}(F_s)} f(l) \geq \min_{l \in \mathscr{N}(F) - s} f(l) = \mathrm{Ind}_f F - f(s).$$

2. We have $\mathscr{N}(F+G) \subset \mathrm{conv}(\mathscr{N}(F) \cup \mathscr{N}(G))$. Hence

$$\mathrm{Ind}_f (F+G) \geq \min_{l \in \mathrm{conv}(\mathscr{N}(F) \cup \mathscr{N}(G))} f(l) = \min(\mathrm{Ind}_f F, \mathrm{Ind}_f G).$$

3. Finally,

$$\begin{aligned} \mathrm{Ind}_f FG &= \min_{l \in \mathscr{N}(FG)} f(l) = \min_{l \in \mathscr{N}(F) + \mathscr{N}(G)} f(l) \\ &= \min_{\substack{\alpha \in \mathscr{N}(F) \\ \beta \in \mathscr{N}(G)}} f(\alpha + \beta) = \mathrm{Ind}_f F + \mathrm{Ind}_f G. \quad \triangleright \end{aligned}$$

Let $y \in \mathscr{R}^m$ and $r \in \mathbf{N}^m$. *The index of the polynomial F at the point y with respect to the set r* is defined to be the index $\mathrm{Ind}(F, y, r) = \mathrm{Ind}_f F(\cdot + y)$ of the polynomial

$$F(x+y) = \sum_s F_s(y) x^s$$

with respect to the functional $f(l) = l_1/r_1 + \cdots + l_m/r_m$. Lemma 4.3 remains valid for $\mathrm{Ind}(\cdot, y, r)$.

If the index of a polynomial is large, this means that many initial Taylor coefficients of the polynomial, in a direction that is easily determined, are equal to zero.

5. We select a set of k polynomials over the field $\mathbf{C}$:

$$\varphi(x) = (\varphi_1(x), \ldots, \varphi_k(x)), \qquad x = (x_1, \ldots, x_m);$$

also, let $i(0), i(1), \ldots, i(k-1)$ be m-dimensional multi-indices such that $|i(\nu)| \leq \nu$, $\nu = 0, \ldots, k-1$.

DEFINITION. The *Wronskian* of the system φ is

$$W = \begin{vmatrix} D^{i(0)} \varphi_1 & \cdots & D^{i(0)} \varphi_k \\ \dots & \dots & \dots \\ D^{i(k-1)} \varphi_1 & \cdots & D^{i(k-1)} \varphi_k \end{vmatrix}.$$

For a function of one variable, it is evident that all Wronskians are zero except perhaps one, for which $i(\nu) = \nu$. In the multidimensional case $(m > 1)$ there may be many nonzero Wronskians.

LEMMA 4.4. *A system* φ *is linearly dependent over* $\mathbf{C}$ *if and only if all its Wronskians vanish.*

PROOF. The proof requires only the part of the lemma that asserts the linear dependence of φ, i.e., the existence of complex numbers $\lambda_1, \dots, \lambda_k$, not all zero, for which

$$\lambda_1\varphi_1(x) + \cdots + \lambda_k\varphi_k(x) \equiv 0.$$

We shall prove the lemma for $m = 1$, not only for polynomials, but also for arbitrary functions $\varphi_1, \dots, \varphi_k$ that are analytic in a domain $\Omega \subset \mathbf{C}$. Let

$$W = \begin{vmatrix} \varphi_1 & \cdots & \varphi_k \\ \varphi_1' & \cdots & \varphi_k' \\ \cdots & \cdots & \cdots \\ \varphi_1^{(k-1)} & \cdots & \varphi_k^{(k-1)} \end{vmatrix} = 0, \tag{4.9}$$

where (without loss of generality) $k > 1$, and

$$\widetilde{W} = \begin{vmatrix} \varphi_1 & \cdots & \varphi_{k-1} \\ \cdots & \cdots & \cdots \\ \varphi_1^{(k-2)} & \cdots & \varphi_{k-1}^{(k-2)} \end{vmatrix} \neq 0.$$

Then the determinant of the system of linear equations

$$\begin{cases} G_1\varphi_1 + \cdots + G_{k-1}\varphi_{k-1} = \varphi_k \\ G_1\varphi_1' + \cdots + G_{k-1}\varphi_{k-1}' = \varphi_k' \\ \cdots\cdots\cdots\cdots\cdots\cdots \\ G_1\varphi_1^{(k-2)} + \cdots + G_{k-1}\varphi_{k-1}^{(k-2)} = \varphi_k^{(k-2)} \end{cases} \tag{4.10}$$

is different from zero at some point of Ω and therefore in some domain $\Omega' \subset \Omega$, and the system can be solved for $G_1, \dots, G_{k-1}$. The functions G_j are analytic in Ω'. If we multiply the first $k-1$ columns of the matrix (4.9) by $G_1, \dots, G_{k-1}$ and subtract them from the last column, we find that in Ω'

$$\widetilde{W} \cdot (\varphi_k^{(k-1)} - G_1\varphi_1^{(k-1)} - \cdots - G_{k-1}\varphi_{k-1}^{(k-1)}) = 0$$

or

$$G_1\varphi_1^{(k-1)} + \cdots + G_{k-1}\varphi_{k-1}^{(k-1)} = \varphi_k^{(k-1)}. \tag{4.11}$$

By differentiating (4.10) we obtain

$$(G_1\varphi_1^{(t+1)} + \cdots + G_{k-1}\varphi_{k-1}^{(t+1)}) + (G_1'\varphi_1^{(t)} + \cdots + G_{k-1}'\varphi_{k-1}^{(t)}) = \varphi_k^{(t+1)},$$

$t = 0, \dots, k-2$, from which, by (4.10) and (4.11), we have

$$G_1'\varphi_1 + \cdots + G_{k-1}'\varphi_{k-1} = 0$$
$$\cdots\cdots\cdots\cdots\cdots\cdots$$
$$G_1'\varphi_1^{(k-2)} + \cdots + G_{k-1}'\varphi_{k-1}^{(k-2)} = 0.$$

Consequently, $G'_j = 0$ in Ω', G_j is constant, and the system φ is linearly dependent.

We now turn to the case $m > 1$. If

$$\xi(t) = (\xi_1(t), \ldots, \xi_m(t)), \qquad 0 \le t \le 1,$$

is an analytic curve in $\mathbf{R}^m$, i.e., the functions $\xi_j(t)$ are holomorphic in a neighborhood of the line segment $[0, 1]$, and take real values on this segment, we can define a system of analytic functions of one variable,

$$\varphi_{(\xi)}(t) = (\varphi_1(\xi(t)), \ldots, \varphi_k(\xi(t))).$$

Their Wronskian is zero, since it has the form

$$W = \sum W_\nu \xi_1^{(\nu_1)} \cdots \xi_m^{(\nu_m)},$$

where W_ν is the Wronskian of φ. By what was proved above, there are constants $\lambda_1, \ldots, \lambda_k$, depending on the curve ξ, such that

$$\lambda_1\varphi_1(x) + \cdots + \lambda_k\varphi_k(x) = 0 \tag{4.12}$$

on ξ, and we may suppose that

$$\max|\lambda_j| = 1, \qquad j = 1, \ldots, k. \tag{4.13}$$

We choose a sequence $\{\xi^{(n)}\}$ of curves whose points are dense in some open ball B in $\mathbf{R}^m$. Such a sequence exists: it is enough to select a sequence, dense in B, of points and the coordinates of the first n points at which the polynomials are interpolated. Since the sequence of vectors $\lambda^{(n)} = (\lambda_1^{(n)}, \ldots, \lambda_k^{(n)})$ corresponding to the curves $\xi^{(n)}$ is bounded, we can select a convergent subsequence. For its limit, (4.12) is satisfied in every ball B, and naturally so is (4.13), i.e., the system φ is linearly dependent. ▹

6. We now turn to *Roth's fundamental lemma.*

LEMMA 4.5 (Roth). *Let*

$$\omega = \omega_m(\varepsilon) = 24 \cdot 2^{-m}(\varepsilon/12)^{2^{m-1}}, \tag{4.14}$$

where m is a positive integer and $0 < \varepsilon < 1/12$; $S(x_1, \ldots, x_m)$ is a polynomial with integral coefficients, $S \ne 0$, whose degree in x_μ does not exceed $r_\mu \in \mathbf{N}$, and

$$r_{\mu+1} \le \omega r_\mu, \qquad \mu = 1, \ldots, m-1. \tag{4.15}$$

If p_μ and q_μ are relatively prime integers and $q_\mu > 0$,

$$q_\mu^{r_\mu} \ge q_1^{r_1}, \qquad q_\mu^{\omega} \ge 2^{3m}, \qquad \mu = 1, \ldots, m, \tag{4.16}$$

and the height of S satisfies the inequality

$$H(S) \le q_1^{\omega r_1}, \tag{4.17}$$

then its index at the point $(p_1/q_1, \dots, p_m/q_m)$ *with respect to* $r = (r_1, \dots, r_m)$ *does not exceed* ε.

The proof is by induction on $m \in \mathbf{N}$.

Let $m = 1$. We set

$$S(x_1) = (q_1 x_1 - p_1)^t T(x_1), \tag{4.18}$$

where $T(x_1)$ is a polynomial with $T(p_1/q_1) \neq 0$, $t \in \mathbf{Z}_+$. By the corollary of the Gauss lemma, T has integral coefficients. Comparing the leading coefficients on both sides of (4.18) and using (4.17), we get

$$q_1^t \leq H(S) \leq q_1^{\omega r_1} = q_1^{\varepsilon r_1},$$

since $\omega = \omega_1(\varepsilon) = \varepsilon$. Hence $t/r_1 \leq \varepsilon$, and the lemma is established for $m = 1$.

Let us write S in the form

$$S = \sum_{j=1}^{h} \varphi_j(x_1, \dots, x_{m-1}) \psi_j(x_m), \tag{4.19}$$

where φ_j and ψ_j are polynomials with rational coefficients. It is clear that at least one such representation is possible, for example one with $h = r_m + 1$ and $\psi_j(x_m) = x_m^{j-1}$. We select the representation (4.19) that has the smallest h. Let us show that the selected $\{\varphi_j\}$ and $\{\psi_j\}$ form linearly independent systems over R.

In fact, if

$$\begin{gathered} c_1\varphi_1 + \dots + c_h\varphi_h = 0, \\ c_1^2 + \dots + c_h^2 > 0, \end{gathered} \tag{4.20}$$

with real coefficients c_j, the relation (4.20) is possible with rational c_j also. Then one of the functions φ_j can be linearly represented (with rational coefficients) in terms of the others, and we would obtain a representation (4.19) with a smaller h. The linear independence of $\{\psi_j\}$ is proved similarly.

By Lemma 4.4, there exist nonzero Wronskians,

$$U(x_m) = \det(\delta^j \psi_k), \qquad j, k = 1, \dots, h, \tag{4.21}$$

and

$$V(x_1, \dots, x_{m-1}) = \det(\Delta^i \varphi_k), \qquad i, k = 1, \dots, h, \tag{4.22}$$

where

$$\delta^j = \frac{1}{(j-1)!} \frac{d^{j-1}}{dx_m^{j-1}}, \qquad \Delta^i = \frac{1}{\nu_1(i)! \cdots \nu_{m-1}(i)!} \frac{\partial^{\nu_1(i) + \dots + \nu_{m-1}(i)}}{\partial x_1^{\nu_1(i)} \cdots \partial x_{m-1}^{\nu_{m-1}(i)}},$$

$$(\nu_1(i), \dots, \nu_{m-1}(i)) \in \mathbf{Z}_+^{m-1}, \qquad \nu_1(i) + \dots + \nu_{m-1}(i) \leq i - 1.$$

Here $h \leq r_m + 1$, and U and V have rational coefficients.

If we multiply the determinant (4.21) by the transpose of the determinant (4.22), we obtain

$$\begin{aligned} W(x_1, \dots, x_m) &= V(x_1, \dots, x_{m-1})U(x_m) \\ &= \det\left(\Delta^i\delta^j \sum_{k=1}^{h} \varphi_k(x_1, \dots, x_{m-1})\psi_k(x_m)\right) \\ &= \det(\Delta^i\delta^j S(x_1, \dots, x_m)), \qquad i, j = 1, \dots, h. \end{aligned} \tag{4.23}$$

The coefficients of W are integers. Therefore, by the corollary of the Gauss lemma, we may suppose (replacing, if necessary, U by KU and V by $(1/K)V$, where $K \in \mathbf{Q}$, $K > 0$) that the coefficients of U and V in (4.23) are integers.

Let S satisfy the hypotheses of the lemma. By (4.23), $\deg_\mu W \le hr_\mu$; therefore $\deg U \le hr_m$ and $\deg V \le hr_\mu$. For $i, j = 1, \dots, h$, we have

$$H(\Delta^i\delta^j S) \le 2^{|r|}H(S) \le 2^{|r|}q_1^{\omega r_1}$$

(here we used (4.3)). By the formula for the expansion of a determinant, we obtain the inequality

$$H(W) \le h!((r_1+1)\cdots(r_m+1))^h(2^{|r|}q_1^{\omega r_1})^h, \tag{4.24}$$

since the number of coefficients of each polynomial $\Delta^i\delta^j S$ does not exceed $(r_1+1)\cdots(r_m+1)$. If we use the inequalities

$$h! \le 2^{h(h-1)}, \qquad (r_1+1)\cdots(r_m+1) \le 2^{|r|}$$

and (4.15), (4.16), and (4.24), we obtain

$$H(W) \le 2^{h(h-1)}2^{2|r|h}q_1^{\omega r_1 h} \le 2^{3|r|h}q_1^{\omega r_1 h} \le (2^{3m}q_1^{\omega})^{r_1 h} \le q_1^{2\omega r_1 h}$$

(we have also used the fact that $\omega < 1$). Since $H(W) = H(U)H(V)$ (U and V are polynomials in different sets of variables!), we obtain

$$H(U) \le q_1^{2\omega r_1 h}, \qquad H(V) \le q_1^{2\omega r_1 h}.$$

Now it is easy to verify, for V, the hypotheses of the lemma with m replaced by $m-1$, ε by $\varepsilon^2/12$ (since $2\omega_m(\varepsilon) = \omega_{m-1}(\varepsilon^2/12)$), and r_μ by hr_μ, $\mu = 1, \dots, m-1$. By the induction hypothesis, the index of V at the point $(p_1/q_1, \dots, p_{m-1}/q_{m-1})$ with respect to $(hr_1, \dots, hr_{m-1})$ does not exceed $\varepsilon^2/12$. We can show similarly that the index of U at p_m/q_m with respect to hr_m does not exceed $\varepsilon^2/12$.

We now apply Lemma 4.3. If

$$y = (p_1/q_1, \dots, p_m/q_m) \quad \text{and} \quad r = (r_1, \dots, r_m),$$

we have

$$\begin{aligned} \operatorname{Ind}(W, y, r) &= \operatorname{Ind}(U, y, r) + \operatorname{Ind}(V, y, r) \\ &\le h\varepsilon^2/12 + h\varepsilon^2/12 = h\varepsilon^2/6. \end{aligned}$$

Let $\theta = \mathrm{Ind}(S, y, r)$. For $i_1 + \cdots + i_{m-1} \le h - 1$, we have

$$\begin{aligned}
&\mathrm{Ind}(S_{(i_1, \ldots, i_{m-1}, j)}, y, r) \\
&\quad \ge \theta - (i_1/r_1 + \cdots + i_{m-1}/r_{m-1} + j/r_m) \\
&\quad \ge \theta - (i_1 + \cdots + i_{m-1})/r_{m-1} - j/r_m \\
&\quad \ge \theta - (h-1)/r_{m-1} - j/r_m \ge \theta - r_m/r_{m-1} - j/r_m \\
&\quad \ge \theta - \omega - j/r_m \ge \theta - \varepsilon^2/24 - j/r_m.
\end{aligned}$$

From the formula for the complete expansion of a determinant, it follows that

$$\begin{aligned}
\mathrm{Ind}(W, y, r) &\ge \sum_{j=0}^{h-1} \max(\theta - \varepsilon^2/24 - j/r_m, 0) \\
&\ge -h\varepsilon^2/24 + \sum_{j=0}^{h-1} \max(\theta - j/r_m, 0).
\end{aligned}$$

Consequently we have the inequalities

$$\frac{1}{h}\sum_{j=0}^{h-1} \max(\theta - j/r_m, 0) \le \frac{\varepsilon^2}{6} + \frac{\varepsilon^2}{24} < \frac{\varepsilon^2}{4}.$$

Let us show that this implies the required inequality $\theta < \varepsilon$. In fact, if $\theta \ge (h-1)/r_m$, we have

$$\frac{\varepsilon^2}{4} > \frac{1}{h}\sum_{j=0}^{h-1}(\theta - j/r_m) = \theta - \frac{1}{2}\frac{h-1}{r_m} \ge \frac{1}{2}\theta.$$

Hence $\theta < \varepsilon$. If also $\theta < (h-1)/r_m$, we have

$$\frac{1}{h}\sum_{j=0}^{[\theta r_m]}(\theta - j/r_m) \ge \frac{1}{h}([\theta r_m] + 1)\cdot\frac{1}{2}\theta \ge \frac{\theta^2 r_m}{2h} \ge \frac{\theta^2}{4}.$$

($[\cdot]$ denotes the integral part.) Hence $\theta < \varepsilon$. ▹

REMARKS

1. Concerning convex sets, see, for example, [55, Chapter 1].

2. Let $\|F\| = \exp(-\mathrm{Ind}_f F)$, where F is a polynomial in several variables. By Lemma 4.3, $\|FG\| = \|F\| \cdot \|G\|$, $\|F + G\| \le \max(\|F\|, \|G\|)$; thus $\|F\| = 0 \Leftrightarrow F = 0$.

Therefore $\|F/G\| \overset{\text{def}}{=} \|F\|/\|G\|$ is a non-Archimedean norm in the field of rational functions of several variables.

3. For various proofs of the lemma on Wronskians, see [48], [117], [108]. The same references serve for Roth's fundamental lemma and Roth's theorem.

4. The Newton polygon for the multidimensional case appears to have been considered for the first time by A. D. Bryuno.

PROBLEMS

1. If $A = \{x_1, \ldots, x_n\}$ is a finite set of points in a linear space,

$$\operatorname{conv} A = \{\lambda_1 x_1 + \cdots + \lambda_n x_n : \lambda_1, \ldots, \lambda_n \ge 0, \lambda_1 + \cdots + \lambda_n = 1\}.$$

2. Let $A \subset \mathbf{R}^m$ be a closed bounded convex set. Show that every point $x \in A$ is representable as a convex combination of at most $m+1$ extreme points of A:

$$x = \lambda_1 x_1 + \cdots + \lambda_\mu x_\mu, \qquad \lambda_1, \ldots, \lambda_\mu > 0, \lambda_1 + \cdots + \lambda_\mu = 1,$$

where $x_1, \ldots, x_\mu \in \mathscr{K}(A)$, $\mu \le m+1$. This implies the theorem about convex sets stated in subsection 2. (See [55, p. 21].)

§5. Roth's theorem

1. We restate Roth's theorem (see §3, subsection 4).

THEOREM (Roth). *Let ξ be an algebraic irrational. Then for every $\delta > 0$ the inequality*

$$\left|\xi - \frac{p}{q}\right| < \frac{1}{q^{2+\delta}} \tag{5.1}$$

is satisfied by only a finite numbers of pairs (p, q), where $p \in \mathbf{Z}$ and $q \in \mathbf{N}$.

It is sufficient to prove the theorem for algebraic integers.

The proof of the following lemma is obvious.

LEMMA 5.1. *Let ξ satisfy the equation*

$$\xi^n + a_{n-1}\xi^{n-1} + \cdots + a_0 = 0$$

with integers $a_{n-1}, \ldots, a_0$ and $a = \max|a_j|$, $j = 0, \ldots, n-1$. Then for every $l \in \mathbf{Z}_+$ we have

$$\xi^l = a_{n-1}^{(l)}\xi^{n-1} + \cdots + a_0, \tag{5.2}$$

where $a_j^{(l)} \in \mathbf{Z}$, $|a_j^{(l)}| \le (a+1)^l$, $j = 0, \ldots, n-1$.

2. In the following lemma we use *Dirichlet's principle.*

LEMMA 5.2. *Let there be given M linear forms*

$$L_i(z) = a_{i1}z_1 + \cdots + a_{iN}z_N, \qquad i = 1, \ldots, M,$$

in N variables $z = (z_1, \ldots, z_N)$, $N > M$,

$$a_{ij} \in \mathbf{Z}, \qquad |a_{ij}| \le A, \qquad i = 1, \ldots, M, \ j = 1, \ldots, N.$$

Then there exist integers $z_1^0, \ldots, z_N^0$, not all zero, such that

$$L_i(z^0) = 0, \qquad i = 1, \ldots, M, \ z^0 = (z_1^0, \ldots, z_N^0),$$
$$|z_j^0| \le (NA)^{M/(N-M)}, \qquad j = 1, \ldots, N.$$

PROOF. We set $K = [(NA)^{M/(N-M)}]$ (integral part). Then $NA < (K+1)^{(N-M)/M}$. If

$$0 \le z_j \le K, \qquad j = 1, \ldots, N, \tag{5.3}$$

then

$$-B_iK \le L_i(z) \le C_iK, \qquad i = 1, \ldots, M, \tag{5.4}$$

where B_i is the sum of the moduli of the negative numbers in the set a_{ij}, $j = 1, \ldots, N$, and C_i is the sum of the positive numbers. Clearly $B_i + C_i \leq NA$, $i = 1, \ldots, M$. There are $(K+1)^N$ vectors z that satisfy (5.3). At the same time, there are at most

$$(NAK+1)^M \leq (NA)^M (K+1)^M < (K+1)^N$$

vectors $(L_1(z), \ldots, L_M(z))$ that satisfy (5.4). Consequently there exist two unequal vectors z^1 and z^2 that satisfy (5.3) and have the property that $L_i(z^1) = L_i(z^2)$, $i = 1, \ldots, m$. The vector $z^0 = z^1 - z^2$ is the required vector. ▷

LEMMA 5.3. *Let $r_1, \ldots, r_m \in \mathbf{N}$. If $I_m(\lambda)$ is the number of integral points in the simplex $0 \leq i_k \leq r_k$, $k = 1, \ldots, m$,*

$$i_1/r_1 + \cdots + i_m/r_m \leq (m - \lambda)/2, \tag{5.5}$$

then

$$I_m(\lambda) \leq \frac{\sqrt{2m}}{\lambda}(r_1 + 1) \cdots (r_m + 1). \tag{5.6}$$

PROOF. For $\lambda \geq m$ or $\lambda \leq \sqrt{2m}$, the lemma is trivial. In particular, it is valid for $m = 1$. We use induction on m. It follows from (5.5) that

$$\sum_{k=1}^{m-1} \frac{i_k}{r_k} \leq \frac{1}{2}(m - \lambda) - \frac{i_m}{r_m} = \frac{1}{2}\left((m-1) - \left(\lambda - 1 + \frac{2i_m}{r_m}\right)\right).$$

Therefore

$$\begin{aligned} I_m(\lambda) &\leq \sum_{i_m=0}^{r_m} I_{m-1}\left(\lambda - 1 + \frac{2i_m}{r_m}\right) \\ &\leq (r_1 + 1) \cdots (r_{m-1} + 1)\sqrt{\frac{m-1}{2}} \sum_{i=0}^{r_m} \frac{2}{\lambda - 1 + 2i/r_m}. \end{aligned} \tag{5.7}$$

In addition,

$$\begin{aligned} \sum_{i=0}^{r_m} \frac{2}{\lambda - 1 + 2i/r_m} &= \sum_{i=0}^{r_m} \left(\frac{1}{\lambda - 1 + 2i/r_m} + \frac{1}{\lambda + 1 - 2i/r_m}\right) \\ &= \sum_{i=0}^{r_m} \frac{2\lambda}{\lambda^2 - (1 - 2i/r_m)^2} \leq \frac{2\lambda}{\lambda^2 - 1}(r_m + 1). \end{aligned} \tag{5.8}$$

If $\lambda > \sqrt{2m}$, we have

$$\frac{\sqrt{2(m-1)}\lambda}{\lambda^2 - 1} \leq \frac{\sqrt{2m}}{\lambda}. \tag{5.9}$$

Then (5.6) follows from (5.7), (5.8), and (5.9). ▷

LEMMA 5.4. *Let ξ be an algebraic integer of degree n, and let a be its height. Then for any $\varepsilon > 0$, every positive integer $m > 8n^2\varepsilon^{-2}$, and positive integers $r_1, \dots, r_m$, there exists a polynomial $R(x_1, \dots, x_m)$ with integral coefficients, such that*

1) $R \neq 0$, $\deg_j R \leq r_j$, $j = 1, \dots, m$;
2) *The index of R at the point $\xi = (\xi, \dots, \xi)$ with respect to $r = (r_1, \dots, r_m)$ is not less than $m(1-\varepsilon)/2$;*
3) *The height of R satisfies the inequality $H(R) \leq (4(a+1))^{|r|}$.*

PROOF. We write the polynomial

$$R(x) = \sum_{k \leq r} C_k x^k$$

with $N = (r_1+1)\cdots(r_m+1)$ undetermined coefficients. We denote by Ω the set of points $i \in \mathbf{Z}_+^m$, $i \leq r$, $i_1/r_1 + \cdots + i_m/r_m \leq m(1-\varepsilon)/2$; here M is the number of points in Ω. Assumption 2) means that

$$R_i(\xi) = 0 \tag{5.10}$$

for all $i \in \Omega$. We have

$$R_i(\xi) = \sum_{k \leq r} C_k \binom{k_1}{i_1} \cdots \binom{k_m}{i_m} \xi^{|k|-|i|}. \tag{5.11}$$

If we write each power ξ^l in (5.11) as a sum (5.2) and equate the coefficients of ξ^j, $j = 0, \dots, n-1$, to zero, each relation (5.10) contributes n homogeneous linear equations with integral coefficients in the unknowns C_k. The moduli of the coefficients of these equations are bounded by the number

$$\max\left(\binom{k_1}{i_1} \cdots \binom{k_m}{i_m} |a_j^{(l)}|\right) \leq 2^{|r|}(a+1)^{|r|}.$$

If we apply Lemma 5.3 with $\lambda = m\varepsilon$, we obtain $M \leq \sqrt{2m}N/m$, whence $nM \leq N/2$, since $m > 8n^2\varepsilon^{-2}$. By Lemma 5.2, there is a nontrivial set of integers C_k such that (5.10) is satisfied and

$$|C_k| \leq (N(2a+2)^{|r|})^{nM/(N-nM)} \leq N(2a+2)^{|r|} \leq (4a+4)^{|r|}. \quad \triangleright$$

3. If ξ is well approximated by rational numbers, we may determine a lower bound for the index of R by Lemma 5.4, at certain points close to ξ.

LEMMA 5.5. *Let ξ be an algebraic integer of degree n; a its height; $0 < \delta < 1/12$, $0 < \varepsilon < \delta/20$; $m > 8n^2\varepsilon^{-2}$ a positive integer; $q_\mu > 0$, p_μ integers $(\mu = 1, \dots, m)$;*

$$\eta_\mu = p_\mu/q_\mu - \xi, \qquad |\eta_\mu| < q_\mu^{-2-\delta}, \tag{5.12}$$

$$q_\mu^\varepsilon > 64(a+1)(|\xi|+1); \tag{5.13}$$

and let the positive integer r_μ *satisfy the inequalities*

$$r_1 \log q_1 \le r_\mu \log q_\mu \le (1+\varepsilon) r_\mu \log q_\mu \qquad (\mu = 1, \dots, m). \tag{5.14}$$

Then at the point $\zeta = (p_1/q_1, \dots, p_m/q_m)$ *the index of* R *with respect to* $r = (r_1, \dots, r_m)$, *constructed in Lemma* 5.4, *is not less than* $\delta m/8$.

PROOF. Let $T = R_i$ (the derivative), $i \in \mathbf{Z}_+^m$, $i \le r$, and

$$i_1/r_1 + \cdots + i_m/r_m < \delta m/8. \tag{5.15}$$

By Taylor's formula,

$$T(\zeta) = \sum_{j \le r} T_j(\xi) \eta_1^{j_1} \cdots \eta_m^{j_m}. \tag{5.16}$$

Since R has many zero derivatives at ξ, many of the numbers $T_i(\xi) = R_{i+j}(\xi)$ are zero. More precisely, if

$$j_1/r_1 + \cdots + j_m/r_m < m(1 - \delta/3)/2, \tag{5.17}$$

then $T_i(\xi) = 0$. In fact, it follows from (5.15) and (5.17) that

$$(i_1 + j_1)/r_1 + \cdots + (i_m + j_m)/r_m < m(1 - \delta/12)/2 < m(1 - \varepsilon)/2.$$

Then $R_{i+j}(\xi) = 0$ by the construction of R.

All the terms in (5.16) for which (5.17) is satisfied are zero. Let us estimate the other terms. We have (see assertion (3) of Lemma 5.4 and the inequality (4.3))

$$|T_j(\xi)| = |R_{i+j}(\xi)| \le (r_1 + 1) \cdots (r_m + 1)(4a + 4)^{|r|} 2^{|r|} (1 + |\xi|)^{|r|} \le \gamma^{|r|},$$

where $\gamma = 16(a+1)(1+|\xi|)$. As a result, we obtain from (5.16)

$$|T(\zeta)| \le (2\gamma)^{|r|} \max |\eta_1|^{j_1} \cdots |\eta_m|^{j_m}, \tag{5.18}$$

where the maximum is taken over those $j \le r$ for which

$$j_1/r_1 + \cdots + j_m/r_m \ge m(1 - \delta/3)/2. \tag{5.19}$$

Let the maximum be attained at $j = j^*$.

We now estimate the number

$$\kappa = q_1^{r_1} \cdots q_m^{r_m} T(p_1/q_1, \dots, p_m/q_m).$$

It follows from (5.13) that

$$|r| \log(2\gamma) \le \varepsilon(r_1 \log q_1 + \cdots + r_m \log q_m);$$

then by (5.12) and (5.14)

$$\begin{aligned}
\log|\kappa| &\le (1+\varepsilon)(r_1 \log q_1 + \cdots + r_m \log q_m) \\
&\quad + (j_1^* \log|\eta_1| + \cdots + j_m^* \log|\eta_m|) \\
&\le m(1+\varepsilon)^2 r_1 \log q_1 - (2+\delta)(j_1^* \log q_1 + \cdots + j_m^* \log q_m).
\end{aligned} \tag{5.20}$$

From (5.19) and (5.14), we obtain

$$j_1^* \log q_1 + \cdots + j_m^* \log q_m \geq \frac{m}{2}\left(1 - \frac{\delta}{3}\right) r_1 \log q_1. \tag{5.21}$$

Relations (5.20) and (5.21) yield

$$-\log|\kappa| \geq m r_1 \log q_1((1+\delta/2)(1-\delta/3) - (1+\varepsilon)^2).$$

Since $(1+\delta/2)(1-\delta/3) > 1+\delta/8 > (1+\varepsilon)^2$, we have $|\kappa| < 1$ and consequently $T(\zeta) = 0$. ▹

Proof of Roth's theorem. We recall that ξ is an algebraic integer of order $n \geq 2$, a is its height, and $\delta > 0$. We need to show that the inequality (5.1) is satisfied by a finite number of pairs (p, q), where $p \in \mathbf{Z}$, $q \in \mathbf{N}$. Suppose the contrary. We may assume without loss of generality that $\delta < 1/12$. Let $0 < \varepsilon < \delta/20$ and let $m > 8n^2\varepsilon^{-2}$ be a positive integer; we define $\omega = \omega_m(\varepsilon)$ by (4.14). We find solutions $(p_1, q_1), \ldots, (p_m, q_m)$ of (5.1) such that p_μ and q_μ are relatively prime, and

$$q_1^\omega > (4(a+1))^m, \qquad q_\mu^\varepsilon > 64(a+1)(|\xi|+1), \quad \mu = 1, \ldots, m,$$
$$\log q_\mu < \tfrac{1}{2}\omega \log q_{\mu+1}, \qquad \mu = 1, \ldots, m-1.$$

We then fix the point $\zeta = (p_1/q_1, \ldots, p_m/q_m)$. We choose a positive integer r_1 so large that $\varepsilon r_1 \log q_1 \geq \log q_m$, and set

$$r_\mu = \left[\frac{r_1 \log q_1}{\log q_\mu}\right] + 1, \qquad \mu = 2, \ldots, m.$$

Then

$$\begin{aligned} r_1 \log q_1 \leq r_\mu \log q_\mu &\leq r_1 \log q_1 + \log q_\mu \\ &\leq (1+\varepsilon) r_1 \log q_1, \qquad \mu = 1, \ldots, m, \\ \omega r_\mu \geq \omega \frac{r_1 \log q_1}{\log q_\mu} &\geq \frac{2}{\log q_{\mu+1}} r_1 \log q_1 \\ &\geq \frac{r_1 \log q_1}{\log q_{\mu+1}} + 1 \geq r_{\mu+1}, \qquad \mu = 1, \ldots, m-1. \end{aligned}$$

We fix the set $r = (r_1, \ldots, r_m)$.

The assumptions of Lemmas 4.5 and 5.5 are now satisfied. By Lemma 5.5, $\operatorname{Ind}(R, \zeta, r) \geq \delta m/8$ for some nontrivial polynomial R with integral coefficients. On the other hand, $\operatorname{Ind}(R, \zeta, r) \leq \varepsilon < \delta m/8$ by Roth's fundamental lemma. ▹

Remarks

1. See [160] for Roth's original proof. All subsequent proofs, including the one presented here, follow the original proof closely.

2. In [132] there is an estimate of the number of solutions of (5.1).

THEOREM. *If* $\alpha \in A$, $n = \deg \alpha \geq 3$, $0 < \delta < 1/3$, *and* $H = H(\alpha)$ *is the height of* α, *then the number of solutions of the inequality*

$$|\alpha - p/q| < \tfrac{1}{2} q^{-2-\delta}$$

with $p \in \mathbf{Z}$, $q \in \mathbf{N}$, *and* p *and* q *relatively prime, does not exceed*

$$2\delta^{-1} \log C + \exp(70 n^2 \delta^{-2}),$$

where

$$C = 3 + \log(1 + |\alpha|) + 2\log(1 + H).$$

3. W. M. Schmidt [117] obtained a generalization of Roth's theorem to the simultaneous approximation of several numbers.

CHAPTER 2

Padé Approximants and Orthogonal Polynomials

§1. Formal power series

1. We denote by $\mathbf{C}((z))$ the set of formal power series

$$a(z) = \sum_{k=-\infty}^{+\infty} a_k z^k$$

in the variable z with complex coefficients a_k such that only finitely many a_k with negative index k are different from zero. Let

$$b(z) = \sum_{k=-\infty}^{+\infty} b_k z^k \in \mathbf{C}((z)).$$

We define the sum and product of series a and b by the formulas

$$(a+b)(z) = \sum_{k=-\infty}^{+\infty} (a_k + b_k) z^k$$

and

$$(a \cdot b)(z) = \sum_{k=-\infty}^{+\infty} \left(\sum_{l=-\infty}^{+\infty} a_l b_{k-l} \right) z^k, \tag{1.1}$$

respectively. The inner sum in (1.1) has a finite number of terms, and the series $a+b$ and $a \cdot b$ again belong to $\mathbf{C}((z))$.

It is trivial to verify that $\mathbf{C}((z))$ is an Abelian group under addition, and its neutral element is the series 0, all of whose coefficients are zero. Multiplication is commutative and associative, and $\mathbf{C}((z))$ contains an identity element (the constant equal to $1 = 1 \cdot z^0$): $1 \cdot a = a$ for every $a \in \mathbf{C}((z))$. The distributive law is valid.

Let us show that for every $a \in \mathbf{C}((z))$ with $a \neq 0$ there exists an inverse element b such that

$$a \cdot b = 1. \tag{1.2}$$

We write a in the form

$$a(z) = z^n \sum_{k=0}^{\infty} \alpha_k z^k, \tag{1.3}$$

where $\alpha_0 \neq 0$ and n is an integer. We look for a series b of the form

$$b(z) = z^{-n} \sum_{k=0}^{\infty} \beta_k z^k .$$

Then (1.2) leads, by the multiplication rule (1.1), to the following infinite sequence of equations:

$$\begin{aligned} \alpha_0 \beta_0 &= 1 , \\ \alpha_0 \beta_1 + \alpha_1 \beta_0 &= 0 , \\ \alpha_0 \beta_2 + \alpha_1 \beta_1 + \alpha_2 \beta_0 &= 0 , \\ \cdots\cdots\cdots\cdots\cdots\cdots \end{aligned}$$

from which, since $a_0 \neq 0$, we can successively obtain $\beta_0, \beta_1, \beta_2, \ldots$.

Consequently the set of formal power series $\mathbf{C}((z))$ becomes a field. We can define a non-Archimedean norm in this field (see Chapter 1, §1). Indeed, if a is the series (1.3), we have

$$\|a\| = \rho^n , \tag{1.4}$$

where ρ can be taken to be any number on the interval $(0, 1)$, for example, $\rho = 1/e$ $(\|0\| = 0)$. The axioms for a non-Archimedean norm are easily verified for (1.4).

2. Let $\mathbf{C}(z)$ be the field of rational functions in the variable z with complex coefficients. If $a \in \mathbf{C}(z)$, the Laurent series of $a(z)$ (with center at $z = 0$) is an element of the field $\mathbf{C}((z))$. In this sense, $\mathbf{C}(z)$ is a subfield of $\mathbf{C}((z))$.

If a is the series (1.3), its partial sums

$$a_\nu(z) = z^n \sum_{k=0}^{\nu} \alpha_k z^k , \qquad \nu = 0, 1, 2, \ldots ,$$

belong to $\mathbf{C}(z)$ and

$$\|a - a_\nu\| \leq \rho^{n+\nu+1} \to 0$$

as $\nu \to \infty$. Hence $\mathbf{C}(z)$ is dense in $\mathbf{C}((z))$ under the norm (1.4), and $\mathbf{C}((z))$, in turn, is the closure of $\mathbf{C}(z)$ under this norm.

A rational function $b \in \mathbf{C}(z)$ can be written in the form $b(z) = p(z)/q(z)$, where p and q are relatively prime polynomials with $q \neq 0$. We set

$$\deg b = \max(\deg p , \deg q) .$$

The number $\deg b$ is known as the *degree* or the *order* of $b(z)$.

By analogy with the approximation of real numbers by rationals, we can consider the problem of approximating elements of $\mathbf{C}((z))$ by rational functions.

PROBLEM A. *Let* $a \in \mathbf{C}((z))$, *and let*

$$e_n(a) = \inf\{\|a - b\| : b(z) \in \mathbf{C}(z) , \deg b \leq n\}$$

for $n \in \mathbf{Z}_+$. *Find a rational function* $b(z)$ *for which this infimum is attained.*

A precise formulation of the problem is as follows. We must find, among the rational functions $b(z)$, with $\deg b \leq n$ one for which the Laurent expansion, with center at $z = 0$, coincides with the series (1.3) for $k = 0, \ldots, \nu$ with the largest possible ν.

3. We shall solve Problem A, or more precisely, reduce it to a system of linear equations. Without loss of generality, we may restrict ourselves to series $a(z)$ of the form

$$a(z) = \sum_{k=0}^{\infty} a_k z^k, \tag{1.5}$$

i.e., those for which $\|a\| \leq 1$.

PROBLEM B. *Let* $a(z)$ *be a series* (1.5), *with* $n \in \mathbf{Z}_+$. *Find polynomials* $q_n = q$ *and* $p_n = p$ *such that*

$$\deg q \leq n, \quad \deg p \leq n, \qquad q \neq 0,$$

and

$$q(z)a(z) - p(z) = cz^{2n+1} + \cdots, \tag{1.6}$$

i.e., the series (1.6) *starts with terms of degree no less than* $2n + 1$ *in* z.

Let us show that Problem B has a solution. In fact,

$$q(z)a(z) = \sum_{k=0}^{\infty} d_k z^k,$$

where d_k is a linear functional of the coefficients of q. Hence (1.6) is equivalent to the system

$$d_{n+1} = d_{n+2} = \cdots = d_{2n} = 0$$

of n homogeneous linear equations in the $n + 1$ unknown coefficients of q. Such a system always has a nonzero solution. The polynomial $p(z)$ is determined automatically as the corresponding partial sum of the series:

$$p(z) = \sum_{k=0}^{n} d_k z^k.$$

This establishes the existence of a solution.

It is clear that the solution (q_n, p_n) of Problem B is not unique. For example, it can be multiplied by any nonzero number. This nonuniqueness can also be more substantial. Let us show that, nevertheless, the fractions $p(z)/q(z)$ determine the same rational function, no matter which solution (q, p) of Problem B we consider. Let (q', p') and (q'', p'') be two solutions of Problem B. Then

$$\begin{aligned} q'a - p' &= c'z^{2n+1} + \cdots, \\ q''a - p'' &= c''z^{2n+1} + \cdots. \end{aligned} \tag{1.7}$$

If we multiply the first equation in (1.7) by q'' and the second by q', and subtract one from the other, we obtain

$$q'p'' - q''p' = cz^{2n+1} + \cdots . \tag{1.8}$$

Since the left-hand side of (1.8) is a polynomial of degree at most $2n$, we have $q'p'' - q''p' = 0$, or $p'/q' = p''/q''$. ▷

DEFINITION. Let $a(z)$ be the series (1.5), $n \in \mathbf{Z}_+$, and (q_n, p_n) any solution of Problem B for the index n. The rational function

$$\pi_n(f; z) = \pi_n(z) = \frac{p_n(z)}{q_n(z)}$$

is called the *nth diagonal Padé approximant of* $a(z)$. (The origin of the terminology will be explained later; see the Remarks.)

Let us now show that the infimum in Problem A (for the series (1.5)) is attained by the functions $\pi_n = \pi_n(a)$ and only by these. Let $b(z) = p(z)/q(z)$ be any rational function of degree at most n. We write the difference $a - b$ in the form

$$a - b = (a - \pi_n) + (\pi_n - b), \tag{1.9}$$

and $\pi_n = p^*/q^*$ as the quotient of relatively prime polynomials. We call $\lambda = n - \deg \pi_n$ the deficiency of π_n in the class of rational functions of degree at most n. We have

$$\pi_n - b = \frac{p^*q - pq^*}{qq^*}. \tag{1.10}$$

The degree of the numerator in (1.10) does not exceed $2n - \lambda$. Consequently if $b \neq \pi_n$, we have

$$\|\pi_n - b\| \geq \exp(-(2n - \lambda)). \tag{1.11}$$

It is clear that $q^*(0) \neq 0$, and that the pair

$$(z^\lambda q^*(z), z^\lambda p^*(z))$$

is a solution of Problem B for a given n. Consequently

$$a(z) - \pi_n(z) = cz^{2n+1-\lambda} + \cdots ,$$

i.e.,

$$\|a - \pi_n\| \leq \exp(-(2n + 1 - \lambda)). \tag{1.12}$$

It follows from (1.9), (1.11), and (1.12) that when $b \neq \pi_n$

$$\|a - \pi_n\| < \|\pi_n - b\| = \|a - b\|. \quad ▷$$

REMARK

For Padé approximants, see [123], [167], [155], [157], and [171]; see also the original papers by A. Padé [153] and Frobenius [135].

§2. Padé approximants for e^z

We shall find the diagonal Padé approximants for e^z, or, more precisely, their Taylor series with center at $z = 0$. (In this exposition, we follow the book by C. L. Siegel [166].) Generally speaking, the determination of an explicit form and the investigation of Padé approximants is not a simple matter in any specific case, and requires individual guesswork. This lack of a general method gives Padé approximants a distinctive fascination.

The Padé approximants for the exponential function were determined explicitly by Charles Hermite [137] in his proof of the transcendency of e (here it was a question of the still more general technique of simultaneous Padé approximants).

1. Let $n \in \mathbf{Z}_+$,

$$\deg p_n \leq n, \quad \deg q_n \leq n, \qquad q_n \neq 0,$$

and

$$q_n(z)e^z - p_n(z) = c_n z^{2n+1} + \cdots, \tag{2.1}$$

where the constants c_n might, to begin with, be zero. The left-hand side of (2.1) is an entire function, i.e., its Taylor series (written out on the right-hand side) converges to it for every complex number z. (The same remark also applies to the other power series that will be encountered in this section.)

If $f(z)$ is an entire function, we have

$$D(e^z f(z)) = e^z f(z) + e^z f'(z) = e^z (I + D) f(z), \tag{2.2}$$

where $D = d/dz$ is the operator of differentiation and I is the identity operator. From (2.2), we obtain inductively

$$D^m (e^z f(z)) = e^z (I + D)^m f(z), \qquad m \in \mathbf{Z}_+. \tag{2.3}$$

Using this formula, we differentiate (2.1) $n + 1$ times:

$$e^z (I + D)^{n+1} q_n(z) = c_n (2n+1) 2n \cdots (n+1) z^n + \cdots.$$

Since $e^{-z} = 1 - z + \cdots$, we obtain

$$(I + D)^{n+1} q_n(z) = c_n (2n+1) 2n \cdots (n+1) z^n + \cdots. \tag{2.4}$$

The left-hand side of (2.4) is a polynomial of degree at most n. Therefore

$$(I + D)^{n+1} q_n(z) = c_n (2n+1) 2n \cdots (n+1) z^n. \tag{2.5}$$

The formula (2.5), as we shall now show, completely determines the polynomials $q_n(z)$.

On the linear space of polynomials, we define the linear operators

$$A_m T = \sum_{k=0}^{\infty} \binom{-m}{k} D^k T, \qquad m \in \mathbf{Z}_+. \tag{2.6}$$

The sum on the right-hand side of (2.6) is finite for every polynomial T; in fact, if $k > \deg T$, there is no term of order k. By the recursion formula for the binomial coefficients, we obtain

$$\begin{aligned} A_{m+1}(I+D)T &= \sum_{k=0}^{\infty}\binom{-m-1}{k}D^kT + \sum_{k=0}^{\infty}\binom{-m-1}{k}D^{k+1}T \\ &= T + \sum_{k=1}^{\infty}\left(\binom{-m-1}{k}+\binom{-m-1}{k-1}\right)D^kT \\ &= \sum_{k=0}^{\infty}\binom{-m}{k}D^kT = A_mT. \end{aligned}$$

Hence

$$A_{n+1}(I+D)^{n+1}T = A_0T = T,$$

and consequently

$$q_n(z) = c_n(2n+1)2n\cdots(n+1)A_{n+1}z^n. \tag{2.7}$$

Hence $q_n(z)$ is uniquely determined by (2.1) (up to a nonzero numerical multiplier); a nonzero solution exists only when $c_n \neq 0$. Indices $n \in \mathbf{Z}_+$ with this property are said to be *normal.* (Thus we have shown that all indices for e^z are normal.) We take

$$c_n = ((2n+1)2n\cdots(n+1))^{-1}.$$

Then by (2.6) and (2.7)

$$\begin{aligned} q_n(z) &= \sum_{k=0}^{n}\binom{-n-1}{k}D^kz^n \\ &= \sum_{k=0}^{n}(-1)^k\frac{(n+1)(n+2)\cdots(n+k)}{k!}n(n-1)\cdots(n-k+1)z^{n-k}. \end{aligned}$$

To determine $p_n(z)$ we replace z by $(-z)$ in (2.1) and multiply the equation by $(-e^z)$. We obtain $-q_n(-z)+p_n(-z)e^z = c_nz^{2n+1}+\cdots$, from which we have $p_n(z) = q_n(-z)$. We note that the polynomials $p_n(z)$ and $q_n(z)$ have integral coefficients.

Let

$$R_n(z) = q_n(z)e^z - p_n(z). \tag{2.8}$$

It follows from (2.3) and (2.5) that

$$D^{n+1}R_n(z) = z^ne^z. \tag{2.9}$$

Let us set

$$\widetilde{R}(z) = \frac{1}{n!}\int_0^z (z-t)^nt^ne^t\,dt. \tag{2.10}$$

By differentiating (2.10), we also obtain

$$D^{n+1}\widetilde{R}_n(z) = z^ne^z. \tag{2.11}$$

Comparing (2.9) and (2.11), we see that $R_n(z) = \widetilde{R}_n(z) + T(z)$, where T is a polynomial and $\deg T \le n$. Making the change of variable $t \to zt$ in (2.10), we obtain

$$\widetilde{R}_n(z) = \frac{z^{2n+1}}{n!} \int_0^1 t^n(1-t)^n e^{zt}\, dt.$$

Therefore $\widetilde{R}_n(z)$, like $R_n(z)$, has a zero of order $2n+1$ at $z=0$. Hence $T = 0$ and $R_n = \widetilde{R}_n$. Finally,

$$R_n(z) = \frac{z^{2n+1}}{n!} \int_0^1 t^n(1-t)^n e^{zt}\, dt. \tag{2.12}$$

Since $t(1-t) \le 1/4$, we obtain

$$|R_n(z)| \le \frac{1}{n!}\left(\frac{|z|}{2}\right)^{2n} |z| M(z) \tag{2.13}$$

from (2.12), where

$$M(z) = \max(1, \exp(\operatorname{Re} z)).$$

2. THEOREM 2.1. *The number π is irrational.*

PROOF. Suppose the contrary, i.e., that $\pi = l/m$, where l and $m \in \mathbf{N}$. It follows from (2.12) that

$$R_n(\pi i) = (-1)^{n+1} \frac{\pi^{2n+1}}{n!} \int_0^1 t^n(1-t)^n \sin(\pi t)\, dt \neq 0.$$

In (2.8), we take $z = \pi i$ and multiply both sides of the equation by m^n:

$$m^n(-q_n(\pi i) - p_n(\pi i)) = m^n R_n(\pi i). \tag{2.14}$$

On the left side of (2.14) we have a nonzero complex number, $x + i \cdot y$ with integral x and y; and on the right, a number that approaches zero as $n \to \infty$. ▷

A proof similar to that of Theorem 2.1 shows that e^α is irrational for rational $\alpha \neq 0$.

REMARKS

1. In subsection 1 we defined a normal index. There is another (not equivalent) definition of normal indices, which will be used in Chapter 4, §1.

2. The Padé approximants of the exponential function can be used to obtain other arithmetic results similar to Theorem 2.1. For example [166], we have the following theorem.

THEOREM. *If $\alpha \neq 0$ is rational, the number* $\tan\alpha$ *is irrational.*

§3. Padé approximants at infinity

Padé approximants at $z = \infty$ are no different in principle from those at a finite point. However, the change from infinity to, for example, $z = 0$, leads to a reformulation of the theorem, which needs to be made rather carefully.

1. We denote by $\mathbf{C}((1/z))$ the field of formal power series whose elements are obtained from the elements of $\mathbf{C}((z))$ by replacing z by $1/z$. The operations of addition, multiplication, and taking the norm are carried over from one field to the other by using the same replacement. We consider only elements $f(z) \in \mathbf{C}((1/z))$ of the form

$$f(z) = \sum_{k=0}^{\infty} \frac{f_k}{z^{k+1}}. \tag{3.1}$$

PROBLEM C. *Let $f(z)$ be defined by* (3.1), $n \in \mathbf{Z}_+$. *Find a polynomial $Q_n \neq 0$, $\deg Q_n \leq n$, such that*

$$Q_n(z)f(z) - P_n(z) = \frac{c}{z^{n+1}} + \cdots,$$

where $P_n(z)$ is the polynomial part of the series $Q_n(z)f(z)$.

A solution of Problem C exists (see §1) and $\deg P_n \leq n-1$. Pairs (P_n, Q_n) that solve Problem C will be called *nth Padé pairs for $f(z)$.*

PROPOSITION 3.1. *The quotient P_n/Q_n defines the same rational function for every nth Padé pair.*

PROOF. Let (P'_n, Q'_n) and (P''_n, Q''_n) be two nth Padé pairs. Then

$$Q'_n(z)f(z) - P'_n(z) = \frac{c'}{z^{n+1}} + \cdots,$$

$$Q''_n(z)f(z) - P''_n(z) = \frac{c''}{z^{n+1}} + \cdots,$$

from which $Q'_nP''_n - Q''_nP'_n = c/z + \cdots$. The left-hand side of this equation is a polynomial; consequently $Q'_nP''_n - Q''_nP'_n = 0$, i.e., $P'_n/Q'_n = P''_n/Q''_n$. ▷

Proposition 3.1 shows that the following definition is reasonable.

DEFINITION. Let $f(z)$ be the series (3.1), $n \in \mathbf{Z}_+$, and let (P_n, Q_n) be an nth Padé pair for the series f. The rational function $\pi_n(z) = P_n(z)/Q_n(z)$ is called the *nth diagonal Padé approximant of f.*

It is clear that $\pi_n(f(1/z), z) = \pi_n(f(z), 1/z)$.

DEFINITION. If $\deg Q_n = n$ for any nth Padé pair, we say that n is *normal* (for f). The set of normal indices is denoted by $\Lambda = \Lambda(f)$.

The concept of a normal index plays an essential role in the subsequent discussion. We shall establish a determinantal test for normality. Let

$$H_0 = 1,$$

$$H_n = H_n(f) = \begin{vmatrix} f_0 & f_1 & \cdots & f_{n-1} \\ f_1 & f_2 & \cdots & f_n \\ \cdots & \cdots & \cdots & \cdots \\ f_{n-1} & f_n & \cdots & f_{2n-2} \end{vmatrix}, \qquad n = 1, 2, \ldots,$$

be the Hankel determinants constructed for the series (3.1).

PROPOSITION 3.2. $n \in \Lambda \Leftrightarrow H_n \neq 0$.

PROOF. The index $n = 0$ is always normal, and $H_0 = 1 \neq 0$.

For $n > 0$ we write out the explicit system of linear equations for determining the coefficients of the polynomial Q_n. Let

$$Q_n(z) = \sum_{k=0}^{n} x_k z^k ;$$

then the system has the form

$$\begin{array}{llll} x_0 f_0 & + \cdots + x_{n-1} f_{n-1} & + x_n f_n & = 0 \\ x_0 f_1 & + \cdots + x_{n-1} f_n & + x_n f_{n+1} & = 0 \\ \cdots & \cdots & \cdots & \cdots \\ x_0 f_{n-1} & + \cdots + x_{n-1} f_{2n-2} & + x_n f_{2n-1} & = 0 . \end{array} \tag{3.2}$$

If $n \notin \Lambda$, there is a nonzero solution of (3.2) with $x_n = 0$. Hence $H_n = 0$. Let $n \in \Lambda$. Then, by the definition of a normal index, the system (3.2) with $x_n = 0$ has only the trivial solution. Therefore $H_n \neq 0$. ▷

Let us notice some easily verified properties of normal indices. If $n \in \Lambda$, the nth Padé pair (P_n, Q_n) is the only (up to multiplication by a nonzero constant) relatively prime pair of polynomials P_n, Q_n. In addition, $n \in \Lambda \Leftrightarrow \deg \pi_n = n$.

2. The following proposition completely describes the structure of the sequence of diagonal Padé approximants. It turns out that this sequence consists only of approximants corresponding to normal indices.

PROPOSITION 3.3. *Let $n \in \Lambda$, let the integer $\nu > n$, and let $(n, \nu] \cap \Lambda = \varnothing$. Then $\pi_\nu = \pi_n$.*

PROOF. We write π_ν as an irreducible fraction: $\pi_\nu = P/Q$. Let $\deg \pi_\nu = r$. Since $\nu \notin \Lambda$, we have $r < \nu$. We show that r is a normal index. The pair (P, Q) is the rth Padé pair; consequently $\pi_r = \pi_\nu$, $\deg \pi_r = r$, $r \in \Lambda$. It is clear that $r \leq n$. Then (P, Q) is the nth Padé pair: $\pi_n = \pi_\nu$. ▷

In the following proposition, we obtain the exact degree of approximation of a Padé approximant to the series from which it was constructed.

PROPOSITION 3.4. *Let $n_0 < n_1$ be two successive normal indices for f. Then*

$$f(z) - \pi_{n_0}(z) = \frac{c}{z^{n_0+n_1}} + \cdots ,$$

where $c \neq 0$.

PROOF. Let r be a positive integer such that

$$f(z) - \pi_{n_0}(z) = \frac{c}{z^{n_0+r}} + \cdots ,$$

where $c \neq 0$, and let the leading coefficient of Q_{n_0} be 1. Then

$$Q_{n_0}(z) f(z) - P_{n_0}(z) = \frac{c}{z^r} + \cdots .$$

By Proposition 3.3, $\pi_{n-1} = \pi_{n_0}$, and therefore

$$Q_{n_0}(z)f(z) - P_{n_0}(z) = \frac{c_1}{z^{n_1}} + \cdots ,$$

i.e., $r \geq n_1$. On the other hand, if $r > n_1$ the pair (P_{n_0}, Q_{n_0}) must be the n_1th Padé pair, contradicting the normality of n_1. Hence $r = n_1$. ▹

3. As a corollary of the results of this section, we obtain a theorem of Kronecker.

THEOREM 3.1 (Kronecker). *A necessary and sufficient condition for the series $f(z)$ to be the Laurent expansion (with center at infinity) of a rational function is that $H_n(f) = 0$ for all sufficiently large n.*

PROOF. If $f(z)$ is the expansion of a rational function, we have $\pi_n = f$ for all $n \geq \deg f$ and all indices $n > \deg f$ are not normal, so that $H_n = 0$ for $n > \deg f$. Conversely, if Λ is a finite set, we have $\pi_n = \pi_N$ for $n \geq N = \max \Lambda$. Hence the expansion of $f - \pi_N$ cannot begin at a finite index: $f = \pi_N$. ▹

REMARK. It is clear from the proof of the theorem that the index of the last nonzero Hankel determinant $H_n(f)$ is equal to $\deg f$.

PROBLEMS

1. For a series $f(z)$ of the form (3.1), prove the following analog of the proposition in subsection 3 of §1: $\inf \|f - r\|$, where r is a rational function of order not exceeding n, is attained only for the nth diagonal Padé approximant $r = \pi_n$. Here $\|\cdot\|$ is the usual non-Archimedean norm in $\mathbf{C}((1/z))$.

2. The solution of Problem C is unique (up to a nonzero numerical multiple) when n or $n+1$ is a normal index.

3. Generalization of Problem 2. Let L_n consist of the pairs (P_n, Q_n) that are solutions of Problem C, together with the pair $(0, 0)$; then L_n is a linear space over the field $\mathbf{C}$. Find its dimension.

SOLUTION. If $n_0 \leq n < n_1$, where n_0 and n_1 are consecutive normal indices, then

$$\dim L_n = \left[\frac{n_1 - n_0 + 1}{2}\right] - \left[\left|n - \frac{n_0 + n_1 - 1}{2}\right|\right].$$

Here $|\cdot|$ denotes the absolute value, and $[\cdot]$ the integral part, of a number. If $n \geq N = \max \Lambda$, we have $\dim L_n = n - N + 1$.

§4. Continued fractions and Padé approximants

1. Let $f(z)$ be the series (3.1). We write the algorithm for the expansion of $f(z)$ as a continued fraction of functions; it is an exact analog of the algorithm for real numbers (see Chapter 1, §2).

If $f \neq 0$, we have $f(z) = 1/(1/f(z))$, $1/f(z) = a_1(z) + \varphi_1(z)$, where $a_1(z)$ is a polynomial of degree at least 1, and $\varphi_1(z)$ is a series of the form (3.1). In addition, if $\varphi_1 \neq 0$, we obtain, in the same way,

$$f(z) = \cfrac{1}{a_1(z) + \cfrac{1}{a_2(z) + \varphi_2(z)}},$$

where $a_2(z)$ is a polynomial ($\deg a_2 \geq 1$), and $\varphi_2(z)$ is a series of the form (3.1). Continuing this process, we have the following alternative: either $\varphi_N = 0$ for some index N, or $\varphi_n \neq 0$ for all positive integers n. In the first case,

$$f(z) = \cfrac{1}{a_1(z) + \cfrac{1}{a_2(z) + \ddots + \cfrac{1}{a_N(z)}}} \tag{4.1}$$

and $f(z)$ is a rational function (the exact expansion of a rational function in a Laurent series). Conversely, if $f(z)$ is a rational function, it can be expanded as a finite continued fraction (4.1) by the use of the division algorithm in the field $\mathbf{C}[z]$ of polynomials.

Consequently, the second case occurs when f is not a rational function. In this case we formally associate with f the infinite continued fraction

$$f(z) \sim \cfrac{1}{a_1(z) + \cfrac{1}{a_2(z) + \ddots + \cfrac{1}{a_n(z) + \ddots}}} \tag{4.2}$$

2. We denote by

$$\tau_n(z) = \cfrac{1}{a_1(z) + \cfrac{1}{a_2(z) + \ddots + \cfrac{1}{a_n(z)}}}$$

the convergents of the continued fraction (4.2). It is clear that $\tau_n(z)$ admits the representation $\tau_n = A_n/B_n$, where the numerators and denominators of the convergents, for $n \in \mathbf{N}$, can be determined successively from the recurrence relations (see Chapter 1, §2)

$$\begin{aligned} A_{n+1} &= a_{n+1}A_n + A_{n-1}, \\ B_{n+1} &= a_{n+1}B_n + B_{n-1}, \end{aligned} \tag{4.3}$$

with the initial conditions

$$\begin{aligned} A_0 &= 0, \qquad & A_1 &= 1, \\ B_0 &= 1, \qquad & B_1 &= a_1. \end{aligned} \tag{4.4}$$

From (4.3) and (4.4) we obtain the identity

$$B_{n+1}A_n - A_{n+1}B_n = (-1)^{n+1},$$

which implies that A_n and B_n are relatively prime $(n = 0, 1, 2, \dots)$.

The following proposition establishes the convergence, in the norm of $\mathbf{C}((1/z))$, of the continued fraction (4.2) to f, and determines the degree of approximation of the convergents.

PROPOSITION 4.1. *We have*

$$\|f - \tau_n\| \to 0 \qquad (n \to \infty). \tag{4.5}$$

Moreover,

$$f(z) - \tau_n(z) = \frac{C_n}{z^{s_n + s_{n+1}}} + \cdots, \tag{4.6}$$

where $C_n \neq 0$, $s_n = d_1 + \cdots + d_n$, $d_n = \deg a_n$, $n \in \mathbf{N}$.

PROOF. Continued fractions satisfy the equation

$$f(z) = [0; a_1, a_2, \dots, a_{n+1} + \varphi_{n+1}] = \frac{A_{n+1} + A_n\varphi_{n+1}}{B_{n+1} + B_n\varphi_{n+1}}.$$

Hence

$$f - \frac{A_n}{B_n} = \frac{(-1)^n}{B_n B_{n+1}\left(1 + \varphi_{n+1}\frac{B_n}{B_{n+1}}\right)}.$$

It follows easily, by induction from (4.3) and (4.4), that $\deg B_n = s_n$, $n \in \mathbf{N}$. Then

$$\|f - \tau_n\| = \|B_n B_{n+1}\|^{-1} = \exp(-(s_n + s_{n+1})) \to 0. \quad \triangleright$$

Since, for every continued fraction (4.2) (not necessarily obtained from some series by our algorithm)

$$\tau_{n+1} - \tau_n = \frac{(-1)^n}{B_n B_{n+1}}, \qquad n \in \mathbf{Z}_+,$$

where $\tau_0 = 0$, the series

$$\tilde{f} = \sum_{k=0}^{\infty} (\tau_{k+1} - \tau_k)$$

converges (in the form of $\mathbf{C}((1/z))$). If we apply the algorithm described above to the expansion of f, we obtain the original fraction (uniqueness theorem).

3. PROPOSITION 4.2. *In the preceding notation, we have*

$$\Lambda(f) = \{0, s_1, s_2, \dots\}. \tag{4.7}$$

PROOF. Let us denote the set on the right of (4.7) by S. Since $\deg B_n = s_n$, by (4.6) and (4.5) we have that $\tau_n = A_n/B_n$ is the s_nth Padé approximant of f; and since the polynomials A_n and B_n are relatively prime, we have $\deg \tau_n = s_n$, i.e., $s_n \in \Lambda$. Thus $S \subset \Lambda$.

Let us establish the converse inclusion. Let $k \in \Lambda$ and let $\pi_k = P_k/Q_k$ be the kth Padé approximant of f. Then $s_n \le k < s_{n+1}$ $(s_0 = 0)$ for some $n \in \mathbf{Z}_+$ and

$$\|\pi_k - \tau_n\| \le \max(\|f - \tau_n\|, \|f - \pi_k\|) \le \exp(-\min(s_n + s_{n+1}, 2k+1)).$$

Hence

$$\begin{aligned}\|\pi_k B_n - Q_k A_n\| &= \|B_n Q_k\| \, \|\pi_k - \tau_n\| \\ &\le \exp(-\min(s_{n+1} - k, k+1-s_n)) \le e^{-1} < 1.\end{aligned}$$

Therefore $P_k B_n - Q_k A_n = 0$ and consequently $\pi_k = \tau_n$, i.e., $k = s_n$. ▷

REMARK. It is clear from the proof of Proposition 4.2 that every convergent is some Padé approximant of f, and each Padé approximant of this series is a convergent. In particular, the appropriately normalized numerator and denominator of a Padé approximant π_{s_n} with a normal index satisfy the recursion relations (4.3) and (4.4), where $\deg a_n = s_n - s_{n-1}$.

REMARK

On various types of continued fractions, their applications and generalizations, see [171], [157], [98], and also [113], [73], [147], and [41].

PROBLEM

1. Prove the uniqueness theorem for a continued fraction.

§5. Orthogonal polynomials

We now discuss a fundamental special case of the general theory of Padé approximants, when the arguments can be extended further.

1. We shall need the following concepts. Let $s = \{s_n\}_0^\infty$ be a sequence of real numbers.

DEFINITION. We say that a sequence s is *positive* if, for all $n \in \mathbf{Z}_+$, the Hankel form

$$\sum_{i,j=0}^{n} s_{i+j} x_i x_j, \qquad x = (x_0, \dots, x_n) \in \mathbf{R}^{n+1},$$

is strictly positive definite, i.e., takes positive values when $x \ne 0$.

By Sylvester's criterion, this is equivalent to saying that for each $n \in \mathbf{Z}_+$ the nth Hankel determinant is positive:

$$H_n = \begin{vmatrix} s_0 & s_1 & \cdots & s_{n-1} \\ s_1 & s_2 & \cdots & s_n \\ \cdots & \cdots & \cdots & \cdots \\ s_{n-1} & s_n & \cdots & s_{2n-2} \end{vmatrix} > 0.$$

On the complex linear space $\mathbf{C}[\lambda]$ of polynomials, we define a linear functional $\mathfrak{S}$ corresponding to the sequence s of real numbers: if

$$T(\lambda) = t_0 + t_1\lambda + \cdots + t_n\lambda^n \in \mathbf{C}[\lambda],$$

then

$$\mathfrak{S}\{T(\lambda)\} = t_0 s_0 + t_1 s_1 + \cdots + t_n s_n.$$

(This is the general form of a real linear functional on $\mathbf{C}[\lambda]$.)

DEFINITION. A functional $\mathfrak{S}$ is said to be *positive* if, for every polynomial $T(\lambda) \in \mathbf{C}[\lambda]$ such that $T(\lambda) \geq 0$, $-\infty < \lambda < +\infty$, the inequality $\mathfrak{S}\{T(\lambda)\} \geq 0$ is satisfied, and $\mathfrak{S}\{T(\lambda)\} = 0$ only if $T = 0$.

Let us determine the general form of a polynomial that is nonnegative on the real axis.

LEMMA 5.1. *If* $T(\lambda) \in \mathbf{C}[\lambda]$ *and* $T(\lambda) \geq 0$, $\lambda \in \mathbf{R}$, *then* $T(\lambda) = A(\lambda)^2 + B(\lambda)^2$, *where both* A *and* B *have real coefficients.*

PROOF. Since $T(\lambda)$ takes real values for $\lambda \in \mathbf{R}$, it has real coefficients. Therefore if λ_0 is a nonreal root of $T(\lambda)$, the conjugate complex number $\overline{\lambda}_0$ is also a root. Real roots λ_0 have even multiplicity (if this were not the case, $T(\lambda)$ would take both positive and negative values in a neighborhood of λ_0). Consequently, the decomposition of $T(\lambda)$ into linear factors has the form

$$T(\lambda) = c\prod_{k=1}^{n}(\lambda - \lambda_k)(\lambda - \overline{\lambda}_k), \qquad c > 0. \tag{5.1}$$

Let $\lambda_k = a_k + ib_k$ with a_k, $b_k \in \mathbf{R}$. Then

$$(\lambda - \lambda_k)(\lambda - \overline{\lambda}_k) = (\lambda - a_k)^2 + b_k^2. \tag{5.2}$$

If we repeatedly apply the identity

$$(c_1^2 + d_1^2)(c_2^2 + d_2^2) = (c_1c_2 - d_1d_2)^2 + (c_1d_2 + c_2d_1)^2$$

to (5.1) and (5.2), we obtain the conclusion of the lemma. ▷

LEMMA 5.2. *A sequence* s *is positive if and only if the corresponding functional* $\mathfrak{S}$ *is positive.*

PROOF. Let $\mathfrak{S}$ be positive. For any $n \in \mathbf{Z}_+$ and $x = (x_0, \dots, x_n) \in \mathbf{R}^{n+1}$, $x \neq 0$, we consider the polynomial

$$A(\lambda) = \sum_{k=0}^{n} x_k\lambda^k.$$

The polynomial $A(\lambda)^2$ is not zero and not negative on the real axis; consequently

$$\sum_{i,j=0}^{n} s_{i+j}x_ix_j = \mathfrak{S}\{A(\lambda)^2\} > 0,$$

i.e., s is a positive sequence.

Conversely, let s be positive and let $T(\lambda)$ be a polynomial that is nonnegative for $\lambda \in \mathbf{R}$, with $T(\lambda) \not\equiv 0$. By Lemma 5.1,

$$T(\lambda) = A(\lambda)^2 + B(\lambda)^2,$$

where

$$A(\lambda) = \sum_{k=0}^{n} x_k \lambda^k, \qquad x = (x_0, \dots, x_n) \in \mathbf{R}^{n+1},$$

$$B(\lambda) = \sum_{k=0}^{m} y_k \lambda^k, \qquad y = (y_0, \dots, y_m) \in \mathbf{R}^{m+1},$$

and at least one of the vectors x, y is different from zero. Consequently

$$\begin{aligned}\mathfrak{S}\{T(\lambda)\} &= \mathfrak{S}\{A(\lambda)^2\} + \mathfrak{S}\{B(\lambda)^2\} \\ &= \sum_{i,j=0}^{n} s_{i+j} x_i x_j + \sum_{i,j=0}^{m} s_{i+j} y_i y_j > 0,\end{aligned}$$

i.e., $\mathfrak{S}$ is a positive functional. ▷

2. Let us mention some examples of positive sequences. We recall some definitions from measure theory and at the same time comment on the terminology.

The *Borel σ-algebra* (on the line $\mathbf{R}$) is the smallest σ-algebra containing all compact subsets of $\mathbf{R}$. Its elements are called *Borel sets.* A *finite positive Borel measure* (from now on, we shall simply say "a measure") is a σ-additive set function defined on the Borel σ-algebra and taking finite nonnegative values.

A point $\lambda_0 \in \mathbf{R}$ is called a *point of increase* of the measure μ if, for every $\varepsilon > 0$, we have $\mu(\lambda_0 - \varepsilon, \lambda_0 + \varepsilon) > 0$. The set of points of increase of μ is closed; it is called the *spectrum* of μ and is denoted by S_μ.

The *support* Δ_μ of a measure μ is the smallest interval containing S_μ. A support can be a finite interval, $\Delta_\mu = [a, b]$, a semi-infinite interval, $\Delta_\mu = (-\infty, a]$ or $[a, +\infty)$, or the whole real line, $\Delta_\mu = (-\infty, +\infty)$. (If $\mu = 0$, we take $\Delta_\mu = \varnothing$.)

A point $\lambda_0 \in \mathbf{R}$ is a *mass-point* if $\mu\{\lambda_0\} > 0$. The set of such points is at most countable. We denote it by Γ_μ.

For every measure μ we define the measure

$$\mu_d(E) = \sum_{\lambda \in \Gamma_\mu} \mu\{\lambda\} \delta_\lambda(E),$$

where δ_λ is the unit measure concentrated at λ: $\delta_\lambda(E) = 0$ if $\lambda \notin E$, $\delta_\lambda(E) = 1$ if $\lambda \in E$, for every Borel set E. The measure μ_d is the discrete part of μ. It is easy to see that $\mu - \mu_d$ is a measure, whose discrete part is zero. Measures that consist only of discrete parts are called *discrete* or *atomic.*

Further facts of measure theory will be recalled as needed.

Let $F \subset \mathbf{R}$ be an infinite closed set. We denote by $\mathscr{M}(F)$ the class of measures μ that have infinite spectra $S_\mu \subset F$ and have the property that all the integrals

$$\int_F |\lambda|^n \, d\mu(\lambda), \qquad n \in \mathbf{Z}_+,$$

are finite. We use the abbreviation $\mathscr{M} = \mathscr{M}(\mathbf{R})$.

An important property of measures $\mu \in \mathscr{M}$ is the positivity of the sequences s of power moments constructed from them,

$$s_n = \int_{\mathbf{R}} \lambda^n \, d\mu(\lambda), \qquad n \in \mathbf{Z}_+. \tag{5.3}$$

In fact, the functional $\mathfrak{S}$ corresponding to s has the form

$$\mathfrak{S}\{T(\lambda)\} = \int_{\mathbf{R}} T(\lambda)\, d\mu(\lambda), \qquad T(\lambda) \in \mathbf{C}[\lambda],$$

and is evidently positive. As we shall show, (5.3) is the general form of a positive sequence.

3. We consider the problem of constructing Padé approximants for a formal power series

$$f(z) = \sum_{n=0}^{\infty} \frac{s_n}{z^{n+1}}, \tag{5.4}$$

where $s = \{s_n\}_0^\infty$ is a positive sequence. Let $n \in \mathbf{Z}_+$ and let (P_n, Q_n) be an arbitrary nth Padé pair for f. For all $n \in \mathbf{Z}_+$, we have $H_n > 0$, and all indices n are normal. Hence, as we saw, the degree of Q_n is n and Q_n is unique (up to a nonzero numerical multiplier). The polynomials P_n and Q_n are relatively prime. To determine the coefficients of the polynomial

$$Q_n(\lambda) = x_0 + x_1\lambda + \cdots + x_n\lambda^n,$$

we write down a system of equations that they must satisfy, namely

$$\sum_{j=0}^{n} s_{i+j}x_j = 0, \qquad i = 0, \ldots, n-1 \tag{5.5}$$

(see §3). If we set $s_m = \mathfrak{S}\{\lambda^m\}$ in (5.5), we may rewrite the system as

$$\mathfrak{S}\{Q_n(\lambda)\lambda^i\} = 0, \qquad i = 0, \ldots, n-1. \tag{5.6}$$

DEFINITION. A nonzero polynomial Q_n of degree at most n, satisfying (5.6), is called the *nth orthogonal polynomial* (with respect to the positive functional $\mathfrak{S}$ or the positive sequence s, and also with respect to the measure μ, if (5.3) is satisfied).

We have therefore shown that Q_n is the Padé denominator of f when Q_n is the nth orthogonal polynomial with respect to $\mathfrak{S}$. Similarly we can establish a formula for the numerators P_n of the nth Padé pair:

$$P_n(z) = \mathfrak{S}_\lambda \left\{ \frac{Q_n(z) - Q_n(\lambda)}{z - \lambda} \right\}, \qquad n \in \mathbf{Z}_+ \tag{5.7}$$

($\mathfrak{S}_\lambda$ acts on a polynomial as a function of λ). The polynomial (5.7) is usually called a *polynomial of the second kind* for Q_n (relative to $\mathfrak{S}$).

4. Let us note the simplest properties of orthogonal polynomials.

PROPOSITION 5.1. *Zeros of the polynomials Q_n $(n \in \mathbf{Z}_+)$ that are orthogonal with respect to the positive functional $\mathfrak{S}$ are real and simple.*

PROOF. We may suppose that the coefficients of Q_n are real, since the coefficients in (5.5) are real. Let $\lambda_1, \dots, \lambda_k$ be the different real zeros of odd multiplicity of Q_n. Then $0 \leq k \leq n$ and

$$T(\lambda) = \varepsilon Q_n(\lambda)(\lambda - \lambda_1) \cdots (\lambda - \lambda_k) \geq 0, \qquad \lambda \in \mathbf{R},$$

for some choice of $\varepsilon = \pm 1$. In addition, $T(\lambda) \not\equiv 0$. If $k < n$, we obtain $\mathfrak{S}\{T(\lambda)\} = 0$ from the orthogonality relations (5.6), contradicting the positivity of $\mathfrak{S}$. ▷

Proposition 5.1 can be made more precise. Let s be the sequence of power moments of a measure $\mu \in \mathscr{M}$ with support Δ_μ. Then all the zeros of Q_n lie inside Δ_μ.

It follows from the preceding section that Q_n and P_n satisfy a recurrence relation of the form (4.3) with linear polynomials a_n. We can give a direct proof of this relation, and also eliminate the arbitrariness of the normalization of Q_n. Let us suppose that $\{Q_n\}$ are *orthonormal* polynomials, i.e., the leading coefficients α_n of Q_n are positive and the relations

$$\mathfrak{S}\{Q_n(\lambda)Q_m(\lambda)\} = \delta_{n,m}, \qquad n, \ m \in \mathbf{Z}_+,$$

are satisfied, where $\delta_{n,m}$ is the Kronecker symbol. These polynomials $\{Q_n\}$ are uniquely defined.

The sesquilinear form $\mathfrak{S}\{R(\lambda)\overline{S(\lambda)}\}$, $R, \ S \in \mathbf{C}[\lambda]$, satisfies the axioms of a scalar product and consequently induces the structure of a pre-Hilbert space in $\mathbf{C}[\lambda]$. The polynomials $\{Q_n\}$ are orthogonal with respect to this scalar product and can be obtained from the linearly independent system 1, $\lambda, \lambda^2, \dots$ by the Gram-Schmidt orthogonalization process.

PROPOSITION 5.2. *We have the determinantal formula*

$$Q_n(\lambda) = \frac{1}{\sqrt{H_{n+1}H_n}} \begin{vmatrix} s_0 & \cdots & s_n \\ \cdots & \cdots & \cdots \\ s_{n-1} & \cdots & s_{2n-1} \\ 1 & \cdots & \lambda^n \end{vmatrix}, \qquad n \in \mathbf{Z}_+. \tag{5.8}$$

PROOF. Let $Q_n^*(\lambda)$ be the right-hand side of (5.8). Let us show that the polynomials Q_n^* are orthonormal; then, since they are unique, $Q_n = Q_n^*$. If we calculate $\mathfrak{S}\{Q_n^*(\lambda)\lambda^k\}$ for $k < n$, we obtain a vanishing determinant (with two identical rows). The leading coefficient of the polynomial Q_n^* – $\sqrt{H_n/H_{n+1}}$ is positive. Finally,

$$\mathfrak{S}\{Q_n^*(\lambda)^2\} = \frac{H_n}{H_{n+1}H_n} \mathfrak{S} \left\{ \begin{vmatrix} s_0 & \cdots & s_n \\ \cdots & \cdots & \cdots \\ s_{n-1} & \cdots & s_{2n-1} \\ \lambda^n & \cdots & \lambda^{2n} \end{vmatrix} \right\} = 1. \quad ▷$$

Below we shall assume that $s_0 = 1$.

PROPOSITION 5.3. *The orthonormal polynomials Q_n and the polynomials P_n of the second kind satisfy the relations*

$$\lambda P_n(\lambda) = h_n P_{n+1}(\lambda) + v_n P_n(\lambda) + h_{n-1} P_{n-1}(\lambda), \tag{5.9}$$

$$\lambda Q_n(\lambda) = h_n Q_{n+1}(\lambda) + v_n Q_n(\lambda) + h_{n-1} Q_{n-1}(\lambda), \tag{5.10}$$

$n \in \mathbf{Z}_+$, *with the initial conditions*

$$P_{-1} = 1, \qquad P_0 = 0, \tag{5.11}$$

$$Q_{-1} = 0, \qquad Q_0 = 1, \tag{5.12}$$

where

$$\begin{gathered} h_n = \mathfrak{S}\{\lambda Q_n(\lambda) Q_{n+1}(\lambda)\} = \frac{\alpha_n}{\alpha_{n+1}} > 0, \qquad h_{-1} = -1, \\ v_n = \mathfrak{S}\{\lambda Q_n(\lambda)^2\}, \qquad n \in \mathbf{Z}_+. \end{gathered} \tag{5.13}$$

PROOF. Let $Q_{-1} = 0$. By (5.8), we have $Q_0 = 1$. Let $n \in \mathbf{Z}_+$. We expand $\lambda Q_n(\lambda)$ in terms of the basis $Q_0, \dots, Q_{n+1}$:

$$\begin{aligned} \lambda Q_n(\lambda) = h_{n+1}^{(n)} Q_{n+1}(\lambda) + v_n Q_n(\lambda) \\ + h_{n-1}^{(n)} Q_{n-1}(\lambda) + \sum_{k=2}^{n} d_k Q_{n-k}(\lambda). \end{aligned} \tag{5.14}$$

We calculate the coefficients in (5.14) by Fourier's formulas:

$$d_k = \mathfrak{S}\{\lambda Q_n(\lambda) Q_{n-k}(\lambda)\} = 0, \qquad k = 2, \dots, n,$$

since $\deg \lambda Q_{n-k}(\lambda) \le n - 1$.

We define h_n and v_n, $n \in \mathbf{Z}_+$, by (5.13). Then

$$h_{n+1}^{(n)} = \mathfrak{S}\{\lambda Q_n(\lambda) Q_{n+1}(\lambda)\} = h_n, \qquad n = 0, 1, 2, \dots,$$

$$h_{n-1}^{(n)} = \mathfrak{S}\{\lambda Q_n(\lambda) Q_{n-1}(\lambda)\} = h_{n-1}, \qquad n = 1, 2, \dots.$$

This establishes (5.10) and (5.12).

By (5.7), for $n = 1, 2, \dots$, we have

$$\begin{aligned} h_n P_{n+1}(z) + v_n P_n(z) + h_{n-1} P_{n-1}(z) &= \mathfrak{S}_\lambda \left\{ \frac{z Q_n(z) - \lambda Q_n(\lambda)}{z - \lambda} \right\} \\ &= z \mathfrak{S}_\lambda \left\{ \frac{Q_n(z) - Q_n(\lambda)}{z - \lambda} \right\} + \mathfrak{S}\{Q_n(\lambda)\} = z P_n(z). \end{aligned}$$

Evidently $P_0 = 0$ and $Q_1(\lambda) = (\lambda - s_1)/\sqrt{H_2}$. Consequently

$$P_1(\lambda) = \frac{1}{\sqrt{H_2}} \mathfrak{S}\{1\} = \frac{1}{\sqrt{H_2}} = \frac{1}{h_0}.$$

Therefore (5.9) is also valid for $n = 0$ if we set $h_{-1} = -1$ and $P_{-1} = 1$. ▷

DEFINITION. The tridiagonal matrix

$$\mathscr{A} = \begin{pmatrix} v_0 & h_0 & & 0 \\ h_0 & v_1 & h_1 & \\ & h_1 & v_2 & h_2 \\ 0 & \cdot & \cdot & \cdot \end{pmatrix}, \tag{5.15}$$

with real v_n and positive h_n, is called a *Jacobi matrix* or $\mathscr{J}$*-matrix* (in this case we write $\mathscr{A} \in (\mathscr{J})$).

In Proposition 5.3, we actually constructed a mapping that associates a Jacobi matrix with each positive sequence s $(s_0 = 1)$. This is a one-to-one mapping. If we successively solve the relations (5.12) and (5.10) generated by the matrix (5.15), we obtain a sequence of polynomials Q_n of degree n with leading coefficients $\alpha_n = (h_0, \dots, h_{n-1})^{-1}$ $(\alpha_0 = 1)$. In $\mathbf{C}[\lambda]$ we introduce a linear functional defined by the basis vectors:

$$\mathfrak{S}\{Q_0(\lambda)\} = 1, \qquad \mathfrak{S}\{Q_n(\lambda)\} = 0, \qquad n \in \mathbf{N}. \tag{5.16}$$

If $n - m \geq 1$, it follows from (5.10) that $\lambda^m Q_n(\lambda)$ is a linear combination of $Q_j(\lambda)$, $j = n - m, \dots, n + m$, and by (5.16), $\mathfrak{S}\{\lambda^m Q_n(\lambda)\} = 0$. Similarly,

$$\mathfrak{S}\{\alpha_n \lambda^n Q_n(\lambda)\} = \alpha_n h_{n-1} \cdots h_0 \mathfrak{S}\{Q_0(\lambda)\} = 1,$$

from which $\mathfrak{S}\{Q_n(\lambda)^2\} = 1$. Finally,

$$\mathfrak{S}\{Q_n(\lambda) Q_m(\lambda)\} = \delta_{n,m}, \qquad n, m \in \mathbf{Z}_+.$$

In other words, $\mathfrak{S}\{R\overline{S}\}$ is a positive definite sesquilinear form, with respect to which Q_n are orthonormal polynomials.

We shall now see how to replace (5.9)–(5.12) by a different natural normalization. Let us set

$$\begin{aligned} \widetilde{Q}_n &= h_0 \cdots h_{n-1} Q_n, \quad n \in \mathbf{N}, \qquad \widetilde{Q}_0 = Q_0, \\ \widetilde{P}_n &= h_0 \cdots h_{n-1} P_n, \quad n \in \mathbf{N}, \qquad \widetilde{P}_0 = P_0. \end{aligned} \tag{5.17}$$

Then the leading coefficient of $\widetilde{Q}_n$ is unity. If we substitute (5.17) into (5.9) and (5.12), we obtain

$$\begin{aligned} \widetilde{Q}_{n+1}(\lambda) &= (\lambda - v_n)\widetilde{Q}_n(\lambda) - h_{n-1}^2 \widetilde{Q}_{n-1}(\lambda), \\ \widetilde{P}_{n+1}(\lambda) &= (\lambda - v_n)\widetilde{P}_n(\lambda) - h_{n-1}^2 \widetilde{P}_{n-1}(\lambda), \end{aligned} \tag{5.18}$$

for $n \in \mathbf{N}$, with the initial conditions

$$\begin{aligned} \widetilde{P}_0(\lambda) &= 0, \qquad \widetilde{P}_1(\lambda) = 1, \\ \widetilde{Q}_0(\lambda) &= 1, \qquad \widetilde{Q}_1(\lambda) = \lambda - s_1. \end{aligned} \tag{5.19}$$

Equations (5.18) and (5.19) are equivalent to the expansion of $\pi_n = \widetilde{P}_n/\widetilde{Q}_n$ as the continued fraction

$$\pi_n = \cfrac{1}{z - v_0 - \cfrac{h_0^2}{z - v_1 - \cfrac{h_1^2}{z - v_2 - \ddots - \cfrac{h_{n-2}^2}{z - v_{n-1}}}}}.$$

Therefore, with the power series (5.4) we can associate the continued fraction

$$f(z) \sim \cfrac{1}{z - v_0 - \cfrac{h_0^2}{z - v_1 - \cfrac{h_1^2}{z - v_2 - \ddots}}} \tag{5.20}$$

which converges to $f(z)$ in the norm of $\mathbf{C}((1/z))$. This is known as the Chebyshev continued fraction, the study of which first led Chebyshev and Gauss (independently) to the concept of orthogonal polynomials.

5. We continue the study of the properties of orthogonal polynomials.

Since all the zeros of Q_n are simple, the rational function $\pi_n = P_n/Q_n$ has the following partial fraction decomposition:

$$\pi_n(z) = \sum_{j=1}^{n} \frac{\mu_{n,j}}{z - \lambda_{n,j}}. \tag{5.21}$$

Here $\lambda_{n,j}$ are the zeros of $Q_n(\lambda)$, indexed, say, in increasing order, and $\mu_{n,j}$ are the residues:

$$\mu_{n,j} = \operatorname*{res}_{z=\lambda_{n,j}} \pi_n(z) = \frac{P_n(\lambda_{n,j})}{Q_n'(\lambda_{n,j})}, \qquad j = 1, \dots, n. \tag{5.22}$$

These are known as the *Christoffel coefficients.*

PROPOSITION 5.4. *The Gauss-Jacobi quadrature formula*

$$\mathfrak{S}\{T(\lambda)\} = \sum_{j=1}^{n} \mu_{n,j} T(\lambda_{n,j}) \tag{5.23}$$

is valid for every polynomial $T(\lambda)$ *of degree at most* $2n-1$.

PROOF. According to (5.21), the expansion of a rational function $\pi_n(z)$ in a series of powers of $1/z$ has the form

$$\pi_n(z) = \sum_{k=0}^{\infty} \frac{1}{z^{k+1}} \sum_{j=1}^{n} \mu_{n,j} \lambda_{n,j}^k.$$

In addition,

$$f(z) - \pi_n(z) = \frac{c}{z^{2n+1}} + \cdots .$$

Hence, for $k = 0, \dots, 2n-1$ we obtain

$$s_k = \sum_{j=1}^{n} \mu_{n,j} \lambda_{n,j}^k ,$$

i.e., (5.23) is verified for $T(\lambda) = \lambda^k$ with the specified values of k. ▷

We shall need the following properties of the Christoffel coefficients.

PROPOSITION 5.5. *Let* $n \in \mathbf{N}$. *Then*

1) $\mu_{n,1} + \cdots + \mu_{n,n} = 1$,
2) $\mu_{n,j} > 0$, $j = 1, \dots, n$.

PROOF. We obtain 1) by applying the quadrature formula (5.23) to the polynomial $T(\lambda) = 1$. Let

$$l_{n,j}(\lambda) = \frac{Q_n(\lambda)}{Q_n'(\lambda_{n,j})(\lambda - \lambda_{n,j})}, \qquad j = 1, \dots, n.$$

Again applying (5.23), we obtain

$$\mu_{n,j} = \mathfrak{S}\{l_{n,j}(\lambda)^2\} > 0,$$

which proves 2). ▷

PROPOSITION 5.6. *Let* $n \in \mathbf{N}$, $j = 1, \dots, n$. *Then*

$$\frac{1}{\mu_{n,j}} = \sum_{k=0}^{n-1} Q_k(\lambda_{n,j})^2 . \tag{5.24}$$

PROOF. Set

$$U = \begin{pmatrix} \sqrt{\mu_{n,1}} Q_0(\lambda_{n,1}) & \sqrt{\mu_{n,2}} Q_0(\lambda_{n,2}) & \cdots & \sqrt{\mu_{n,n}} Q_0(\lambda_{n,n}) \\ \sqrt{\mu_{n,1}} Q_1(\lambda_{n,1}) & \sqrt{\mu_{n,2}} Q_1(\lambda_{n,2}) & \cdots & \sqrt{\mu_{n,n}} Q_1(\lambda_{n,n}) \\ \sqrt{\mu_{n,1}} Q_{n-1}(\lambda_{n,1}) & \sqrt{\mu_{n,2}} Q_{n-1}(\lambda_{n,2}) & \cdots & \sqrt{\mu_{n,n}} Q_{n-1}(\lambda_{n,n}) \end{pmatrix}.$$

It follows from the quadrature formula and the orthogonality relations that

$$\sum_{j=1}^{n} \sqrt{\mu_{n,j}} Q_i(\lambda_{n,j}) \sqrt{\mu_{n,j}} Q_k(\lambda_{n,j}) = \mathfrak{S}\{Q_i(\lambda) Q_k(\lambda)\} = \delta_{i,k},$$

where $i, k = 0, \dots, n-1$, i.e., U is an orthogonal matrix ($UU^* = I$, the unit matrix). Then $U^*U = I$, i.e.,

$$\sum_{k=0}^{n-1} \sqrt{\mu_{n,i}} Q_k(\lambda_{n,i}) \sqrt{\mu_{n,j}} Q_k(\lambda_{n,j}) = \delta_{i,j}, \tag{5.25}$$

where $i, j = 1, \dots, n$. In particular, (5.25) is equivalent to (5.24) for $i = j$. ▷

A different approach to (5.24) is based on the *Christoffel-Darboux formula*, which is of independent interest.

PROPOSITION 5.7. *For the sequence of orthogonal polynomials, we have the formula*

$$\sum_{k=0}^{n} Q_k(\kappa)Q_k(\lambda) = h_n \frac{Q_{n+1}(\kappa)Q_n(\lambda) - Q_n(\kappa)Q_{n+1}(\lambda)}{\kappa - \lambda}. \tag{5.26}$$

PROOF. We write out the recurrence relations (see (5.10)):

$$\kappa Q_k(\kappa) = h_k Q_{k+1}(\kappa) + v_k Q_k(\kappa) + h_{k-1} Q_{k-1}(\kappa),$$
$$\lambda Q_k(\lambda) = h_k Q_{k+1}(\lambda) + v_k Q_k(\lambda) + h_{k-1} Q_{k-1}(\lambda).$$

Let us multiply the first equation by $Q_k(\lambda)$ and subtract the second, multiplied by $Q_k(\kappa)$:

$$(\kappa - \lambda)Q_k(\kappa)Q_k(\lambda) = \xi_k - \xi_{k-1}, \tag{5.27}$$

where

$$\xi_k = h_k(Q_{k+1}(\kappa)Q_k(\lambda) - Q_k(\kappa)Q_{k+1}(\lambda)).$$

Summing (5.27) for $k = 0, \dots, n$ we obtain

$$(\kappa - \lambda)\sum_{k=0}^{n} Q_k(\kappa)Q_k(\lambda) = \xi_n - \xi_{-1} = \xi_n. \quad \triangleright$$

If we take the limit in (5.26) as $\kappa \to \lambda$, we obtain the following corollary.

COROLLARY. *We have the formula*

$$\sum_{k=0}^{n} Q_k(\lambda)^2 = h_n(Q'_{n+1}(\lambda)Q_n(\lambda) - Q_{n+1}(\lambda)Q'_n(\lambda)). \tag{5.28}$$

COROLLARY. *The polynomials Q_n and Q_{n+1} have no zeros in common; moreover, the zeros of these polynomials are interlaced, i.e., between any two consecutive zeros of Q_{n+1} there is a single zero of Q_n $(n \in \mathbf{N})$.*

PROOF. If λ_0 is a zero of Q_{n+1}, then according to (5.28) we have

$$Q'_{n+1}(\lambda_0)Q_n(\lambda_0) > 0. \tag{5.29}$$

Let λ' and λ'' be two successive zeros of Q_{n+1}. Then $Q'_{n+1}(\lambda')Q'_{n+1}(\lambda'') < 0$. By (5.29), $Q_n(\lambda')Q_n(\lambda'') < 0$ and consequently Q_n has at least one zero on the interval (λ', λ''). Since there are n such intervals, Q_n has a single zero on each one. $\triangleright$

PROPOSITION 5.8. *We have the formula*

$$h_n(P_{n+1}(\lambda)Q_n(\lambda) - P_n(\lambda)Q_{n+1}(\lambda)) = 1. \tag{5.30}$$

PROOF. Let us denote the left-hand side of (5.30) by η_n. If we express $P_{n+1}(\lambda)$ and $Q_{n+1}(\lambda)$ in (5.30) by means of the recurrence formulas (5.9) and (5.10), we obtain $\eta_n = \eta_{n-1}$, $n \in \mathbf{Z}_+$. Consequently $\eta_n = \eta_{-1} = 1$. $\triangleright$

COROLLARY. *The zeros of $Q_n(\lambda)$ and $P_n(\lambda)$ are interlaced; that is, between any two successive zeros of $Q_n(\lambda)$ there is precisely one zero of $P_n(\lambda)$ ($n = 2, 3, \ldots$).*

PROOF. Let $n \in \mathbf{N}$ and $Q_{n+1}(\lambda) = 0$. Then by (5.30)

$$P_{n+1}(\lambda_0)Q_n(\lambda_0) > 0. \tag{5.31}$$

Let λ' and λ'' be two successive zeros of Q_{n+1}. We have $Q_n(\lambda')Q_n(\lambda'') < 0$. Then, by (5.31), $P_{n+1}(\lambda')P_{n+1}(\lambda'') < 0$. Consequently P_{n+1} has at least one zero on the interval (λ', λ''). ▹

This corollary can also be deduced from (5.21) and the positivity of the Christoffel coefficients.

If we take account of (5.22), (5.28), and (5.30), we obtain

$$\mu_{n,j} = \frac{P_n(\lambda_{n,j})}{Q'_n(\lambda_{n,j})} = \frac{h_{n-1}P_n(\lambda_{n,j})Q_{n-1}(\lambda_{n,j})}{h_{n-1}Q'_n(\lambda_{n,j})Q_{n-1}(\lambda_{n,j})} = \left(\sum_{k=0}^{n-1} Q_k(\lambda_{n,j})^2\right)^{-1}.$$

We again obtain (5.24).

REMARKS

1. On positive sequences, see [8], [11], and [55].
2. For measure theory, see the textbooks [51] and [64], and the monograph [110].
3. The term "support of a measure" is often used to denote its spectrum. We are using the term in a different sense.
4. The book [87] is a basic monograph on orthogonal polynomials.
5. For recurrence relations connected with orthogonal polynomials (considered as an independent subject for research), see [7].
6. For Chebyshev's work, see [114].

PROBLEMS

1. Prove that $T(\lambda) = A(\lambda)^2 + \lambda B(\lambda)^2$, where A and B are real polynomials, is the general form of polynomials that are nonnegative on $[0, +\infty)$. *Hint*: Use induction on the degree of T.

2. Prove that $T(\lambda) = \lambda A(\lambda)^2 + (1-\lambda)B(\lambda)^2$ and $T(\lambda) = A(\lambda)^2 + \lambda(1-\lambda)B(\lambda)^2$, where A and B are real polynomials, is the general form of polynomials of odd and even degrees, respectively, that are positive on $[0, 1]$. *Hint*: Make the change of variable $\lambda \to \lambda/(1+\lambda)$ and use the preceding problem.

3. Prove the following formula for the Hankel determinants constructed from the moments of the measure μ:

$$H_n = \frac{1}{n!}\int_{-\infty}^{+\infty} \cdots \int_{-\infty}^{+\infty} W(t_1, \ldots, t_n)^2 \, d\mu(t_1)\cdots d\mu(t_n),$$

where

$$W(t_1, \ldots, t_n) = \begin{vmatrix} 1 & 1 & \cdots & 1 \\ t_1 & t_2 & \cdots & t_n \\ \cdots & \cdots & \cdots & \cdots \\ t_1^{n-1} & t_2^{n-1} & \cdots & t_n^{n-1} \end{vmatrix}$$

is a Vandermonde determinant.

§6. Markov's theorem

As we have seen, the nth diagonal Padé approximant π_n of the formal power series (3.1) is its best approximation in the class of rational functions of degree at most n (best with respect to the norm $\mathbf{C}((1/z))$), and the sequence

$\{\pi_n\}$ converges to this series as $n \to \infty$, in the same norm. Even these very simple properties of Padé approximants make it possible to obtain nontrivial results on formal power series. One such result is the theorem of Kronecker that was proved in §3. Later (in §7) we shall discuss the application of Padé approximants to the moment problem. However, these approximants are chiefly of interest as a tool for approximating analytic functions, generalizing such classical analytic machinery as Taylor (or Laurent) series. Although we know that the domain of convergence of a Taylor series is the largest disk of holomorphy of the function that is being approximated, it is not possible to obtain such a simple result for Padé approximants. The study of their convergence encounters essential difficulties. One of the first, and up to now one of the few, general results on convergence was obtained by A. A. Markov.

1. Let μ be a measure with the compact support $\Delta = \Delta_\mu = [a, b]$. Then all its power moments $\{s_n\}_0^\infty$ exist. The function

$$\hat{\mu}(z) = \int_\Delta \frac{d\mu(\lambda)}{z - \lambda} \tag{6.1}$$

is well defined and holomorphic in the domain $\overline{\mathbf{C}} \backslash \Delta$ (even outside the spectrum S_μ) and $\hat{\mu}(\infty) = 0$; (6.1) is known as the *Markov* function of μ.

We expand the Cauchy kernel in a series

$$\frac{1}{z - \lambda} = \sum_{n=0}^{\infty} \frac{\lambda^n}{z^{n+1}};$$

this series converges absolutely and uniformly for $\lambda \in \Delta$ and for z belonging to a compact subset (in the topology of $\overline{\mathbf{C}}$) of the domain $|z| > R = \max(|a|, |b|)$. If we integrate the series termwise with respect to μ, we obtain

$$\hat{\mu}(z) = \sum_{n=0}^{\infty} \frac{s_n}{z^{n+1}}, \qquad |z| > R. \tag{6.2}$$

Thus the power series on the right-hand side of (6.2) is the Laurent series, centered at infinity, for $\hat{\mu}(z)$, and converges to it in the specified domain. We denote the diagonal Padé approximants of (6.2) by π_n.

THEOREM 6.1 (A. A. Markov). *Let μ have the compact support Δ and let $\hat{\mu}$ be its Markov function. Then*

$$\pi_n(z) \rightrightarrows \hat{\mu}(z), \qquad n \to \infty,$$

uniformly in the domain $\overline{\mathbf{C}} \backslash \Delta$.

REMARK. If μ has a finite number of points of increase, $\hat{\mu}$ is a rational function; in that case, the theorem is trivial. Therefore we shall prove the theorem for the significant case, when S_μ is infinite. We may suppose, without loss of generality, that $\mu(\Delta) = 1$.

By convergence (or boundedness) uniformly inside a domain, we shall understand uniform convergence (or boundedness) on every compact subdomain of that domain.

Since the approximants π_n in Theorem 6.1 converge to $\hat{\mu}$ in the domain $\overline{\mathbf{C}}\backslash\Delta$, which is larger than the domain of convergence of the Laurent series, they solve the problem of analytic continuation of the Laurent series into this domain.

PROOF OF THE THEOREM. The rational functions π_n have their poles on the interval Δ and are therefore holomorphic in $G = \overline{\mathbf{C}}\backslash\Delta$.

Let us prove that the sequence $\{\pi_n\}$ is uniformly bounded inside G. Let K be a compact subset of G, and let

$$\eta = \operatorname{dist}(K, \Delta) = \min\{|z-\lambda|: z \in K, \lambda \in \Delta\} > 0$$

be the distance between K and Δ. Then, by the formula (5.21) and Proposition 5.5, we obtain, for every $z \in K$,

$$|\pi_n(z)| = \left|\sum_{j=1}^{n} \frac{\mu_{n,j}}{z-\lambda_{n,j}}\right| \le \sum_{j=1}^{n} \frac{\mu_{n,j}}{|z-\lambda_{n,j}|} \le \frac{1}{\eta}\sum_{j=1}^{n}\mu_{n,j} = \frac{1}{\eta}. \tag{6.3}$$

This establishes the boundedness of $\{\pi_n\}$.

By Montel's theorem, the boundedness of the sequence inside G implies that it is precompact in the space of functions that are holomorphic in G (with the topology of convergence inside G). Consequently, in order to prove that $\{\pi_n\}$ converges to $\hat{\mu}$, it is enough to establish that $\hat{\mu}$ is the only limit point of this sequence. Let g be any limit point of $\{\pi_n\}$, i.e., let g be the limit of some subsequence $\{\pi_{n_k}\}$. For any $j \in \mathbf{Z}_+$ and for all sufficiently large indices k, the coefficient of $1/z^{j+1}$ in the Laurent series of π_{n_k}, centered at infinity, is equal to s_j. Consequently the Laurent series of g coincides with (6.2), and $g = \hat{\mu}$. ▷

2. Let us present another proof of Markov's theorem.

For $n \in \mathbf{N}$, let

$$\mu_n = \sum_{j=1}^{n} \mu_{n,j}\,\delta_{\lambda_{n,j}} \tag{6.4}$$

be a measure with finitely many points $\lambda_{n,1}, \dots, \lambda_{n,n}$ of increase (these are the poles of π_n) belonging to Δ, at which masses $\mu_{n,1}, \dots, \mu_{n,n}$ are concentrated ($\mu_{n,j}$ are the Christoffel coefficients), and

$$\mu_n(\Delta) = \int_\Delta d\mu_n(\lambda) = 1.$$

The function π_n can be expressed in the form

$$\pi_n(z) = \hat{\mu}_n(z) = \int_\Delta \frac{d\mu_n(\lambda)}{z-\lambda}, \tag{6.5}$$

and the quadrature formula (5.23) in the form

$$\int_{\Delta} T(\lambda)\, d\mu(\lambda) = \int_{\Delta} T(\lambda)\, d\mu_n(\lambda) \tag{6.6}$$

(T is an arbitrary polynomial of degree at most $2n-1$).

We denote by $C(\Delta)$ the Banach space of complex-valued functions that are defined and continuous on the interval Δ, with the norm defined by the formula

$$\|\varphi\|_{\Delta} = \max\{|\varphi(\lambda)| : \lambda \in \Delta\}$$

(the uniform norm). For a measure ν on Δ we shall use the notation

$$\nu(\varphi) = \int_{\Delta} \varphi(\lambda)\, d\nu(\lambda), \qquad \varphi \in C(\Delta).$$

Let us show that the sequence of measures μ_n converges weakly to μ as $n \to \infty$, i.e., for every function $\varphi \in C(\Delta)$ we have $\mu_n(\varphi) \to \mu(\varphi)$ as $n \to \infty$. We fix φ and select an arbitrary positive number ε. By Weierstrass's theorem, the set of polynomials is dense in $C(\Delta)$ (by the compactness of Δ); consequently there is a polynomial T such that $\|\varphi - T\|_{\Delta} < \varepsilon$. We have

$$|\mu_n(\varphi) - \mu(\varphi)| \le |\mu_n(\varphi - T)| + |\mu(\varphi - T)| + |\mu_n(T) - \mu(T)|.$$

Since μ_n and μ are unit measures, we have

$$|\mu_n(\varphi - T)| \le \|\varphi - T\|_{\Delta} < \varepsilon,$$
$$|\mu(\varphi - T)| \le \|\varphi - T\|_{\Delta} < \varepsilon.$$

According to (6.6), for sufficiently large n we have $\mu_n(T) - \mu(T) = 0$, so that $|\mu_n(\varphi) - \mu(\varphi)| < 2\varepsilon$.

For $z \in G$, the function $\varphi_z(\lambda) = 1/(z-\lambda)$, $\lambda \in \Delta$, is continuous on Δ. Therefore

$$\hat{\mu}_n(z) = \mu_n(\varphi_z) \to \mu(\varphi_z) = \hat{\mu}(z) \tag{6.7}$$

as $n \to \infty$. By (6.5), this means that $\pi_n(z) \to \hat{\mu}(z)$. That the limit (6.7) is uniform for $z \in K$, where K is compact in G, follows from the compactness of the family $\Phi_K = \{\varphi_z : z \in K\}$ in the space $C(\Delta)$. Let $\eta = \operatorname{dist}(K, \Delta) > 0$, and select $\varepsilon > 0$. Let $Z = \{z_m\}_1^M$ be an ε-net on K. If $z', z'' \in K$, $|z' - z''| < \varepsilon$, we have

$$|\varphi_z(\lambda) - \varphi_{z''}(\lambda)| = \left| \frac{z'' - z'}{(z'-\lambda)(z''-\lambda)} \right| < \frac{\varepsilon}{\eta^2}, \qquad \lambda \in \Delta.$$

Consequently, $\{\varphi_{z_M}\}_1^M$ is an $\varepsilon\eta^{-2}$-net on Φ_K. By what was proved above, for every m, $1 \le m \le M$, there is a number N_m such that for all $n > N_m$ the inequality

$$|\mu_n(\varphi_{z_m}) - \mu(\varphi_{z_m})| < \varepsilon$$

is satisfied. Set $N = \max\{N_m : m = 1, \dots, M\}$. For $n > N$ and $\varphi \in \Phi_K$, there are numbers m such that $\|\varphi - \varphi_{z_m}\|_\Delta < \varepsilon\eta^{-2}$. Consequently

$$\begin{aligned}|\mu_n(\varphi) - \mu(\varphi)| &\le |\mu_n(\varphi - \varphi_{z_m})| + |\mu(\varphi - \varphi_{z_m})| + |\mu_n(\varphi_{z_m}) - \mu(\varphi_{z_m})| \\ &\le 2\|\varphi - \varphi_{z_m}\|_\Delta + \varepsilon < 2\varepsilon\eta^{-2} + \varepsilon. \quad \triangleright\end{aligned}$$

3. As a complement to Markov's theorem, we shall find an estimate for the degree of convergence of the diagonal Padé approximants to the Markov function.

Let ψ_Δ map the domain $G = \overline{\mathbf{C}}\backslash\Delta$ conformally on the exterior of the unit disk so that $\psi_\Delta(\infty) = \infty$ and $\psi'_\Delta(\infty) > 0$ (this function exists, by the Riemann mapping theorem). Then

$$\psi_\Delta(z) = \psi\left(\frac{z - (b+a)/2}{(b-a)/2}\right),$$

where $\Delta = [a, b]$, and $\psi = \psi_{[-1,1]}$ is the inverse of the Zhukovskiĭ function: $\psi(z) = z + \sqrt{z^2 - 1}$, with $\sqrt{z^2 - 1} > 0$ for $z > 1$. Let

$$L_R = \{z \in \mathbf{C} : |\psi_\Delta(z)| = R\}, \qquad R > 1,$$

be a level line of $|\psi_\Delta|$. This is an ellipse with foci at a and b and the sum of its semi-axes equal to $\frac{1}{2}(b-a)R$. Every point $z \in G$ lies outside an ellipse L_R, $R > 1$. We shall, as usual, denote by $\|\varphi\|_{L_R}$ the maximum modulus of φ on the compact set L_R.

THEOREM 6.2. *We have the inequality*

$$\overline{\lim_{n\to\infty}} \|\hat{\mu} - \pi_n\|_{L_R}^{1/n} < \frac{1}{R^2}, \qquad R > 1. \tag{6.8}$$

PROOF. The function $\hat{\mu} - \pi_n$ is holomorphic in G and has a zero of order $2n+1$ at infinity. The function ψ_Δ is holomorphic in $G\backslash\{\infty\}$ and has a simple pole at infinity. Consequently $(\hat{\mu} - \pi_n)\psi_\Delta^{2n}$ is holomorphic in G.

Let us fix $R > 1$ and take $\delta \in (1, R)$. If we use (6.3) (and a similar inequality for $\hat{\mu}$), we find that the inequality

$$|(\hat{\mu} - \pi_n)(z)(\psi_\Delta(z))^{2n}| \le \frac{2}{c(\delta)}\delta^{2n} \tag{6.9}$$

is satisfied for all $z \in L_\delta$, where

$$c(\delta) = \operatorname{dist}(L_\delta, \Delta) = \frac{1}{2}\left(\sqrt{R} - \frac{1}{\sqrt{R}}\right)^2 \frac{b-a}{2} > 0.$$

By the maximum principle for analytic functions, (6.9) is valid for all z outside L_δ and, in particular, for all $z \in L_R$, i.e.,

$$\|\hat{\mu} - \pi_n\|_{L_R} \le \frac{2}{c(\delta)}\left(\frac{\delta}{R}\right)^{2n}. \tag{6.10}$$

If we take nth roots on both sides of (6.10) and then take limits as $n \to \infty$, we obtain

$$\varlimsup_{n\to\infty} \|\hat{\mu} - \pi_n\|_{L_R}^{1/n} \le \left(\frac{\delta}{R}\right)^2 . \tag{6.11}$$

Since $\delta \in (1, R)$ is arbitrary, (6.8) follows from (6.11). ▷

REMARKS

1. Many papers discuss generalizations of Markov's theorem. These generalizations have been connected, in particular, with the passage to multipoint Padé approximants, and with replacing $\hat{\mu}(z)$ by $\hat{\mu}(z) + R(z)$, where R is a rational function. A few such papers are [33], [35], [36], and [82].

2. In this book we consider mainly the diagonal Padé approximants; hence we have omitted the theory of the rows of the Padé table and their generalizations. In this connection, we mention the following papers: [34], [19], [101], [102], [162], and [1], as well as the first papers by Hadamard [136] and Montessus de Ballore [150].

3. We do not consider complex measures and similar questions. We mention only [152] and [158].

PROBLEMS

1. Prove that

1) if $d\mu(x) = dx$, $-1 < x < 1$, then

$$\hat{\mu}(z) = \ln\left(\frac{z-1}{z+1}\right), \qquad z \in \overline{\mathbf{C}}\setminus[-1, 1],$$

with the principal branch of the logarithm;

2) if

$$d\mu(x) = \frac{1}{\pi}\frac{dx}{\sqrt{1-x^2}}, \qquad -1 < x < 1,$$

then

$$\hat{\mu}(z) = \frac{1}{\sqrt{z^2-1}}, \qquad z \in \overline{\mathbf{C}}\setminus[-1, 1]$$

with the principal branch;

3) if

$$d\mu(x) = \frac{1}{\mathrm{B}(a, b-a)} x^{a-1}(1-x)^{b-a-1}\, dx, \qquad 0 < x < 1,$$

where $b > a > 0$ and B is the beta function, then

$$\hat{\mu}(z) = \frac{1}{z}\, {}_1F_1\left(1, a, b; \frac{1}{z}\right), \qquad z \in \overline{\mathbf{C}}\setminus[0, 1],$$

where

$${}_1F_1(\alpha, \beta, \gamma; x) = \sum_{n=0}^{\infty} \frac{(\alpha)_n(\beta)_n}{(\gamma)_n}\frac{x^n}{n!}$$

is the hypergeometric function; here $(\alpha)_n = \alpha(\alpha+1)\cdots(\alpha+n-1)$ and $(\alpha)_0 = 1$;

4) if

$$d\mu(x) = \frac{1}{\Gamma(\alpha)}\left(\ln\frac{1}{x}\right)^{\alpha-1} dx, \qquad 0 < x < 1,$$

where $\alpha > 0$ and Γ is the gamma function, then

$$\hat{\mu}(z) = L_\alpha\left(\frac{1}{z}\right), \qquad z \in \overline{\mathbf{C}}\setminus[0, 1],$$

where

$$L_\alpha(x) = \sum_{n=1}^{\infty} \frac{x^n}{n^\alpha}$$

is the polylogarithmic function.

§7. The moment problem and extremal properties of orthogonal polynomials

1. Let F be an infinite closed subset of $\mathbf{R}$. Let there be given a sequence $s = \{s_n\}_0^\infty$ of real numbers. It is required to find a measure $\mu \in \mathscr{M}(F)$ (see §5, subsection 2, for notation) such that

$$s_n = \int_F \lambda^n \, d\mu(\lambda), \qquad n \in \mathbf{Z}_+ .$$

This is known as the *power moment problem* (for F). We distinguish the three classical cases: 1) $F = (-\infty, \infty)$, *the Hamburger moment problem*; 2) $F = [0, \infty)$, *the Stieltjes moment problem*; and 3) $F = [0, 1]$, *the Hausdorff moment problem.*

Two questions arise in connection with solutions of moment problems:

A) the question of the existence of a solution (solvability problem);
B) the question of the uniqueness of the solution (determination problem).

We restrict ourselves to the first problem.

THEOREM 7.1 (Hamburger). *A necessary and sufficient condition for the Hamburger moment problem to have a solution is that the sequence s is positive.*

PROOF. The necessity was proved in §5, subsection 2. Let us prove the sufficiency. Let Q_n be the polynomials orthogonal with respect to s, $\lambda_{n,j}$ the zeros of Q_n, and $\mu_{n,j}$ the Christoffel coefficients; μ_n is defined in (6.4). For $f \in \mathbf{Z}_+$, the moments

$$\int_{-\infty}^{+\infty} \lambda^j \, d\mu_n(\lambda) \tag{7.1}$$

are equal to s_j for all sufficiently large n; in particular, the sequence (7.1) is bounded for all j. By the theorem on the weak compactness of balls in the dual space, we can extract from the sequence $\{\mu_n\}$ of measures a subsequence $\{\mu_{n_k}\}$ that converges weakly to a measure μ. This means that for every function φ that is continuous on $\mathbf{R}$ and has limits as $\lambda \to \pm\infty$,

$$\int_{-\infty}^{+\infty} \varphi(\lambda) \, d\mu_{n_k}(\lambda) \to \int_{-\infty}^{+\infty} \varphi(\lambda) \, d\mu(\lambda), \qquad k \to \infty. \tag{7.2}$$

In addition, we may take $\varphi(\lambda) = \lambda^j$ for every $j \in \mathbf{Z}_+$. Then s_j is the jth power moment of μ. It is clear that μ has an infinite number of points of increase, since otherwise the Hankel determinants H_n would vanish for all large n, by Kronecker's theorem. ▷

2. In §6 we associated an arbitrary measure μ with compact support with a Markov function $\hat{\mu}(z)$ whose Laurent series with center at infinity is

$$f(z) = \sum_{n=0}^{\infty} \frac{s_n}{z^{n+1}}, \tag{7.3}$$

where

$$s_n = \int_{-\infty}^{+\infty} \lambda^n \, d\mu(\lambda), \qquad n \in \mathbf{Z}_+.$$

Let $\mu \in \mathscr{M}$; the support of μ can be unbounded. Then the function

$$\hat{\mu}(z) = \int_{-\infty}^{+\infty} \frac{d\mu(\lambda)}{z - \lambda} \tag{7.4}$$

is defined and holomorphic in the upper halfplane $\operatorname{Im} z > 0$. This function is the *Cauchy transform* of μ. We note that $\operatorname{Im} \hat{\mu}(z) < 0$ for $\operatorname{Im} z > 0$. By the Stieltjes inversion formula, the Cauchy transform establishes a one-to-one correspondence between the measures μ of class $\mathscr{M}$ and the functions of the form (7.4). In particular, it follows that the Hausdorff moment problem is always determined, i.e., it has a solution, and the solution must be unique. In fact, for all measures μ that are solutions of the Hausdorff moment problem, the Markov functions $\hat{\mu}$ coincide, since they have the same Laurent expansion; consequently the solutions themselves coincide.

If μ has an unbounded support, the series (7.3) diverges for every finite z. However, in this case it provides a means of approximating $\hat{\mu}$; in fact its asymptotic expansion, as $z \to \infty$, in every sector of the form

$$\varepsilon < \arg z < \pi - \varepsilon, \qquad 0 < \varepsilon < \pi/2, \tag{7.5}$$

i.e., for every $N \in \mathbf{Z}_+$, is

$$\hat{\mu}(z) - \sum_{n=0}^{N-1} \frac{s_n}{z^{n+1}} = O(1/z^{N+1})$$

as $z \to \infty$ in any sector (7.5) (here O depends on ε). For a proof of this assertion, we write the Cauchy kernel in the form

$$\frac{1}{z - \lambda} = \sum_{n=0}^{N-1} \frac{\lambda^n}{z^{n+1}} + \left(\frac{\lambda}{z}\right)^N \frac{1}{z - \lambda}$$

and integrate both sides of this equation with respect to μ. We obtain the required result by using the inequality $|z - \lambda| \geq |z| \sin \varepsilon$, which holds for all z in the sector (7.5).

Let $\{Q_n\}_0^\infty$ be the polynomials orthonormal with respect to a measure $\mu \in \mathscr{M}$, $\{P_n\}_0^\infty$ the polynomials of the second kind defined by (5.7), and $\hat{\mu}$ the Cauchy transform of μ. The functions

$$R_n(z) = Q_n(z)\hat{\mu}(z) - P_n(z), \qquad n \in \mathbf{Z}_+, \tag{7.6}$$

which are holomorphic in the upper halfplane, will be called *functions of the second kind.*

By (7.6) and (5.7), we have

$$R_n(z) = \int_{-\infty}^{+\infty} \frac{Q_n(\lambda)}{z-\lambda}\, d\mu(\lambda), \qquad n \in \mathbf{Z}_+. \tag{7.7}$$

3. THEOREM 7.2. *A sufficient condition for the Hamburger moment problem to be determined is that the series*

$$\sum_{k=0}^{\infty} |Q_k(z)|^2 = \infty \tag{7.8}$$

diverges for some z *with* $\operatorname{Im} z > 0$; *and a necessary condition is that the series* (7.8) *diverges for every* z *with* $\operatorname{Im} z > 0$.

Let us establish the sufficiency of (7.8). We first prove a lemma.

LEMMA 7.1. *Let* $\mu \in \mathscr{M}$, *let* $\{R_k\}_0^\infty$ *be the functions of the second kind, and let* $\operatorname{Im} z > 0$. *Then*

$$\sum_{k=0}^{\infty} |R_k(z)|^2 < +\infty.$$

PROOF. The function $\varphi(\lambda) = 1/(z-\lambda)$, $\lambda \in \mathbf{R}$, is continuous and bounded on the real axis; consequently it belongs to the Hilbert space L_μ^2. The sequence $\{Q_k\}_0^\infty$ is an orthonormal sequence in this space (not necessarily complete). By (7.7), the numbers $R_k(z)$ are the Fourier coefficients of φ in this system. Consequently, if we apply Bessel's inequality, we obtain

$$\sum_{k=0}^{\infty} |R_k(z)|^2 \le \int_{-\infty}^{+\infty} \frac{d\mu(\lambda)}{|z-\lambda|^2} < +\infty. \quad \triangleright$$

COROLLARY. *Let* $\operatorname{Im} z > 0$. *Then the series*

$$\sum_{k=0}^{\infty} |Q_k(z)|^2, \qquad \sum_{k=0}^{\infty} |P_k(z)|^2$$

converge (*or diverge*) *simultaneously.*

PROOF. By (7.6), each of the three sequences $\{Q_k(z)\}_0^\infty$, $\{P_k(z)\}_0^\infty$, and $\{R_k(z)\}_0^\infty$ is a linear combination of the other two, and the last one belongs to the Hilbert space l_2 (by Lemma 7.1). $\triangleright$

Let s be a positive sequence, and $\mathfrak{M}(s)$ the set of solutions of the corresponding Hamburger moment problem. Let us set

$$K(z) = \{\hat{\mu}(z) \colon \mu \in \mathfrak{M}(s)\}, \qquad \operatorname{Im} z > 0. \tag{7.9}$$

We shall not investigate the structure of these sets (in fact the $K(z)$ are disks). We merely note that if, for every z in the upper halfplane, the set (7.9) consists of a single point, the moment problem is determined. In fact, let μ', μ'' be solutions of the moment problem. Then $\hat{\mu}'(z) = \hat{\mu}''(z)$, $\operatorname{Im} z > 0$. Therefore, by the inversion formula, we obtain $\mu' = \mu''$.

LEMMA 7.2. *If the moment problem is not determined, there is a* z *with* $\operatorname{Im} z > 0$ *such that*

$$\sum_{k=0}^{\infty} |Q_k(z)|^2 < +\infty, \qquad \sum_{k=0}^{\infty} |P_k(z)|^2 < +\infty.$$

PROOF. By the above remark, there is a z in the upper halfplane such that the set $K(z)$ contains two different points ω_1 and ω_2. By Lemma 7.1, the sequences

$$\{\omega_j Q_k(z) + P_k(z)\}_0^{\infty}, \qquad j = 1, 2, \tag{7.10}$$

belong to l_2. The sequences $\{Q_k(z)\}_0^{\infty}$ and $\{P_k(z)\}_0^{\infty}$ are linear combinations of the sequences (7.10), and therefore also belong to l_2. ▷

LEMMA 7.3. *Let* $c = \{c_n\}_0^{\infty} \in l_2$; *let* $a_{n,k} \in \mathbf{C}$, $k = 0, \dots, n-1$, $n = 1, 2, \dots$, *have the property that*

$$\sum_{n=1}^{\infty} \sum_{k=0}^{n-1} |a_{n,k}|^2 < +\infty;$$

and let the sequence $\eta = \{\eta_n\}_0^{\infty}$ *be defined by the recurrence formulas*

$$\eta_n = c_n + \zeta \sum_{k=0}^{n-1} a_{n,k} \eta_k, \qquad n \in \mathbf{Z}_+, \tag{7.11}$$

where ζ *is a complex number. Then* $\eta \in l_2$.

PROOF. We write c and η as infinite columns. Let

$$A = \begin{pmatrix} 0 & 0 & 0 & \cdots \\ a_{1,0} & 0 & 0 & \cdots \\ a_{2,0} & a_{2,1} & 0 & \cdots \\ \cdots & \cdots & \cdots & \cdots \end{pmatrix}$$

be an infinite matrix. Then, in matrix form, (7.11) can be expressed in the following way:

$$\eta = c + \zeta A \eta. \tag{7.12}$$

The matrix A generates a bounded operator on l_2. (In what follows, a matrix and the corresponding operator will be denoted by the same letter.) If we multiply (7.12) on the left by A, we obtain

$$A\eta = Ac + \zeta A^2 \eta.$$

We now substitute this into (7.12):

$$\eta = c + \zeta Ac + \zeta^2 A^2 \eta.$$

By induction, we obtain the formula

$$\eta - (\zeta A)^{n+1} \eta = \left(\sum_{k=0}^{n} (\zeta A)^k \right) c. \tag{7.13}$$

Let us estimate $\|A^k\|$ (here $\|\cdot\|$ denotes the norm of an operator in l_2). Let A_m be the matrix obtained from A by deleting the first m rows and the first $m-1$ columns. Note that the first k rows of A^k are zero. Let $\widetilde{A}^k$ be obtained from A^k by deleting these rows. Then it is easily verified, by induction on k, that

$$\widetilde{A}^k = A_k \cdots A_2 A_1 .$$

Consequently

$$\|A^k\| = \|\widetilde{A}^k\| \le \|A_k\| \cdots \|A_2\| \, \|A_1\| .$$

We set

$$\alpha_n = \sum_{k=0}^{n-1} |a_{n,k}|^2 , \qquad r_m = \sum_{n=m}^{\infty} \alpha_n .$$

By the hypotheses of the lemma, r_m is the remainder of a convergent series; consequently, $r_m \to 0$. Since the square of the norm of the operator on l_2 that is generated by the matrix does not exceed the sum of the squares of its elements, we have $\|A_m\|^2 \le r_m$ and

$$\|A^k\|^2 \le r_k \cdots r_2 r_1 .$$

The series

$$\sum_{k=0}^{\infty} |\zeta|^k r_1 \cdots r_k$$

converges, by D'Alembert's test; consequently the series

$$\sum_{k=0}^{\infty} (\zeta A)^k = (I - \zeta A)^{-1}$$

converges in the operator norm (I denotes the identity operator). If we use this formula and the fact that the mth component of the column $(\zeta A)^{n+1}\eta$ is zero for $n \ge m$, we can take the limit as $n \to \infty$ in (7.13). We obtain

$$\eta = (I - \zeta A)^{-1} c \in l_2 . \quad \triangleright$$

LEMMA 7.4. *If the series*

$$\sum_{k=0}^{\infty} |Q_k(z)|^2 < +\infty \tag{7.14}$$

converges for some $z = z_0$ *with* $\operatorname{Im} z_0 > 0$, *then it converges for every complex* z.

PROOF. We have

$$\frac{Q_n(z) - Q_n(z_0)}{z - z_0} = \sum_{k=0}^{n-1} a_{n,k} Q_k(z), \qquad n \in \mathbf{Z}_+ , \tag{7.15}$$

or equivalently

$$Q_n(z) = Q_n(z_0) + (z - z_0)\sum_{k=0}^{n-1} a_{n,k} Q_k(z). \tag{7.16}$$

The coefficients in (7.15) are determined by the Fourier formulas

$$\begin{aligned} a_{n,k} &= \int_{-\infty}^{+\infty} \frac{Q_n(\lambda) - Q_n(z_0)}{\lambda - z_0} Q_k(\lambda)\, d\mu(\lambda) \\ &= Q_k(z_0) \int_{-\infty}^{+\infty} \frac{Q_n(\lambda) - Q_n(z_0)}{\lambda - z_0}\, d\mu(\lambda) \\ &\quad + \int_{-\infty}^{+\infty} (Q_n(\lambda) - Q_n(z_0)) \frac{Q_k(\lambda) - Q_k(z_0)}{\lambda - z_0}\, d\mu(\lambda) \\ &= Q_k(z_0)P_n(z_0) - Q_n(z_0)P_k(z_0), \end{aligned}$$

since Q_n is orthogonal to polynomials of lower degree. Then

$$\begin{aligned} \sum_{n=1}^{\infty}\sum_{k=0}^{n-1} |a_{n,k}|^2 &\le 2\left(\sum_{n=1}^{\infty}\sum_{k=0}^{n-1} |Q_k(z_0)|^2 |P_n(z_0)|^2 + \sum_{n=1}^{\infty}\sum_{k=0}^{n-1} |Q_n(z_0)|^2 |P_k(z_0)|^2\right) \\ &\le 4\sum_{n=0}^{\infty} |Q_n(z_0)|^2 \sum_{k=0}^{\infty} P_k(z_0)|^2 < +\infty. \end{aligned}$$

Consequently, if we apply Lemma 7.3 to (7.16), we obtain the result. ▹

PROOF OF THE SUFFICIENCY OF THE HYPOTHESIS OF THEOREM 7.2. Suppose, on the contrary, that the moment problem is undetermined. Then, by Lemma 7.2, the series (7.14) converges for some $z = z_0$ in the upper half-plane. Hence, by Lemma 7.4, it will converge for all complex z. But this contradicts the hypothesis of the theorem. ▹

4. To establish the necessity of the hypothesis of Theorem 7.2, we need some extremal properties of orthogonal polynomials.

We first establish a simple but fundamental extremal property that is inherent in all orthonormal systems.

PROPOSITION 7.1. *Let $\mu \in \mathcal{M}$, and let T run through the set of polynomials of degree $n \in \mathbf{Z}_+$ with unit leading coefficient. Then*

$$\inf_T \int_{-\infty}^{+\infty} |T(\lambda)|^2\, d\mu(\lambda) = \alpha_n^{-2}, \tag{7.17}$$

where the infimum in (7.17) *is attained only for $T(\lambda) = \frac{1}{\alpha_n} Q_n(\lambda) = \lambda^n + \cdots$.*

PROOF. We expand T in terms of the orthonormal system $\{Q_k\}$:

$$T = \frac{1}{\alpha_n} Q_n + \sum_{k=0}^{n-1} a_k Q_k.$$

By the Parseval equality we have

$$\int_{-\infty}^{+\infty} |T(\lambda)|^2 \, d\mu(\lambda) = \frac{1}{\alpha_n^2} + \sum_{k=0}^{n-1} |a_k|^2. \tag{7.18}$$

The quantity (7.18) attains a minimum when $a_k = 0$, $k = 0, \dots, n-1$. ▷

We now consider a more complicated extremal problem.

PROPOSITION 7.2. *Let $\mu \in \mathscr{M}$, and let T run through the set of polynomials of degree at most $n \in \mathbf{Z}_+$ for which $T(z) = 1$, for a given complex number z. Then*

$$\inf_T \int_{-\infty}^{+\infty} |T(\lambda)|^2 \, d\mu(\lambda) = \rho_n(z), \tag{7.19}$$

where

$$\rho_n(z) = \left(\sum_{k=0}^{n} |Q_k(z)|^2 \right)^{-1},$$

and the infimum is attained in (7.19) *only for $T = T_z$, where*

$$T_z(\lambda) = \frac{\sum_{k=0}^{n} \overline{Q_k(z)} Q_k(\lambda)}{\sum_{k=0}^{n} |Q_k(z)|^2}.$$

PROOF. We expand T in terms of the orthonormal system $\{Q_k\}$:

$$T = \sum_{k=0}^{n} b_k Q_k.$$

By the Parseval equality

$$t \equiv \int_{-\infty}^{+\infty} |T(\lambda)|^2 \, d\mu(\lambda) = \sum_{k=0}^{n} |b_k|^2.$$

By the Cauchy-Bunyakovskiĭ inequality

$$1 = \sum_{k=0}^{n} b_k Q_k(z) \le \left(\sum_{k=0}^{n} |b_k|^2 \right)^{1/2} \left(\sum_{k=0}^{n} |Q_k(z)|^2 \right)^{1/2}, \tag{7.20}$$

i.e.,

$$t \ge \rho_n(z), \tag{7.21}$$

with equality in (7.20), and therefore in (7.21), only for $b_k = \rho_n(z)\overline{Q_k(z)}$, $k = 0, \dots, n$. ▷

We now denote by $\mathfrak{M}_n(s)$, $n \in \mathbf{Z}_+$, the set of solutions of the truncated moment problem, that is, the set of the measures μ for which

$$\int_{-\infty}^{+\infty} \lambda^k \, d\mu(\lambda) = s_k, \qquad k = 0, \dots, n.$$

REMARK. Since everything in Proposition 7.2 depends only on the moments $s_0, \dots, s_{2n}$, the proposition remains valid if μ is any measure in $\mathfrak{M}_{2n}(s)$.

LEMMA 7.5. *Let* $x \in \mathbf{R}$, $n \in \mathbf{Z}_+$. *Then*

$$\sup\{\mu\{x\} : \mu \in \mathfrak{M}_{2n}(s)\} \le \rho_n(x).$$

PROOF. According to Proposition 7.2 and the subsequent remark, for every measure $\mu \in \mathfrak{M}_{2n}(s)$ we have

$$\rho_n(x) = \int_{-\infty}^{+\infty} |T_x(\lambda)|^2 \, d\mu(\lambda) \ge \mu\{x\}. \quad \triangleright$$

LEMMA 7.6. *Let* $x \in \mathbf{R}$, $n \in \mathbf{N}$. *Then*

$$\sup\{\mu\{x\} : \mu \in \mathfrak{M}_{2n-1}(s)\} \ge \rho_n(x). \tag{7.22}$$

PROOF. Let us suppose that x is a zero of Q_n. We define the measure μ_n by (6.4). We have $\mu_n \in \mathfrak{M}_{2n-1}(s)$ by the quadrature formula (5.23). By (5.4),

$$\mu_n\{x\} = \rho_{n-1}(x) = \rho_n(x).$$

The inequality (7.22) follows.

Now let $Q_n(x) \neq 0$. We define the polynomial

$$Q^*_{n+1}(\lambda) = \begin{vmatrix} s_0 & \cdots & s_{n+1} \\ \cdots & \cdots & \cdots \\ s_n & \cdots & s^*_{2n+1} \\ 1 & \cdots & \lambda^{n+1} \end{vmatrix},$$

where $s_0, \dots, s_{2n}$ are the given moments, and s^*_{2n+1} is a parameter which we find from the condition $Q^*_{n+1}(x) = 0$; we can do this since Q^*_{n+1} depends linearly on s^*_{2n+1}. The polynomial Q^*_{n+1} satisfies the same conditions as the $(n+1)$th orthogonal polynomial Q_{n+1}. In particular, it has $n+1$ simple real zeros $\lambda^*_{n,j}$, $j = 1, \dots, n+1$, and the measure

$$\mu^*_{n+1} = \sum_{j=1}^{n+1} \rho_n(\lambda^*_{n,j}) \delta_{\lambda^*_{n,j}}$$

belongs to $\mathfrak{M}_{2n}(s) \subset \mathfrak{M}_{2n-1}(s)$. Hence (7.22) again follows. $\triangleright$

PROPOSITION 7.3. *The maximum mass that can be concentrated at a given point* $x \in \mathbf{R}$ *in a mass distribution, corresponding to a solution of the moment problem, is equal to*

$$\rho(x) = \left(\sum_{n=0}^{\infty} |Q_n(x)|^2 \right)^{-1}.$$

PROOF. We denote the maximum mass by $M(x)$. It follows from Lemma 7.5 that $M(x) \le \rho(x)$. By Lemma 7.6, there is a sequence of measures $\nu_k \in \mathfrak{M}_{2n_k-1}(s)$, $n_1 < n_2 < \cdots$, that converges weakly to a solution ν of the moment problem and has the property that

$$\lim_{k \to \infty} \nu_k\{x\} \ge \rho(x).$$

Then for all $\alpha < x$ and $\beta > x$ that do not belong to Γ_ν, the set of points of concentration of mass of ν, we have

$$\nu(\alpha, \beta) = \lim_{k\to\infty} \nu_k(\alpha, \beta) \geq \rho(x).$$

From this, since α and β are arbitrary,

$$M(x) \geq \rho(x). \quad \rhd$$

PROOF OF THE NECESSITY OF THE HYPOTHESIS OF THEOREM 7.2. Let there exist a point z_0 in the upper halfplane for which $\rho(z_0) > 0$. Then, by Lemma 7.4, $\rho(x) > 0$ for all $x \in \mathbf{R}$. If the moment problem is determined, and μ is its solution, then by Proposition 7.3 we have $\mu\{x\} = \rho(x) > 0$, $x \in \mathbf{R}$, i.e., $\Gamma_\mu = \mathbf{R}$, contradicting the countability of Γ_μ. $\rhd$

5. We present some simple sufficient conditions for the moment problem to be determined, in terms of the moments themselves and of the coefficients of the recursive conditions.

THEOREM 7.3 (Carleman). *If*

$$\sum_{n=0}^{\infty} s_{2n}^{-1/2n} = \infty, \tag{7.23}$$

or

$$\sum_{n=0}^{\infty} \frac{1}{h_n} = \infty, \tag{7.24}$$

the moment problem is determined.

PROOF. Let the moment problem be undetermined. Then by Theorem 7.2, for every $\lambda \in \mathbf{C}$ the sequences $\{Q_n(\lambda)\}_0^\infty$ and $\{P_n(\lambda)\}_0^\infty$ belong to l_2. Consequently, by (5.30), the sequence $\{1/h_n\}_0^\infty$ belongs to l_1. This proves the sufficiency of (7.24).

Moreover, if we apply the Cauchy-Bunyakovskiĭ inequality we obtain

$$\begin{aligned} h_0 h_1 \cdots h_{n-1} &= \frac{1}{\alpha_n} = \int_{-\infty}^{+\infty} \lambda^n Q_n(\lambda)\, d\mu(\lambda) \\ &\leq \left(\int_{-\infty}^{+\infty} \lambda^{2n}\, d\mu(\lambda)\right)^{1/2} \left(\int_{-\infty}^{+\infty} Q_n(\lambda)^2\, d\mu(\lambda)\right)^{1/2} = \sqrt{s_{2n}}. \end{aligned}$$

Hence, by Carleman's inequality (see the Remarks) we obtain

$$\sum_{n=0}^{\infty} s_{2n}^{-1/2n} \leq \sum_{n=0}^{\infty} (h_0 \cdots h_{n-1})^{-1/n} \leq e \sum_{n=0}^{\infty} \frac{1}{h_n} < \infty.$$

This proves the sufficiency of (7.23). $\rhd$

REMARKS

1. For detailed presentations of moment problems, see [8], also [11] and [55].
2. CARLEMAN'S INEQUALITY. *Let* $u_1, u_2, \ldots$ *be positive numbers; then*

$$\sum_{n=1}^{\infty} (u_1 \cdots u_n)^{1/n} \leq e \sum_{n=1}^{\infty} u_n.$$

PROOF. Set

$$c_n = \frac{(n+1)^n}{n^{n-1}} < en, \qquad n = 1, 2, \ldots.$$

Using the inequality between the arithmetic and geometric means, we obtain

$$\sum_{n=1}^{\infty}(u_1 \cdots u_n)^{1/n} = \sum_{n=1}^{\infty} \frac{(c_1 u_1 \cdots c_n u_n)^{1/n}}{n+1} \leq \sum_{n=1}^{\infty} \frac{c_1 u_1 + \cdots + c_n u_n}{n(n+1)}$$
$$= \sum_{k=1}^{\infty} \sum_{n=k}^{\infty} \frac{c_k u_k}{n(n+1)} < e \sum_{k=1}^{\infty} u_k. \quad \triangleright$$

PROBLEMS

1. Prove (7.2) for polynomials.

Hint. Divide the real line into the parts $(-\infty, -N)$, $(-N, N)$, and (N, ∞), and apply the inequalities

$$\int_N^{\infty} \lambda^j \, d\mu_{n_k}(\lambda) \leq \frac{1}{N^{j+2}} \int_N^{\infty} \lambda^{2j+2} \, d\mu_{n_k}(\lambda) \leq \frac{s_{2j+2}}{N^{j+2}} \to 0$$

for $N \to \infty$, where $k > k(j)$.

2. Prove the Stieltjes-Perron inversion formula: *if*

$$f(z) = \int_{-\infty}^{+\infty} \frac{d\mu(\lambda)}{\lambda - z}, \qquad \operatorname{Im} z > 0, \tag{$*$}$$

where μ *is a finite positive measure on* $\mathbf{R}$, *then for all* a *and* b *with* $a < b$ *and* $a, b \notin \Gamma_\mu$,

$$\mu(a, b) = \lim_{\varepsilon \to 0+} \frac{1}{\pi} \int_a^b \operatorname{Im} f(x + i\varepsilon)\, dx. \tag{$**$}$$

Hint. Substitute $(*)$ into $(**)$, change the order of integration, and take the limit under the integral sign.

3. Let a sequence of measures μ_n on $\mathbf{R}$ converge weakly to μ for $a < b$, and let a and b not be mass concentration points for all the measures; then $\mu_n(a, b) \to \mu(a, b)$ as $n \to \infty$.

§8. Orthogonal polynomials and operator theory

1. In studying orthogonal polynomials in §5, we introduced the concept of Jacobi matrices, i.e., the matrices

$$\mathscr{A} = \begin{pmatrix} v_0 & h_0 & & 0 \\ h_0 & v_1 & h_1 & \\ & h_1 & v_2 & h_2 \\ 0 & \cdot & \cdot & \cdot \end{pmatrix}, \tag{8.1}$$

where the v_n are real and the h_n are positive numbers. Such a matrix determines the solvability of the Hamburger moment problem.

DEFINITION. A matrix $\mathscr{A}$ has *type* D ($\mathscr{A} \in$ (D)) if the moment problem is determined; *type* C ($\mathscr{A} \in$ (C)) in the opposite case.

(Let us explain the new terminology. If $\mathscr{A} \in$ (D), then (see §7, subsection 3) the set $K(z)$ is a point ("Dot"); if $\mathscr{A} \in$ (C), $K(z)$ is a disk ("Circle").)

If Q_n are orthonormal polynomials associated with $\mathscr{A}$, then $\mathscr{A} \in$ (D) is equivalent to the divergence of the series

$$\sum_{n=0}^{\infty} |Q_n(\lambda)|^2 \tag{8.2}$$

for $\operatorname{Im}\lambda > 0$; and $\mathscr{A} \in (\mathrm{C})$ to the convergence of the series.

We introduce an operator A, corresponding to the matrix $\mathscr{A}$, on the Hilbert space l_2. Let $\{e_n\}_0^\infty$ be the usual basis in l_2; $\mathscr{D}_0$ the dense linear manifold of essentially finite sequences (that is, having only finitely many nonzero terms); A_0 the linear operator, with domain $\mathscr{D}_0$,

$$A_0 e_n = h_n e_{n+1} + v_n e_n + h_{n-1} e_{n-1}, \qquad n \in \mathbf{Z}_+, \ e_{-1} = 0.$$

It follows from the symmetry of $\mathscr{A}$ that A_0 is a symmetric operator, i.e.,

$$(A_0 x, y) = (x, A_0 y), \qquad x, y \in \mathscr{D}_0.$$

Consequently A_0 has A as its closure; A is a symmetric closed linear operator with domain $\mathscr{D}(A) \supset \mathscr{D}_0$. The operator A is bounded if $\mathscr{A}$ is bounded:

$$\sup |v_n| < +\infty, \quad \sup h_n < +\infty \qquad (n \in \mathbf{Z}_+).$$

Operators corresponding to Jacobi matrices are discrete analogs of differential operators of the second order. In particular, they appear in the approximation of differential operators by difference operators.

2. Let us find the operator A^* adjoint to A. We shall use the fact that $A^* = A_0^*$. Recall that $g \in \mathscr{D}(A^*)$ if there exists $g^* \in l_2$ such that

$$(A_0 x, g) = (x, g^*), \qquad x \in \mathscr{D}_0, \tag{8.3}$$

and then $g^* = A^* g$. Let

$$g = (g_0, g_1, \dots, g_n, \dots) \in l_2. \tag{8.4}$$

If we set $x = e_m$ in (8.3), $m \in \mathbf{Z}_+$, we find

$$g_m^* = h_m g_{m+1} + v_m g_m + h_{m-1} g_{m-1}, \qquad n \in \mathbf{Z}_+, \ g_{-1} = 0. \tag{8.5}$$

Thus $\mathscr{D}(A^*)$ consists of the $g \in l_2$ for which

$$\sum_{m=0}^{\infty} |g_m^*|^2 < +\infty,$$

and

$$A^* g = (g_0^*, \dots, g_m^*, \dots),$$

where g_m^* is given by (8.5).

Let us find the deficiency index of A, i.e., the dimension of the kernel of the operator $(A^* - \lambda I)$, $\operatorname{Im}\lambda \neq 0$. The equation $A^* g = \lambda g$, where g is the sequence (8.4), leads to the relations

$$\lambda g_n = h_n g_{n+1} + v_n g_n + h_{n-1} g_{n-1}, \qquad n \in \mathbf{Z}_+, \ g_{-1} = 0. \tag{8.6}$$

The solutions of (8.6) are proportional to the sequence $\{Q_n\}$ of orthonormal polynomials associated with $\mathscr{A}$. The conditions $g \neq 0$, $g \in l_2$, are therefore equivalent to the convergence of the series (8.2). Thus we have established the following proposition.

PROPOSITION 8.1. *The operator* A *has deficiency index* $(0, 0)$ (*i.e.,* A *is selfadjoint*) *when* $\mathscr{A} \in (\mathrm{D})$, *and deficiency index* $(1, 1)$ *when* $\mathscr{A} \in (\mathrm{C})$.

3. Let A be any linear operator with domain $\mathscr{D}(A)$ in the Hilbert (complex infinite-dimensional separable) space H.

DEFINITION. We say that A is a *Lebesgue* operator if there is a *cyclic element* $\varphi \in H$ for it, i.e., for $k \in \mathbf{Z}_+$ we have $A^k\varphi \in \mathscr{D}(A)$ and the closure of the linear span of φ, $A\varphi$, $A^2\varphi, \ldots$ coincides with H.

PROPOSITION 8.2. *The class of bounded selfadjoint Lebesgue operators coincides with the set of operators generated by bounded Jacobi matrices. Here unitarily equivalent operators are identified.*

PROOF. If A is a bounded selfadjoint Lebesgue operator and φ is a cyclic vector in it, then φ, $A\varphi$, $A^2\varphi, \ldots$ form a linearly independent system. If we orthogonalize it by the Schmidt process, we obtain an orthonormal basis of H:

$$e_n = \beta_{n,0}\varphi + \beta_{n,1}A\varphi + \cdots + \beta_{n,n}A^n\varphi, \qquad n \in \mathbf{Z}_+, \tag{8.7}$$

where we may suppose that $\beta_{n,n} > 0$. Let us find the matrix of the operator A in this basis. According to (8.7),

$$Ae_n = h_{n,0}e_0 + \cdots + h_{n,n+1}e_{n+1},$$

and by the symmetry of A

$$h_{n,j} = (Ae_n, e_j) = (e_n, Ae_j) = 0, \qquad j \le n-2,$$

since for these values of j the vector Ae_j belongs to the linear span of $e_0, \ldots, e_{j+1}$, and therefore is orthogonal to e_n. If we introduce

$$h_n = (Ae_n, e_{n+1}) = \beta_{n,n}/\beta_{n+1,n+1} > 0,$$
$$v_n = (Ae_n, e_n) \in \mathbf{R}, \qquad n \in \mathbf{Z}_+,$$

and take account of $h_{n,n+1} = h_n$, $h_{n,n} = v_n$, and $h_{n,n-1} = h_{n-1}$, we see that the matrix of A coincides with (8.1).

The converse inclusion is trivial. ▷

Generalizing these arguments, we can establish a similar result for unbounded operators (Stone's theorem).

4. We can find the spectrum $S(A)$ of the selfadjoint operator A (corresponding to the $\mathscr{J}$-matrix $\mathscr{A} \in (\mathrm{D})$) by using the connection of these operators with the Hamburger moment problem. Let us recall the basic spectral theorem for selfadjoint operators.

Let H be a Hilbert space. A *resolution of the identity*, or *spectral family*, in H is a family of bounded operators E_λ, $\lambda \in \mathbf{R}$, such that

1) E_λ are orthogonal projectors, i.e., $E_\lambda^* = E_\lambda$, $E_\lambda^2 = E_\lambda$;

2) E_λ are nondecreasing operator functions of λ, i.e., $\lambda_1 < \lambda_2$ implies that $E_{\lambda_2} - E_{\lambda_1}$ is a positive definite operator; this property can be reformulated as follows:

$$E_{\lambda_1} E_{\lambda_2} = E_{\min(\lambda_1, \lambda_2)};$$

3) the family E_λ is left-continuous in the strong operator topology, i.e., $\|E_\lambda x - E_\kappa x\| \to 0$ with $\kappa \to \lambda$, $\kappa > \lambda$ $(x \in H)$;

4) $E_\lambda \to 0$ as $\lambda \to \infty$, $E_\lambda \to I$ as $\lambda \to +\infty$ (in the strong operator topology).

The *basic spectral theorem* states that for a selfadjoint operator A on H there is a unique resolution of the identity E_λ for which

$$A = \int_{-\infty}^{+\infty} \lambda \, dE_\lambda . \tag{8.8}$$

The domain $\mathscr{D}(A)$ of A consists of the elements $f \in H$ for which the integral

$$\int_{-\infty}^{+\infty} \lambda^2 \, d(E_\lambda f, f)$$

converges, and (8.8) means that, for every $f \in \mathscr{D}(A)$,

$$Af = \lim_{\substack{a \to -\infty \\ b \to +\infty}} \int_a^b \lambda \, dE_\lambda f,$$

where the integral over $[a, b]$ is the limit of the Riemann-Stieltjes sums for the integral (all limits are taken in the norm of H).

If A is a bounded selfadjoint operator, the integral (8.8) exists as the uniform limit of the Riemann-Stieltjes sums, where

$$E_\lambda = \begin{cases} 0 & \text{if } \lambda < \inf S(A), \\ I & \text{if } \lambda \geq \sup S(A). \end{cases}$$

Let us apply the spectral theorem to the selfadjoint operator A corresponding to the $\mathscr{J}$-matrix $\mathscr{A} \in (\mathrm{D})$. We set

$$\mu(\lambda) = (E_\lambda e_0, e_0), \qquad \lambda \in \mathbf{R}.$$

It follows from the properties of the spectral family E_λ that μ is a nondecreasing right-continuous function, $\mu(\lambda) \to 0$ as $\lambda \to -\infty$, and $\mu(\lambda) \to 1$ as $\lambda \to +\infty$, i.e., μ is the distribution function of a probability measure. It follows from (8.8) that the moments of this measure are

$$\int_{-\infty}^{+\infty} \lambda^k \, d\mu(\lambda) = (A^k e_0, e_0), \qquad k \in \mathbf{Z}_+.$$

On the Hilbert space L^2_μ, we define the operator

$$U \colon L^2_\mu \to H,$$

which maps a function $\mathfrak{f} \in L^2_\mu$ to the vector

$$f = \mathfrak{f}(A)e_0 \equiv \int_{-\infty}^{+\infty} \mathfrak{f}(\lambda)\, dE_\lambda e_0, \tag{8.9}$$

where the integral is understood as the limit in the norm of H of the Lebesgue-Stieltjes sums. Let us show that U is isometric, i.e.,

$$(U\mathfrak{f}, U\mathfrak{g}) = \int_{-\infty}^{+\infty} \mathfrak{f}(\lambda)\overline{\mathfrak{g}(\lambda)}\, d\mu(\lambda) \tag{8.10}$$

for arbitrary $\mathfrak{f}$, $\mathfrak{g} \in L^2_\mu$. Since the continuous bounded functions form a dense set in L^2_μ, it is sufficient to verify (8.10) for these. Evaluating the integral (8.9) as the limit of the Riemann-Stieltjes sums, we obtain

$$\begin{aligned}(U\mathfrak{f}, U\mathfrak{g}) &= \lim \left(\sum_{j=1}^{n} \mathfrak{f}(\lambda_j)(E_{\lambda_j} - E_{\lambda_{j-1}})e_0, \sum_{k=1}^{n} \mathfrak{g}(\lambda_k)(E_{\lambda_k} - E_{\lambda_{k-1}})e_0 \right) \\ &= \lim \sum_{j,k=1}^{n} \mathfrak{f}(\lambda_j)\overline{\mathfrak{g}(\lambda_k)}((E_{\lambda_k} - E_{\lambda_{k-1}})(E_{\lambda_j} - E_{\lambda_{j-1}})e_0, e_0) \\ &= \lim \sum_{j=1}^{n} \mathfrak{f}(\lambda_j)\overline{\mathfrak{g}(\lambda_j)}(\mu(\lambda_j) - \mu(\lambda_{j-1})) = \int_{-\infty}^{+\infty} \mathfrak{f}(\lambda)\overline{\mathfrak{g}(\lambda)}\, d\mu(\lambda)\end{aligned}$$

(here $-\infty < \lambda_0 < \lambda_1 < \cdots < \lambda_n < +\infty$ is a partition of the real line), since for $k \neq j$ the projections $E_{\lambda_k} - E_{\lambda_{k-1}}$ and $E_{\lambda_j} - E_{\lambda_{j-1}}$ are orthogonal.

Since U carries the polynomials to the set $\langle A^n e_0, n \in \mathbf{Z}_+ \rangle$, which is dense in H, i.e., the image of U coincides with H, it follows that U is a unitary operator. In particular, the polynomials are dense in L^2_μ.

We note a special case of (8.9):

$$\hat{\mu}(z) = (R_z e_0, e_0), \qquad \operatorname{Im} z \neq 0,$$

where $R_z = (zI - A)^{-1}$ is the resolvent of A.

Let $\{Q_k(\lambda)\}_0^\infty$ be the polynomials associated with the $\mathscr{J}$-matrix $\mathscr{A}$. Let us show that

$$e_k = Q_k(A)e_0, \qquad k \in \mathbf{Z}_+. \tag{8.11}$$

We apply induction on k. For $k = 0$, (8.11) is valid. Let (8.11) have been established for $k = 0, \ldots, n$. Then, by the definition of A, we have

$$AQ_n(A)e_0 = Ae_n = h_n e_{n+1} + v_n e_n + h_{n-1}e_{n-1}. \tag{8.12}$$

On the other hand, it follows from the recurrence formula for Q_n that

$$\begin{aligned} AQ_n(A) &= h_n Q_{n+1}(A) + v_n Q_n(A) + h_{n-1}Q_{n-1}(A), \\ AQ_n(A)e_0 &= h_n Q_{n+1}(A)e_0 + v_n e_n + h_{n-1}e_{n-1}. \end{aligned} \tag{8.13}$$

If we compare (8.12) and (8.13), we obtain (8.11) for $k = n + 1$. ▷

Let us note that (8.11) is equivalent to

$$U\colon Q_k(\lambda) \mapsto e_k\,, \qquad k \in \mathbf{Z}_+\,.$$

Consequently the polynomials $\{Q_k\}$ are orthonormal with respect to μ, and this measure is the unique solution of the moment problem associated with the $\mathscr{J}$-matrix $\mathscr{A}$.

In the space L^2_μ, we consider the operator of multiplying a function by its argument:

$$(\Lambda\mathfrak{f})(\lambda) = \lambda\mathfrak{f}(\lambda)\,, \qquad \lambda \in \mathbf{R}\,. \tag{8.14}$$

Its domain $\mathscr{D}(\Lambda)$ consists of the functions $\mathfrak{f}$ for which the integral

$$\int_{-\infty}^{+\infty} \lambda^2 |\mathfrak{f}(\lambda)|^2\, d\mu(\lambda) < +\infty\,.$$

Then Λ is a selfadjoint operator, and its matrix in the basis $\{Q_k(\lambda)\}_0^\infty$ coincides with $\mathscr{A}$, i.e., the operators A and Λ are unitarily equivalent: $A = U\Lambda U^{-1}$.

5. Since the spectra of unitarily equivalent operators are the same, and the spectrum of Λ is the set of points of increase of μ, we have $S(A) = S_\mu$. Orthogonal polynomials can be used to determine the spectrum of A, at least approximately. Since the measures μ_n (see (6.4)) converge weakly to μ, it follows that $S(A)$ is approximated, in some sense, by the zeros of the Q_n. More precisely, between two consecutive zeros of Q_n there is at least one point of increase of μ.

To determine the spectrum S_μ we may apply the Men′shov-Rademacher theorem.

THEOREM 8.1. *Let (X, Σ, μ) be a space with the probability measure μ, and let $\{\varphi_n(x)\}_1^\infty$ be any orthonormal system in $L^2(X, \Sigma, \mu)$. If $\{a_n\}$ is a sequence of real numbers such that*

$$\sum_{n=1}^\infty a_n^2 \log^2 n < +\infty\,,$$

then the series

$$\sum_{n=1}^\infty a_n \varphi_n(x)$$

converges μ-almost everywhere on X.

We precede the proof by the following lemma.

LEMMA 8.1. *Let $\varphi_1, \dots, \varphi_N$ be an orthonormal system in $L^2(X, \Sigma, \mu)$, and $a_1, \dots, a_N$ real numbers, $N \geq 2$. Let us set*

$$S^*(x) = \sup_{1 \leq n \leq N} \left| \sum_{\nu=1}^n a_\nu \varphi_\nu(x) \right|\,.$$

Then

$$\int_X (S^*(x))^2\, d\mu(x) \le C \log^2 N \sum_{\nu=1}^{N} a_\nu^2,$$

where C is an absolute constant.

PROOF. Let $2^{k-1} < N \le 2^k$. Extending, if necessary, the sequence $\{a_\nu\}_1^N$ by zeros, we may suppose that $N = 2^k$. In the sum

$$a_1\varphi_1 + \cdots + a_N\varphi_N$$

we insert brackets in the following way:

$$(k+1)\begin{cases} [a_1\varphi_1 + \cdots + a_N\varphi_N] \\ [a_1\varphi_1 + \cdots + a_{N/2}\varphi_{N/2}] + [a_{N/2+1}\varphi_{N/2+1} + \cdots + a_N\varphi_N] \\ [a_1\varphi_1 + \cdots + a_{N/4}\varphi_{N/4}] \\ \quad + \cdots + [a_{3N/4+1}\varphi_{3N/4+1} + \cdots + a_N\varphi_N] \\ \cdots\cdots\cdots\cdots\cdots\cdots\cdots\cdots\cdots\cdots\cdots\cdots \\ [a_1\varphi_1] + [a_2\varphi_2] + \cdots + [a_N\varphi_N]. \end{cases} \tag{8.15}$$

Every partial sum can be written in the form

$$S_n(x) = [\ ]_0 + [\ ]_1 + \cdots + [\ ]_k,$$

where $[\ ]_j$ is one of the brackets in the jth row of (8.15), or a zero term. This decomposition depends on n. It follows that

$$|S_n(x)| \le |[\ ]_0| + |[\ ]_1| + \cdots + |[\ ]_k| \le \sqrt{k+1}([\ ]_0^2 + [\ ]_1^2 + \cdots + [\ ]_k^2)^{1/2}.$$

Therefore

$$(S^*(x))^2 \le (k+1) \cdot [\textit{sum of squares of the brackets } (8.15)].$$

If we integrate this inequality, we obtain

$$\int_X (S^*(x))^2\, d\mu(x) \le (k+1)[(k+1)(a_1^2 + \cdots + a_N^2)]. \quad \triangleright$$

PROOF OF THEOREM 8.1. We have

$$\sum_{k=1}^{\infty} \int_X |S_{2^{k+1}}(x) - S_{2^k}(x)|\, d\mu(x) \le \sum_{k=1}^{\infty} \left(\int_X (S_{2^{k+1}} - S_{2^k})^2\, d\mu \right)^{1/2}$$

$$= \sum_{k=1}^{\infty} \left(\sum_{\nu=2^k+1}^{2^{k+1}} a_\nu^2 \right)^{1/2} = \sum_{k=1}^{\infty} \frac{1}{k} \left(\sum_{\nu=2^k+1}^{2^{k+1}} a_\nu^2 k^2 \right)^{1/2}$$

$$\le \left(\sum_{k=1}^{\infty} \frac{1}{k^2} \right)^{1/2} \left(\sum_{k=1}^{\infty} \sum_{\nu=2^k+1}^{2^{k+1}} a_\nu^2 \log_2^2 \nu \right)^{1/2} < +\infty.$$

Consequently, by B. Levi's theorem,

$$\sum_{k=1}^{\infty} |S_{2^{k+1}}(x) - S_{2^k}(x)| < +\infty \quad \mu\text{-a.e.} \tag{8.16}$$

Let us set

$$\delta_k(x) = \sup_{2^k < n \le 2^{k+1}} \left| \sum_{\nu=2^k+1}^{n} a_\nu \varphi_\nu(x) \right| .$$

By Lemma 8.1,

$$\int_X \delta_k^2(x)\, d\mu(x) \le C \log^2 2^k \sum_{\nu=2^k+1}^{2^{k+1}} a_\nu^2 \le C \sum_{\nu=2^k+1}^{2^{k+1}} a_\nu^2 \log^2 \nu .$$

Therefore

$$\sum_{k=1}^{\infty} \int_X \delta_k^2(x)\, d\mu(x) < +\infty ,$$

and by B. Levi's theorem

$$\sum_{k=1}^{\infty} \delta_k^2(x) < +\infty \quad \mu\text{-a.e.}$$

In particular, $\delta_k(x) \to 0$ μ-a.e. Hence, if we take (8.16) into account, the theorem follows. ▹

PROPOSITION 8.3. 1) *For every* $\varepsilon > 0$ *and almost every point* λ (*with respect to* μ) *we have the inequality*

$$Q_n(\lambda) = o(\sqrt{n} \log^{3/2+\varepsilon} n) , \qquad n \to \infty . \tag{8.17}$$

2) *If* $S(A)$ *is compact, and* $\lambda_0 \notin S(A)$, *we have*

$$\varlimsup_{n\to\infty} |Q_n(\lambda_0)|^{1/n} > 1 . \tag{8.18}$$

PROOF. Set

$$a_n = (\sqrt{n} \log^{3/2+\varepsilon} n)^{-1} , \qquad n \ge 2 .$$

Then

$$\sum_{n=2}^{\infty} a_n^2 \log^2 n < +\infty .$$

Consequently, by the Men′shov-Rademacher theorem, the series

$$\sum_{n=2}^{\infty} a_n Q_n(\lambda)$$

converges for μ-a.e. λ, and (8.17) follows.

If $\lambda_0 \notin S(A)$, some neighborhood of λ_0 is disjoint from $S(A)$. It follows from this and from the boundedness of the spectrum that there is a quadratic polynomial φ such that $\varphi(\lambda_0) = 1$ and

$$q \equiv \sup\{|\varphi(\lambda)| : \lambda \in S(A)\} < 1 .$$

By (7.19), we have

$$\rho_{2n}(\lambda_0) \le \int_{S(A)} |\varphi(\lambda)|^{2n}\, d\mu(\lambda) \le q^{2n} ,$$

i.e.,

$$\sum_{k=0}^{2n} |Q_k(\lambda_0)|^2 \geq q^{-2n}, \tag{8.19}$$

and now (8.18) follows from (8.19) and simple inequalities. ▷

Points of the spectrum of an operator can be distinguished by the rate of growth of the polynomials $Q_n(\lambda)$: on the spectrum they cannot grow rapidly, whereas outside they grow like a geometric progression (for bounded operators).

Finally, we describe the set of eigenvalues of A; we denote it by $S_d(A)$ and call it the *discrete spectrum.*

PROPOSITION 8.4. *We have*

$$S_d(A) = \Gamma_\mu; \tag{8.20}$$

in addition,

$$\frac{1}{\mu\{s\}} = \sum_{n=0}^{\infty} Q_n(s)^2, \qquad s \in \Gamma_\mu. \tag{8.21}$$

PROOF. Let $s \in \mathbf{R}$. The solutions of the equation

$$\lambda\mathfrak{f}(\lambda) = s\mathfrak{f}(\lambda) \quad (\text{for } \mu\text{-a.e. } \lambda)$$

are proportional to the function

$$\mathfrak{f}_s(\lambda) = \begin{cases} 1 & \text{if } \lambda = s, \\ 0 & \text{if } \lambda \neq s, \end{cases}$$

which is not equivalent to zero if and only if $s \in \Gamma_\mu$. This proves (8.20).

Formula (8.21) was essentially proved in Proposition 7.3. However, it can also be verified directly. Let $s \in \Gamma_\mu$ and let

$$\mathfrak{a}_n = \int_{-\infty}^{+\infty} \mathfrak{f}_s(\lambda) Q_n(\lambda)\, d\mu(\lambda) = Q_n(s)\mu\{s\}$$

be the Fourier coefficients of $\mathfrak{f}_s$. By the Parseval equality

$$\sum_{n=0}^{\infty} \mathfrak{a}_n^2 = \sum_{n=0}^{\infty} Q_n(s)^2 \mu\{s\}^2 = \int_{-\infty}^{+\infty} \mathfrak{f}_s(\lambda)^2\, d\mu(\lambda) = \mu\{s\},$$

which is equivalent to (8.21). ▷

REMARK

For operator theory see, for example, the textbooks and monographs [51], [10], and [85]. The connection of operator theory with moment problems is discussed in [8].

PROBLEMS

1. An orthogonal projection P in the Hilbert space H is an operator $Px = x_1$, where $x = x_1 + x_2$; $x_1 \in H_1$, $x_2 \in H_2$; and H_1 and H_2 are orthogonal subspaces of H whose sum is H. Show that P is an orthogonal projection only if $P^2 = P$ and $P^* = P$.

2. Let P and Q be orthogonal projections; prove that $P \leq Q$, i.e., that the operator $Q - P$ is positive definite if and only if $PQ = QP = P$.

3. Prove that the spectrum of the operator (8.14) is equal to S_μ.

4. Let $\{a_n\}_1^\infty$ be positive numbers. If

$$\sum_{k=1}^{n} a_k \geq \delta^n > 1, \qquad n = 1, 2, \ldots,$$

then

$$\overline{\lim_{n\to\infty}}\, a_n^{1/n} > 1.$$

5. In the space l_2, define an operator A by the formula

$$A\colon (x_0, x_1, x_2, \ldots, x_n, \ldots) \mapsto (0, x_1, 2x_2, \ldots, nx_n, \ldots);$$

its domain consists of the $x \in l_2$ for which

$$\sum_{n=1}^{\infty} n^2 |x_n|^2 < +\infty.$$

Prove that A is a selfadjoint Lebesgue operator.

6. Chebyshev-Markov-Stieltjes separation theorem. *Let $\mu \in \mathscr{M}$ (see §7, subsection 2), $n = 2, 3, \ldots$; then*

$$\mu(-\infty, \lambda_i] < \sum_{j=1}^{i} \rho_{n-1}(\lambda_j) < \mu(-\infty, \lambda_{i+1}), \qquad i = 1, \ldots, n-1, \tag{$*$}$$

where $\lambda_1, \ldots, \lambda_n$ are the zeros of the nth polynomial Q_n orthogonal with respect to the measure μ; the numbers $\rho_{n-1}(\lambda)$ were defined in Proposition 7.2.

Hint. There is a unique polynomial $R(\lambda)$ of degree $2n-2$ with the properties that $R(\lambda_j) = 1$, $j = 1, \ldots, i$; $R(\lambda_j) = 0$, $j = i+1, \ldots, n$; $R'(\lambda_j) = 0$, $j \neq i$. In fact, these conditions reduce to a system of $2n-1$ linear equations, and the corresponding homogeneous system has only the trivial solution. Then $R(\lambda) \geq 0$, $\lambda \in \mathbf{R}$; $R(\lambda) \geq 1$, $\lambda \leq \lambda_i$. If we apply the quadratic formula to $R(\lambda)$, we obtain the left inequality in $(*)$. The right inequality can be proved similarly.

7. Stone's theorem. *An operator A is a selfadjoint Lebesgue operator if and only if A is unitarily equivalent to an operator in l_2 that corresponds to a $\mathscr{J}$-matrix of type* D. *(See* [8].*)*

Hint. The sufficiency has already been established. Let us prove the necessity. Let ψ be a cyclic vector of A. Set $\varphi = \exp(-A^2)\psi$; then φ is also a cyclic vector. Starting from φ, as in subsection 3, construct an orthonormal basis in which the matrix A serves as the $\mathscr{J}$-matrix $\mathscr{A}$. Let $\widetilde{A}$ be the corresponding operator. Show that $\widetilde{A}$ is selfadjoint; then $\mathscr{A} \in (\mathrm{D})$ and $\widetilde{A} = A$ (unitary equivalence). Thus we need to prove that the image of $(\widetilde{A} - iI)$ coincides with l_2. Suppose the contrary; then there is an $x \in H$, $x \neq 0$, such that $((A - iI)e_n, x) = 0$, $n \in \mathbf{Z}_+$. If we pass to the space L^2_μ using the operator U (see subsection 4), we obtain a contradiction.

8. In $L^2(0, 1)$, define the operator $Af(x) = -if'(x)$ on absolutely continuous functions with derivatives in $L^2(0, 1)$, for which $f(0) = f(1)$. Is A a selfadjoint Lebesgue operator?

§9. Some theorems on the spectra of operators

1. Let H be Hilbert space and let A be a bounded operator on H. In §8 we defined $S_d(A)$ to be the discrete spectrum of A, i.e., the set of its eigenvalues. Let us write $S_c(A) = S(A)\backslash S_d(A)$; we call this set the continuous spectrum of A. In this subsection we study the question of how the continuous spectrum of an operator changes under compact perturbations. Before we state the next theorem, which was obtained by H. Weyl, we recall a classical result which is equivalent to the Fredholm alternative: if C

is a compact operator on H, we have $S(C)\backslash\{0\} \subset S_d(C)$, and for every $\varepsilon > 0$ the sum of the dimensions of those eigenspaces that correspond to eigenvalues λ with $|\lambda| > \varepsilon$ is finite.

THEOREM 9.1 (H. Weyl). *Let A be a bounded operator, and C a compact operator, on H. Then*

$$S_c(A) \subset S(A+C).$$

PROOF. Suppose the contrary, i.e., that $\lambda_0 \in S_c(A)$ and the operator $(A + C - \lambda_0 I)$ is invertible. We have

$$A - \lambda_0 I = (A + C - \lambda_0 I) - C = (A + C - \lambda_0 I)(I - (A + C - \lambda_0 I)^{-1} C).$$

The operator

$$F = (A + C - \lambda_0 I)^{-1} C$$

is compact. Consequently either 1 is an eigenvalue of F (but then λ_0 is an eigenvalue of A), or the operator $I - F$ is invertible (but then $A - \lambda_0 I$ is invertible). ▷

The situation in which we will use the Weyl theorem will consist in comparing the spectra of two bounded selfadjoint operators A and B for which $A-B$ is a compact operator. We notice that for a selfadjoint operator A, the set $S_d(A)$ is at most countably infinite, since the eigenspaces corresponding to different eigenvalues are orthogonal, and H is separable. In this case the Weyl theorem asserts that $S(A) \Delta S(B)$ is at most countably infinite (here Δ denotes the symmetric difference). However, the theorem says nothing about the spectrum of B in the resolvent domain of A (i.e., in the complement of $S(A)$). In this domain, of course, there are, by the Weyl theorem, at most countably many points of $S(B)$. The following proposition supplements the Weyl theorem.

PROPOSITION 9.1. *Let A and B be bounded operators on H such that $A - B$ is a compact operator and B is a selfadjoint operator. Let there be no points of the spectrum of A in the domain $\mathscr{G} \subset \mathbf{C}$. Then $S_d(B)$ has no limit points in $\mathscr{G}$.*

PROOF. Suppose the contrary. Let $\lambda_n \in S_d(B)$, $n \in \mathbf{N}$, $\lambda_n \to \lambda_0 \in \mathscr{G}$ as $n \to \infty$, and $\lambda_n \neq \lambda_0$. Let g_n be the eigenvector corresponding to λ_n:

$$Bg_n = \lambda_n g_n, \qquad \|g_n\| = 1, \qquad n \in \mathbf{N}, \tag{9.1}$$

or

$$(B - A)g_n = (\lambda_n I - A)g_n, \qquad n \in \mathbf{N}. \tag{9.2}$$

We select a convergent subsequence $(B - A)g_{n_k}$, $k \in \mathbf{N}$. Then by (9.2) g_{n_k} converges since $\|(\lambda I - A)^{-1}\| \leq \text{const}$ for all λ in a neighborhood of λ_0. Let $g_{n_k} \to g_0$ as $k \to \infty$. Taking the limit in (9.1) over the subsequence n_k, we obtain $Bg_0 = \lambda_0 g_0$, $\|g_0\| = 1$. Since λ_0 is an eigenvalue of B different from λ_{n_k}, by the selfadjointness of B we have

$$(g_{n_k}, g_0) = 0, \qquad k \in \mathbf{N}. \tag{9.3}$$

Taking the limit in (9.3) as $k \to \infty$, we obtain $\|g_0\| = 0$. This contradiction establishes the proposition. ▷

REMARK. If we do not require the operator B to be selfadjoint, we can construct an example in which a compact (and even finite-dimensional) component essentially "destroys" the resolvent domain of the operator (see [111]).

2. Before we apply the results of subsection 1 to the study of the spectra of bounded selfadjoint operators that correspond to Jacobi matrices, we consider the simplest case of matrices with constant coefficients, when v_n and h_n are independent of n. In this case, we may suppose, without loss of generality, that $v_n = 0$, $h_n = 1/2$, and $n \in \mathbf{Z}_+$, i.e.,

$$\mathscr{A} = \begin{pmatrix} 0 & 1/2 & & & \mathbf{0} \\ 1/2 & 0 & 1/2 & & \\ & 1/2 & 0 & 1/2 & \\ & \mathbf{0} & \cdot & \cdot & \cdot \end{pmatrix}, \tag{9.4}$$

and that the operator A in l_2 that corresponds to this matrix acts on basis vectors as follows:

$$Ae_n = \tfrac{1}{2}(e_{n+1} + e_{n-1}), \qquad n \in \mathbf{Z}_+,\ e_{-1} = 0.$$

(For arbitrary matrices $\widetilde{\mathscr{A}}$ with the constants $v_n = v$, $h_n = h \neq 0$, we have $\mathscr{A} = (\widetilde{\mathscr{A}} - vI)/2h$. The spectra of the corresponding operators are easily obtained successively.) We calculate the polynomials $Q_n(\lambda)$ associated with the matrix (9.4), i.e., we solve the recurrence relation

$$u_{n+1} - 2\lambda u_n + u_{n-1} = 0, \qquad n \in \mathbf{Z}_+, \tag{9.5}$$

with the initial conditions

$$u_{-1} = 0, \qquad u_0 = 1. \tag{9.6}$$

To do this, we form the characteristic equation

$$t^2 - 2\lambda t + 1 = 0.$$

Its roots are

$$t_\pm = \lambda \pm \sqrt{\lambda^2 - 1}$$

(we may take either sign). Consequently, the general solution of (9.5) (for $\lambda \neq \pm 1$) has the form

$$u_n = C_+(\lambda + \sqrt{\lambda^2 - 1})^n + C_-(\lambda - \sqrt{\lambda^2 - 1})^n.$$

Determining the constants $C_\pm$ from the initial conditions (9.6), we finally obtain

$$Q_n(\lambda) = \frac{1}{2\sqrt{\lambda^2-1}}((\lambda+\sqrt{\lambda^2-1})^{n+1} - (\lambda-\sqrt{\lambda^2-1})^{n+1}), \qquad n \in \mathbf{Z}_+.$$

If $\lambda \in [-1, 1]$, we obtain, by setting $\lambda = \cos\theta$, $0 \le \theta \le \pi$, the de Moivre-Laplace formula

$$\begin{aligned} Q_n(\cos\theta) &= \frac{1}{2i\sin\theta}((\cos\theta + i\sin\theta)^{n+1} - (\cos\theta - i\sin\theta)^{n+1}) \\ &= \frac{\sin((n+1)\theta)}{\sin\theta}. \end{aligned} \tag{9.7}$$

It follows from this formula and Proposition 8.3, 2) that $[-1, 1] \subset S(A)$, since the sequence (9.7) is bounded for each θ, $0 < \theta < \pi$. If $\lambda \notin [-1, 1]$ we have

$$Q_n(\lambda) = \frac{(\lambda+\sqrt{\lambda^2-1})^{n+1}}{2\sqrt{\lambda^2-1}}(1+o(1)), \qquad n \to \infty,$$

where we take the branch of the square root for which

$$|\lambda + \sqrt{\lambda^2-1}| > 1, \qquad \lambda \notin [-1, 1].$$

Therefore, by Proposition 8.3, 1), all points outside $[-1, 1]$ are points of the resolvent set. Thus

$$S(A) = [-1, 1].$$

Let us find the Markov (resolvent) function of A. By (5.20), in our case

$$\hat{\mu}(z) = \cfrac{1}{z - \cfrac{1/4}{z - \cfrac{1/4}{z - \ddots}}} \tag{9.8}$$

By Markov's theorem, the fraction (9.8) converges outside $[-1, 1]$, and we obtain the equation

$$\hat{\mu}(z) = \left(z - \tfrac{1}{4}\hat{\mu}(z)\right)^{-1}, \qquad z \notin [-1, 1].$$

Hence

$$\hat{\mu}(z) = 2(z - \sqrt{z^2-1}), \qquad z \in \overline{\mathbf{C}} \backslash [-1, 1].$$

To determine the branch of the square root, we start from the expansion

$$\hat{\mu}(z) \sim 1/z, \qquad z \to \infty.$$

Hence $\sqrt{z^2-1} > 0$ for $z > 1$.

Let γ be any simple rectifiable Jordan contour that goes around the interval $[-1, 1]$ clockwise. Then, for every z outside γ, we have

$$\hat{\mu}(z) = \frac{1}{2\pi i}\int_\gamma \frac{\hat{\mu}(x)}{x-z}\,dx$$

by the Cauchy formula. Taking the limit as γ shrinks to the interval $[-1, 1]$, and calculating the limiting values of $\hat{\mu}$ on the upper and lower sides of the interval, we obtain, for every z outside $[-1, 1]$,

$$\begin{aligned}\hat{\mu}(z) &= \frac{1}{2\pi i}\left(\int_{-1}^{1}\frac{2(x-i\sqrt{1-x^2})}{x-z}\,dx - \int_{-1}^{1}\frac{2(x+i\sqrt{1-x^2})}{x-z}\,dx\right)\\ &= \frac{2}{\pi}\int_{-1}^{1}\frac{\sqrt{1-x^2}}{z-x}\,dx,\end{aligned}$$

where $\sqrt{1-x^2}$ is the positive root. Hence we find the spectral measure for A:

$$d\mu(x) = \frac{2}{\pi}\sqrt{1-x^2}\,dx, \qquad -1 < x < 1.$$

PROPOSITION 9.2. *Let B be the operator corresponding to a Jacobi matrix* (8.1) *for which*

$$\lim_{n\to\infty} v_n = 0, \qquad \lim_{n\to\infty} h_n = 1/2.$$

Then the spectrum of B has the form

$$S(B) = [-1, 1] \cup S,$$

where S is a finite or countable set of real numbers outside $[-1, 1]$, *which can accumulate only at the endpoints of this interval.*

PROOF. Let A be the operator corresponding to the matrix (9.4). Then the operator $B - A$ will be completely continuous, and Proposition 9.2 follows from Theorem 9.1, Proposition 9.1, and the results of the present subsection. ▷

REMARKS

1. The term "continuous spectrum" is used in various senses in other books. There are other classifications of the points of the spectrum.
2. We have proved only the simplest version of the Weyl theorem.
3. Here we have not considered the question of when perturbation of an operator reveals only a finite number of points of the spectrum in the resolvent domain of the unperturbed operator. For a discussion of this question and a bibliography, see [71].
4. For the connection between orthogonal polynomials and scattering theory, see [71] and [6]. The book [15] may be useful.
5. We have also not touched on the connection of orthogonal polynomials with symmetric operators of deficiency index (m, m). Such operators lead to matrix orthogonal polynomials; see [53], [54], [52], and [7].

PROBLEMS

1. Find the measure corresponding to the Jacobi matrix

$$\begin{pmatrix} 0 & 1/\sqrt{2} & 0 & 0 & \cdots \\ 1/\sqrt{2} & 0 & 1/2 & 0 & \cdots \\ 0 & 1/2 & 0 & 1/2 & \cdots \\ 0 & 0 & 1/2 & 0 & \cdots \\ \cdots & \cdots & \cdots & \cdots & \cdots \end{pmatrix}.$$

2. An operator A on l_2 is defined by the formula

$$A: (x_0, x_1, \dots, x_n, \dots) \mapsto (a_0 x_0, a_1 x_1, \dots, a_n x_n, \dots),$$

where $\{a_n\}$ is a given bounded sequence of complex numbers. Prove that A is a compact operator if and only if $a_n \to 0$ as $n \to \infty$.

§10. Finite and semi-infinite Toda and Langmuir lattices

In this section we study problems that arise from investigations in theoretical physics. The Toda and Langmuir lattices that are considered below are studied here, not in their basic formulation (lattices infinite in both directions) but in a simpler situation that is, nevertheless, useful for understanding the nature of the processes that arise here. This case was first investigated by J. Moser [151].

1. Let us state the problems for Toda and Langmuir lattices. We begin with points $\{x_n\}$ for which

$$x_{n-1} < x_n, \qquad n \in \mathbf{Z}. \tag{10.1}$$

We suppose that there is a force that pushes on consecutive points and decreases according to the exponential law, so that x_n is acted on by a force

$$F_n = e^{-(x_n - x_{n-1})} - e^{-(x_{n+1} - x_n)}, \qquad n \in \mathbf{Z}.$$

Newton's second law yields the system of differential equations

$$\ddot{x}_n = e^{-(x_n - x_{n-1})} - e^{-(x_{n+1} - x_n)}, \qquad n \in \mathbf{Z}, \tag{10.2}$$

which is called the *Toda lattice*. At $t = 0$, we have the initial data

$$x_n(0) = \kappa_n, \qquad \dot{x}_n(0) = \eta_n, \qquad n \in \mathbf{Z}.$$

The solution of the system (10.2) is described by the physical process as long as (10.1) is satisfied. However, we shall consider arbitrary solutions of the system.

Let us consider a finite Toda lattice:

$$\begin{aligned}
\ddot{x}_1 &= -e^{-(x_2 - x_1)}, \\
\ddot{x}_n &= e^{-(x_n - x_{n-1})} - e^{-(x_{n+1} - x_n)}, \qquad n = 2, \dots, N-1, \\
\ddot{x}_N &= e^{-(x_N - x_{N-1})}, \\
x_n(0) &= \kappa_n, \qquad \dot{x}_n(0) = \eta_n, \qquad n = 1, \dots, N,
\end{aligned}$$

where $N \geq 2$. After the change of notation

$$\begin{aligned}
v_n &= \dot{x}_n, \qquad n = 1, \dots, N, \\
c_n &= e^{-(x_{n+1} - x_n)}, \qquad n = 1, \dots, N-1,
\end{aligned}$$

the lattice takes the form

$$\begin{aligned}
\dot{v}_1 &= -c_1, \\
\dot{v}_n &= c_{n-1} - c_n, \qquad n = 2, \dots, N-1, \\
\dot{v}_N &= c_{N-1},
\end{aligned} \tag{10.3}$$

$$
\begin{aligned}
\dot{c}_n &= c_n(v_n - v_{n+1}), \qquad n = 1, \dots, N-1, \\
v_n(0) &= \eta_n, \qquad n = 1, \dots, N; \\
c_n(0) &= \gamma_n, \qquad n = 1, \dots, N-1.
\end{aligned}
\tag{10.4}
$$

The *two-sided infinite Langmuir lattice* is the system of equations

$$\dot{c}_n = c_n(c_{n+1} - c_{n-1}), \qquad n \in \mathbf{Z}.$$

By a *finite Langmuir lattice*, we mean the finite system

$$
\begin{aligned}
\dot{c}_1 &= c_1 c_2, \\
\dot{c}_n &= c_n(c_{n+1} - c_{n-1}), \qquad n = 2, \dots, N-1, \\
\dot{c}_N &= -c_N c_{N-1},
\end{aligned}
\tag{10.5}
$$

$$c_n(0) = \gamma_n > 0, \qquad n = 1, \dots, N. \tag{10.6}$$

We seek solutions of the lattices under the condition that $c_n(t) > 0$.

Semi-infinite Toda and Langmuir lattices are defined similarly.

2. We now solve the finite Langmuir lattice. In $\mathbf{R}^{n+1}$, we take any orthonormal basis $e_0, \dots, e_N$, and define two operators on this space, whose matrices have the following forms in the selected basis:

$$
L(t) = \begin{pmatrix}
0 & \sqrt{c_1} & & & & \\
\sqrt{c_1} & 0 & \sqrt{c_2} & & \mathbf{0} & \\
& \sqrt{c_2} & 0 & \sqrt{c_3} & & \\
& & & \ddots & & \\
& \mathbf{0} & & \sqrt{c_{N-1}} & 0 & \sqrt{c_N} \\
& & & & \sqrt{c_N} & 0
\end{pmatrix}
\tag{10.7}
$$

and

$$
2M(t) = \begin{pmatrix}
0 & 0 & \sqrt{c_1c_2} & & & & & & \mathbf{0} \\
0 & 0 & 0 & \sqrt{c_2c_3} & & & & & \\
-\sqrt{c_1c_2} & 0 & 0 & 0 & \sqrt{c_3c_4} & & & & \\
& -\sqrt{c_2c_3} & 0 & 0 & 0 & \sqrt{c_4c_5} & & & \\
& & \cdot & \cdot & \cdot & \cdot & \cdot & & \\
& & & -\sqrt{c_{N-3}c_{N-2}} & 0 & 0 & 0 & \sqrt{c_{N-1}c_N} \\
& \mathbf{0} & & & -\sqrt{c_{N-2}c_{N-1}} & 0 & 0 & 0 \\
& & & & & -\sqrt{c_{N-1}c_N} & 0 & 0
\end{pmatrix}
$$

where $c_n = c_n(t)$, which appears in the definitions of $L(t)$ and $M(t)$, are solutions of the Langmuir lattice. A direct verification shows that the pair L, M is a *Lax pair*, i.e., that

$$\dot{L} = [M, L] \equiv ML - LM. \tag{10.8}$$

This verification is completely technical, and we shall omit it.

The operator L (for a given t) is described in the basis $\{e_n\}$ by a finite Jacobi matrix. Consequently, it has a simple spectrum $\lambda_0, \dots, \lambda_N$ and a corresponding system of eigenvectors $f_0, \dots, f_N$, where we suppose that

$\|f_n\| = 1$, $n = 0, \dots, N$. It follows from (10.8) that the spectrum $S(L)$ is independent of t. Let us prove this. Differentiating the relation $Lf_n = \lambda_n f_n$, we obtain

$$\dot{L}f_n + L\dot{f}_n = \dot{\lambda}_n f_n + \lambda_n \dot{f}_n.$$

Taking account of (10.8),

$$\lambda_n M f_n - LMf_n + L\dot{f}_n = \dot{\lambda}_n f_n + \lambda_n \dot{f}_n.$$

Taking the scalar product with f_n, we have

$$\lambda_n (Mf_n, f_n) - (LMf_n, f_n) + (L\dot{f}_n, f_n) = \dot{\lambda}_n,$$

where we have used the relation

$$(\dot{f}_n, f_n) = \frac{1}{2}\frac{d}{dt}(f_n, f_n) = 0.$$

If we use the symmetry of L and the skew-symmetry of M, we obtain

$$\dot{\lambda}_n = -(Mf_n, Lf_n) + (\dot{f}_n, Lf_n) = 0. \quad \triangleright$$

Let

$$\hat{\mu}(z) = ((z - L)^{-1} e_0, e_0) = \sum_{n=0}^{N} \frac{m_n(t)}{z - \lambda_n}$$

be the resolvent function of $L(t)$. Let us find the time dependence of the Christoffel coefficients $m_n(t)$. This will yield a solution of the integration problem, since the spectrum $\{\lambda_n\}$ and the functions $m_n(t)$ completely determine $L(t)$.

Let $Q_0, \dots, Q_N$ be polynomials associated with the matrix $L = L(t)$, depending on the parameter t. Set

$$g_n = \sum_{k=0}^{N} Q_k(\lambda_n) e_k, \qquad n = 0, \dots, N.$$

Then

$$Lg_n = \lambda_n g_n, \tag{10.9}$$

i.e., $\{g_n\}$ is an eigenvector of L, and, as we know (see (5.24)),

$$\frac{1}{m_n(t)} = \sum_{k=0}^{N} Q_k(\lambda_n)^2. \tag{10.10}$$

If we differentiate (10.9) with respect to t, and use (10.8), we obtain

$$L(\dot{g}_n - Mg_n) = \lambda_n(\dot{g}_n - Mg_n),$$

i.e., $(\dot{g}_n - Mg_n)$ is an eigenvector of L with eigenvalues λ_n. Since the eigensubspaces of L are one-dimensional, we have

$$\dot{g}_n - Mg_n = p_n g_n. \tag{10.11}$$

Taking the scalar product of (10.11) with g_n and using the skew-symmetry of M, we obtain

$$\frac{1}{2}\frac{d}{dt}(g_n, g_n) = p_n(g_n, g_n).$$

According to (10.10), $(g_n, g_n) = 1/m_n(t)$. Consequently

$$p_n = \frac{1}{2}\frac{d}{dt}\left(\frac{1}{m_n(t)}\right)\Big/\left(\frac{1}{m_n(t)}\right). \qquad (10.12)$$

Let us write (10.11) for the first coordinate:

$$p_n = -\tfrac{1}{2}\sqrt{c_1c_2}\,Q_2(\lambda_n),$$

since $Q_0 \equiv 1$. Calculating Q_2 from the recurrence formula, we obtain $Q_2(\lambda) = (\lambda^2 - c_1)/\sqrt{c_1c_2}$. Consequently

$$p_n = -\tfrac{1}{2}\lambda_n^2 + \tfrac{1}{2}c_1.$$

If we substitute this value into (10.12), we obtain

$$\ln\frac{1}{m_n(t)} = -\lambda_n^2 t + \int_0^t c_1(\tau)\,d\tau + \ln\frac{1}{m_n},$$

where $m_n = m_n(0)$. Or,

$$\frac{1}{m_n(t)} = \frac{1}{m_n}S(t)e^{-\lambda_n^2 t}, \qquad n = 0, \dots, N,$$

where $S(t)$ is independent of n.

Thus we have

$$\hat{\mu}(z) = \frac{1}{S(t)}\sum_{n=0}^{N}\frac{m_n e^{\lambda_n^2 t}}{z - \lambda_n}.$$

From the relation

$$\hat{\mu}(z) \sim \frac{1}{z}, \qquad z \to \infty,$$

we find

$$S(t) = \sum_{n=0}^{N} m_n e^{\lambda_n^2 t}.$$

Finally, we notice that

$$\hat{\mu}(z) = \cfrac{1}{z - \cfrac{c_1}{z - \cfrac{c_2}{z - \ddots - \frac{c_N}{z}}}}.$$

Consequently we have proved the following theorem.

THEOREM 10.1. *All the solutions of the finite Langmuir lattice* (10.5) *are obtained from the relation*

$$\cfrac{1}{z - \cfrac{c_1}{z} - \cfrac{\ddots}{- \cfrac{c_N}{z}}} = \frac{\sum_{n=0}^{N} m_n e^{\lambda_n^2 t}/(z - \lambda_n)}{\sum_{n=0}^{N} m_n e^{\lambda_n^2 t}}, \tag{10.13}$$

where the time-independent numbers λ_n *and* m_n $(m_0 + \cdots + m_N = 1)$ *are determined by the initial data* (10.6) *from the same relation* (10.3).

The following theorem is obtained by a simple limit process.

THEOREM 10.2. *The solution of the semi-infinite Langmuir lattice*

$$\begin{aligned} \dot{c}_1 &= c_1 c_2, \\ \dot{c}_2 &= c_2(c_3 - c_1), \\ &\cdots\cdots\cdots\cdots \\ \dot{c}_n &= c_n(c_{n+1} - c_{n-1}), \\ &\cdots\cdots\cdots\cdots \end{aligned}$$

with bounded initial data

$$c_n(0) = \gamma_n > 0, \quad \sup \gamma_n < +\infty, \qquad n \in \mathbf{N},$$

is obtained from the relations

$$\cfrac{1}{z - \cfrac{c_1}{z - \cfrac{c_2}{z - \ddots}}} = \frac{\int_{S_\mu} e^{\lambda^2 t}\, d\mu(\lambda)/(z - \lambda)}{\int_{S_\mu} e^{\lambda^2 t}\, d\mu(\lambda)}, \tag{10.14}$$

where the spectrum S_μ *and the measure* μ *are independent of time and are uniquely determined by the initial data from the relation* (10.14).

3. The solutions of finite and semi-infinite Toda lattices can be carried out in precisely the same way as above. Here we set

$$L(t) = \begin{pmatrix} v_1 & \sqrt{c_1} & & & & \\ \sqrt{c_1} & v_2 & \sqrt{c_2} & & 0 & \\ & \sqrt{c_2} & v_3 & \sqrt{c_3} & & \\ & & \cdot & \cdot & \cdot & \\ & 0 & & \sqrt{c_{N-2}} & v_{N-1} & \sqrt{c_{N-1}} \\ & & & & \sqrt{c_{N-1}} & v_n \end{pmatrix}$$

and

$$M(t) = \frac{1}{2}\begin{pmatrix} 0 & -\sqrt{c_1} & & & & \\ \sqrt{c_1} & 0 & -\sqrt{c_2} & & 0 & \\ & \sqrt{c_2} & 0 & -\sqrt{c_3} & & \\ & & \cdot & \cdot & \cdot & \\ & 0 & & \sqrt{c_{N-2}} & 0 & -\sqrt{c_{N-1}} \\ & & & & -\sqrt{c_{N-1}} & 0 \end{pmatrix}.$$

If $\{c_k, v_k\}$ form a Toda lattice, L and M are a Lax pair:

$$\dot{L} = ML - LM.$$

The converse is also valid.

The following theorem is proved similarly to Theorem 10.1. Theorem 10.4 is obtained by passage to the limit.

We state the final results.

THEOREM 10.3. *The solutions of the finite Toda lattice* (10.3) *are given by*

$$\cfrac{1}{z - v_1 - \cfrac{c_1}{z - v_2 - \ddots - \cfrac{c_{N-1}}{z - v_N}}} = \frac{\sum_{n=1}^N m_n e^{-\lambda_n t}/(z - \lambda_n)}{\sum_{n=1}^N m_n e^{-\lambda_n t}}, \tag{10.15}$$

where the time-independent numbers λ_n *and* m_n $(m_1 + \cdots + m_N = 1)$ *are obtained from the initial data* (10.4) *by* (10.15).

THEOREM 10.4. *The solution of the semi-infinite Toda lattice*

$$\begin{aligned} \dot{v}_1 &= -c_1, \\ \dot{v}_2 &= c_1 - c_2, \\ &\cdots\cdots\cdots\cdots \\ \dot{v}_n &= c_{n-1} - c_n, \\ &\cdots\cdots\cdots\cdots \\ \dot{c}_1 &= c_1(v_1 - v_2), \\ &\cdots\cdots\cdots\cdots \\ \dot{c}_n &= c_n(v_n - v_{n+1}) \end{aligned}$$

with bounded initial data

$$v_n(0) = \eta_n, \quad c_n(0) = \gamma_n > 0, \quad \sup\{|\eta_n|, \gamma_n\} < +\infty,$$

is obtained from

$$\cfrac{1}{z - v_1 - \cfrac{c_1}{z - v_2 - \ddots}} = \frac{\int_{S_\mu} e^{-\lambda t}\, d\mu(\lambda)/(z - \lambda)}{\int_{S_\mu} e^{-\lambda t}\, d\mu(\lambda)}, \tag{10.16}$$

where the spectrum S_μ *and the measure* μ *are independent of time and are uniquely determined by the data* (10.16).

CHAPTER 3

Asymptotic properties of orthogonal polynomials

§1. Polynomials orthogonal on the circle

1. Let $\boldsymbol{T} = \{z \in \mathbf{C}\colon |z| = 1\}$ be the unit circle, and σ a measure on $\boldsymbol{T}$ (finite, positive, Borel), with an infinite spectrum; then we can construct an (infinite-dimensional) Hilbert space L^2_σ that consists of (equivalence classes of) functions $f(z)$ that are bounded on $\boldsymbol{T}$ and have finite norm

$$\|f\| = \left(\frac{1}{2\pi}\int_{\boldsymbol{T}} |f(z)|^2\, d\sigma(z)\right)^{1/2} < +\infty,$$

generated by the scalar product

$$(f, g) = \frac{1}{2\pi}\int_{\boldsymbol{T}} f(z)\overline{g(z)}\, d\sigma(z). \tag{1.1}$$

If we apply the Gram-Schmidt orthogonalization process to the linearly independent system of powers $1, z, \dots, z^n, \dots$ using the scalar product (1.1), we obtain a uniquely determined sequence of polynomials $\varphi_0(z)$, $\varphi_1(z), \dots$, $\varphi_n(z), \dots$ for which

(a) $\deg \varphi_n = n$, $n \in \mathbf{Z}_+$;

(b) the leading coefficient of $\varphi_n(z) = \alpha_n z^n + \cdots$ is positive: $\alpha_n > 0$, $n \in \mathbf{Z}_+$;

(c) for all m, $n \in \mathbf{Z}_+$,

$$\frac{1}{2\pi}\int_{\boldsymbol{T}} \varphi_n(z)\overline{\varphi_m(z)}\, d\sigma(z) = \begin{cases} 0 & \text{if } n \neq m, \\ 1 & \text{if } n = m. \end{cases} \tag{1.2}$$

Definition. The polynomials $\{\varphi_n\}_0^\infty$ that satisfy conditions (a)–(c) are called *polynomials orthogonal on the circle* (with respect to σ).

2. We shall use the following notation. If $R(z) = \sum c_j z^j$ is a polynomial,

$$\overline{R}(z) = \overline{R(\overline{z})} = \sum \overline{c}_j z^j.$$

In addition, if R is a polynomial and $\deg R = n$,

$$R^*(z) = z^n \overline{R}(1/z)$$

(polynomial of degree at most n). We shall often use the relation $\overline{z} = 1/z$ if $z \in \boldsymbol{T}$.

Let us find the coefficients $\beta_k^{(n)}$ in the expansion

$$\varphi_n^*(z) = \sum_{k=0}^{n} \beta_k^{(n)} \varphi_k(z), \qquad n \in \mathbf{Z}_+.$$

They can be calculated by the Fourier formula

$$\beta_k^{(n)} = \frac{1}{2\pi} \int_T z^n \overline{\varphi}_n(1/z) \overline{\varphi_k(z)}\, d\sigma(z).$$

Hence

$$\begin{aligned}\overline{\beta}_k^{(n)} &= \frac{1}{2\pi} \int_T \frac{1}{z^n} \varphi_n(z) \varphi_k(z)\, d\sigma(z) \\ &= \frac{1}{2\pi} \int_T \varphi_n(z) \left(\frac{\varphi_k(0)}{z^n} + \frac{h_1}{z^{n-1}} + \cdots + h_n \right) d\sigma(z);\end{aligned}$$

$h_1, h_2, \ldots,$ here and later, are constants (different each time). If we use the orthogonality relations (1.2), we obtain

$$\begin{aligned}\overline{\beta}_k^{(n)} &= \varphi_k(0) \frac{1}{2\pi} \int_T \frac{\varphi_n(z)}{z^n}\, d\sigma(z) \\ &= \frac{\varphi_k(0)}{\alpha_n} \frac{1}{2\pi} \int_T \varphi_n(z) \left(\frac{\alpha_n}{z^n} + \frac{h_1}{z^{n-1}} + \cdots + h_n \right) d\sigma(z) \\ &= \frac{\varphi_k(0)}{\alpha_n} \frac{1}{2\pi} \int_T \varphi_n(z) \overline{\varphi_n(z)}\, d\sigma(z) = \frac{\varphi_k(0)}{\alpha_n}.\end{aligned}$$

Therefore

$$\alpha_n \varphi_n^*(z) = \sum_{k=0}^{n} \overline{\varphi_k(0)} \varphi_k(z), \qquad n \in \mathbf{Z}_+. \tag{1.3}$$

It follows from (1.3) that

$$\alpha_{n+1} \varphi_{n+1}^*(z) - \alpha_n \varphi_n^*(z) = \overline{\varphi_{n+1}(0)} \varphi_{n+1}(z), \qquad n \in \mathbf{Z}_+.$$

If we set $z = 0$ here, and use $\varphi_n^*(0) = \alpha_n$, we obtain

$$\alpha_{n+1}^2 - \alpha_n^2 = |\varphi_{n+1}(0)|^2, \qquad n \in \mathbf{Z}_+.$$

Hence

$$0 < \alpha_n \leq \alpha_{n+1}, \qquad n \in \mathbf{Z}_+, \tag{1.4}$$

i.e., the positive numbers α_n form a nondecreasing sequence.

3. We now find the coefficients $\mu_k^{(n)}$ of the expansion

$$\frac{\varphi_n(z) - \alpha_n z^n}{z^{n-1}} = \sum_{k=0}^{n-1} \mu_k^{(n)} \overline{\varphi}_k \left(\frac{1}{z} \right), \qquad n \in \mathbf{N}.$$

We have

$$\begin{aligned}\mu_k^{(n)} &= \frac{1}{2\pi} \int_T \frac{\varphi_n(z) - \alpha_n z^n}{z^{n-1}} \varphi_k(z)\, d\sigma(z) \\ &= \frac{1}{2\pi} \int_T \varphi_n(z) \frac{\varphi_k(z)}{z^{n-1}}\, d\sigma(z) - \alpha_n \frac{1}{2\pi} \int_T z \varphi_k(z)\, d\sigma(z).\end{aligned} \tag{1.5}$$

The first integral in (1.5) is zero, by the orthogonality relation (1.2). Let us denote the second integral by λ_k, $k \in \mathbf{Z}_+$. Then $\mu_k^{(n)} = -\alpha_n\lambda_k$. Consequently

$$\varphi_n(z) = \alpha_n\left(z^n - z^{n-1}\sum_{k=0}^{n-1}\lambda_k\overline{\varphi}_k\left(\frac{1}{z}\right)\right), \qquad n \in \mathbf{Z}_+.$$

Hence we obtain the relation

$$\frac{\varphi_{n+1}(z)}{\alpha_{n+1}} - \frac{z\varphi_n(z)}{\alpha_n} = -\lambda_n\varphi_n^*(z), \qquad n \in \mathbf{Z}_+.$$

If we set $z = 0$ here, we obtain

$$\lambda_n = -\frac{\varphi_{n+1}(0)}{\alpha_n\alpha_{n+1}}, \qquad n \in \mathbf{Z}_+.$$

Therefore we have arrived at the recursion relation

$$\alpha_n\varphi_{n+1}(z) = \alpha_{n+1}z\varphi_n(z) + \varphi_{n+1}(0)\varphi_n^*(z), \qquad n \in \mathbf{Z}_+. \tag{1.6}$$

Let

$$\Phi_n(z) = \frac{\varphi_n(z)}{\alpha_n} = z^n + \cdots$$

be orthogonal polynomials with leading coefficient unity, and

$$a_n = \overline{-\Phi_{n+1}(0)},$$

where $n \in \mathbf{Z}_+$. The numbers a_n, called *cyclic parameters*, play a fundamental role in the theory of polynomials orthogonal on the circle.

For the polynomials Φ_n the recursion relation (1.6) takes the form

$$\Phi_{n+1}(z) = z\Phi(z) - \overline{a}_n\Phi_n^*(z), \qquad n \in \mathbf{Z}_+. \tag{1.7}$$

If we replace z by $1/\overline{z}$ in (1.7), take the complex conjugate, and multiply the resulting equation by z^{n+1}, we obtain a second recursion relation

$$\Phi_{n+1}^*(z) = \Phi_n^*(z) - a_nz\Phi_n^*(z), \qquad n \in \mathbf{Z}_+. \tag{1.8}$$

4. Let $h(z)$ be a polynomial with no zeros in the disk $|z| \le 1$. We introduce the measure

$$d\sigma(z) = \frac{|dz|}{|h(z)|^2}.$$

(We suppose, without loss of generality, that $h(0) > 0$.) As G. Szegő showed, in this case we can find the orthogonal polynomials φ_n successively, starting with index $m = \deg h$. For this purpose, we set

$$\varphi_n(z) = z^{n-m}h^*(z), \qquad n = m, m+1, \dots.$$

We have

$$\begin{aligned}\frac{1}{2\pi}\int_T \varphi_n(z)z^{-\nu}\,d\sigma(z) &= \frac{1}{2\pi}\int_T z^{n-\nu}\overline{h}(1/z)\,d\sigma(z)\\ &= \frac{1}{2\pi}\int_T z^{n-\nu}\overline{h}(1/z)\frac{|dz|}{h(z)\overline{h}(1/z)} = \frac{1}{2\pi}\int_T \frac{z^{n-\nu}}{h(z)}|dz|.\end{aligned}$$

Let us note that $|dz| = dz/(iz)$ for $z \in \boldsymbol{T}$, since

$$z = e^{i\theta}, \quad \theta \in \mathbf{R}; \qquad dz = ie^{i\theta}\, d\theta, \qquad |dz| = d\theta.$$

Therefore if $\nu + 1 \le n$, by Cauchy's theorem we have

$$\frac{1}{2\pi}\int_{\boldsymbol{T}} \varphi_n(z) z^{-\nu}\, d\sigma(z) = \frac{1}{2\pi i}\int_{\boldsymbol{T}} \frac{z^{n-\nu-1}}{h(z)}\, dz = 0,$$

since the integrand is a function that is holomorphic in the closed unit disk. In addition,

$$\|\varphi_n\|^2 = \frac{1}{2\pi}\int_{\boldsymbol{T}} |h(z)|^2 \frac{|dz|}{|h(z)|^2} = 1. \quad \triangleright$$

The simplest system of polynomials orthogonal on the circle is the system of powers,

$$\varphi_n(z) = z^n, \qquad n \in \mathbf{Z}_+,$$

corresponding to the case $h(z) \equiv 1$.

REMARK.

On polynomials orthogonal on the circle, see [87], [31].

§2. An extremal problem and the Szegő condition

1. We shall use the following notation:

$$\ln^+ x = \begin{cases} \ln x & \text{if } x \ge 1, \\ 0 & \text{if } 0 < x < 1, \end{cases}$$
$$\ln^- x = \begin{cases} 0 & \text{if } x \ge 1, \\ -\ln x & \text{if } 0 < x < 1. \end{cases}$$

Then

$$\ln x = \ln^+ x - \ln^- x, \quad |\ln x| = \ln^+ x + \ln^- x, \quad \ln^\pm x \le |\ln x|.$$

DEFINITION. We say that a function p (nonnegative, and summable on $[0, 2\pi]$ with respect to Lebesgue measure) satisfies the *Szegő condition* if

$$\int_0^{2\pi} \ln p(\theta)\, d\theta > -\infty. \tag{2.1}$$

In particular, a necessary condition for the validity of (2.1) is that $p(\theta) > 0$ on a set of full measure. We note that (2.1) is equivalent to the summability of $\ln p(\theta)$ or of $\ln^- p(\theta)$, since if p is summable the integral

$$\int_0^{2\pi} \ln^+ p(\theta)\, d\theta \le \int_0^{2\pi} p(\theta)\, d\theta < +\infty$$

converges.

DEFINITION. The *geometric mean* of a nonnegative measurable function p such that

$$\int_0^{2\pi} \ln^+ p(\theta)\, d\theta < +\infty$$

is the number

$$\mathfrak{G}(p) = \exp\left\{\frac{1}{2\pi}\int_0^{2\pi} \ln p(\theta)\, d\theta\right\}.$$

Notice that $\mathfrak{G}(p) > 0$ if (2.1) is satisfied, and $\mathfrak{G}(p) = 0$ otherwise. We have the following elementary properties:

$$\mathfrak{G}(pq) = \mathfrak{G}(p)\mathfrak{G}(q), \qquad \mathfrak{G}(1/p) = 1/\mathfrak{G}(p) \qquad (\mathfrak{G}(p) \neq 0).$$

We shall also need the inequality between the arithmetic and geometric means:

$$\frac{1}{2\pi}\int_0^{2\pi} \exp\{f(\theta)\}\, d\theta \geq \exp\left\{\frac{1}{2\pi}\int_0^{2\pi} f(\theta)\, d\theta\right\}, \tag{2.2}$$

which is valid for all real summable functions f. More generally,

$$\int_\Delta \exp\{f(\theta)\}\, d\mu(\theta) \geq \exp\left\{\int_\Delta f(\theta)\, d\mu(\theta)\right\} \tag{2.3}$$

for every probability measure μ on an interval Δ and for every real function f that is summable with respect to μ. (In essence, inequalities (2.2) and (2.3) illustrate the convexity of the exponential function. For a measure μ with finite spectrum, (2.3) is an elementary inequality which can be proved by induction; for general μ this can be done by an easy limiting process.)

2. Let σ be a measure on $\boldsymbol{T}$; $\{\varphi_n\}$ polynomials orthogonal on the circle with respect to σ; and α_n the leading coefficient of φ_n. Then we see easily that

$$\frac{1}{\alpha_n^2} = \min \frac{1}{2\pi}\int_{\boldsymbol{T}} |R(z)|^2\, d\sigma(z), \tag{2.4}$$

where the minimum is taken over all polynomials R of degree n with unit leading coefficient and the minimum is attained for, and only for, the polynomial $R = \Phi_n$.

Set

$$d\sigma(e^{i\theta}) = p(\theta)\, d\theta + d\nu(e^{i\theta}), \tag{2.5}$$

where p is a nonnegative function, summable on $[0, 2\pi]$, and ν is the sum of the singular and discrete components of σ, so that

$$\frac{d\nu(e^{i\theta})}{d\theta} = 0 \quad \text{for a.e. } \theta \in [0, 2\pi).$$

(Here and later, "a.e." means "almost everywhere [or "for almost every"] with respect to Lebesgue measure.") In what follows we shall denote a singular measure by ν. If

$$\tilde{\sigma}(\theta) = \int_0^\theta d\sigma(e^{i\theta}), \qquad 0 \leq \theta < 2\pi,$$

is a distribution function, then $\tilde{\sigma}'(\theta) = p(\theta)$ for a.e. θ. We also denote the distribution function of a measure ν by $\tilde{\nu}$. In addition, we say that a measure σ satisfies the Szegő condition if the weight function p satisfies this condition.

By (1.4), the numbers α_n do not decrease, and consequently the limit

$$\mu_0 = \lim_{n\to\infty} \frac{1}{\alpha_n^2}$$

exists. The basic result of the present section is the following theorem.

THEOREM 2.1 (Szegő-Kolmogorov-Kreĭn). *For every measure of the form* (2.5) *the equality*

$$\mu_0 = \mathfrak{G}(p)$$

is satisfied.

There are several proofs of Theorem 2.1. (See [87] and [40].) Here we shall give a direct proof that makes a minimal use of references to other facts.

3. We have the following lemma.

LEMMA 2.1. *For every measure* σ *of the form* (2.5), *the inequality*

$$\mathfrak{G}(p) \le \mu_0 \le \inf_k \left\{ \mathfrak{G}(k)^{-1} \frac{1}{2\pi} \int_0^{2\pi} k(\theta)\, d\tilde{\sigma}(\theta) \right\} \tag{2.6}$$

is satisfied, where the lower bound is taken over the 2π*-periodic positive continuous functions* $k(\theta)$.

PROOF. Since $|R(e^{i\theta})| = |R^*(e^{i\theta})|$ (where R is a polynomial), we have

$$\begin{aligned} \frac{1}{\alpha_n^2} &= \min_R \frac{1}{2\pi} \int_0^{2\pi} |R(e^{i\theta})|^2\, d\tilde{\sigma}(\theta) \\ &\ge \min_R \frac{1}{2\pi} \int_0^{2\pi} |R(e^{i\theta})|^2 p(\theta)\, d\theta = \min_R \frac{1}{2\pi} \int_0^{2\pi} |R^*(e^{i\theta})|^2 p(\theta)\, d\theta, \end{aligned}$$

where R runs through the polynomials of degree n with leading coefficient unity. Notice that then $S = R^*$ runs through the polynomials of degree at most n for which $S(0) = 1$. Hence, by (2.2),

$$\frac{1}{\alpha_n^2} \ge \min_S \exp \frac{1}{2\pi} \int_0^{2\pi} (\ln p(\theta) + 2\ln|S(e^{i\theta})|)\, d\theta.$$

We remark that

$$\frac{1}{2\pi} \int_0^{2\pi} \ln|1 + \beta e^{i\theta}|\, d\theta = \ln^+ |\beta| \ge 0.$$

In fact, first let $|\beta| \le 1$; then, by the mean value theorem for harmonic functions, we obtain

$$\frac{1}{2\pi} \int_0^{2\pi} \ln|1 + \beta e^{i\theta}|\, d\theta = \frac{1}{2\pi} \int_0^{2\pi} \operatorname{Re}\ln(1 + \beta e^{i\theta})\, d\theta = \operatorname{Re}\ln(1 + \beta z)|_{z=0} = 0.$$

If, on the other hand, $|\beta| > 1$, then by what was said above

$$\frac{1}{2\pi}\int_0^{2\pi} \ln|1+\beta e^{i\theta}|\,d\theta = \ln|\beta| + \frac{1}{2\pi}\int_0^{2\pi} \ln\left|1+\frac{e^{i\theta}}{\beta}\right|\,d\theta = \ln|\beta|.$$

If $S(0) = 1$, the polynomial S has the following decomposition into linear factors:

$$S(z) = \prod_j \left(1 - \frac{z}{z_j}\right).$$

Consequently

$$\frac{1}{2\pi}\int_0^{2\pi} \ln|S(e^{i\theta})|\,d\theta = \sum_j \ln^+ \frac{1}{|z_j|} \geq 0.$$

Therefore

$$\frac{1}{\alpha_n^2} \geq \exp\left\{\frac{1}{2\pi}\int_0^{2\pi} \ln p(\theta)\,d\theta\right\} = \mathfrak{G}(p).$$

This completes the proof of the first inequality in (2.6).

Now let $T(\theta)$ be any trigonometric polynomial such that $T(\theta) > 0$ for all θ. By the Riesz-Fejér theorem, we have the representation

$$T(\theta) = |h(e^{i\theta})|^2,$$

where h is an algebraic polynomial with no zeros in the disk $|z| \leq 1$. By the mean value theorem for harmonic functions, we have

$$|h(0)|^2 = \exp\left\{\frac{1}{2\pi}\int_0^{2\pi} \ln|h(e^{i\theta})|^2\,d\theta\right\} = \mathfrak{G}(T).$$

In addition, by (2.4) we have, for $n \geq \deg h$,

$$\mu_0 \leq \frac{1}{\alpha_n^2} \leq \frac{1}{2\pi}\int_0^{2\pi} \left|\frac{h(e^{i\theta})}{h(0)}\right|^2 d\tilde{\sigma}(\theta) = \mathfrak{G}(T)^{-1}\frac{1}{2\pi}\int_0^{2\pi} T(\theta)\,d\tilde{\sigma}(\theta).$$

If we use the Weierstrass theorem on the completeness of the trigonometric polynomials in the space of continuous 2π-periodic functions, we obtain

$$\mu_0 \leq \mathfrak{G}(k)^{-1}\frac{1}{2\pi}\int_0^{2\pi} k(\theta)\,d\tilde{\sigma}(\theta), \tag{2.7}$$

where k is any continuous 2π-periodic positive function. We denote the set of such functions by $\mathscr{K}^+$. The second inequality in (2.6) follows from (2.7). ▷

4. Let us set

$$\mathscr{W}(k) = \mathfrak{G}(k)^{-1}\frac{1}{2\pi}\int_0^{2\pi} k(\theta)\,d\tilde{\sigma}(\theta), \qquad k \in \mathscr{K}^+. \tag{2.8}$$

LEMMA 2.2. *For every measure σ of the form (2.5), we have*

$$\inf \mathscr{W}(k) = \mathfrak{G}(p), \qquad k \in \mathscr{K}^+.$$

PROOF. (Suggested by S. Sh. Galstyan.) In addition to (2.8) we introduce the notation

$$\mathscr{V}(k) = \exp\{\mathscr{T}_1(k)\}\mathscr{T}_2(k), \qquad k \in \mathscr{K}^+,$$

where

$$\mathscr{T}_1(k) = \frac{1}{2\pi}\int_0^{2\pi} \ln\frac{1}{k(\theta)}\,d\theta, \qquad \mathscr{T}_2(k) = \frac{1}{2\pi}\int_0^{2\pi} k(\theta)p(\theta)\,d\theta.$$

It is clear that

$$\mathfrak{G}(p) \le \mathscr{V}(k) \le \mathscr{W}(k), \qquad k \in \mathscr{K}^+. \tag{2.9}$$

Let us prove that

$$v \ge w, \tag{2.10}$$

where $v = \inf\mathscr{V}(k)$, $w = \inf\mathscr{W}(k)$, $k \in \mathscr{K}^+$.

We fix $k \in \mathscr{K}^+$ and choose $\varepsilon > 0$. Then we find a number $\delta > 0$ such that

$$\mathscr{V}(k) + \varepsilon > \exp\{\mathscr{T}_1(k) + 2\delta\}(\mathscr{T}_2(k) + 4\delta). \tag{2.11}$$

We select an $M > 1$ such that

$$\frac{1}{M} < k(\theta) < M \tag{2.12}$$

for all $\theta \in I = [0, 2\pi)$, and

$$\frac{1}{2\pi M}\tilde{\nu}(I) < \delta. \tag{2.13}$$

It follows from the absolute continuity of the Lebesgue integral, and the singular nature of the measure $\tilde{\nu}$, that there are disjoint sets F_1 and F_2, closed in I, such that

$$\frac{1}{2\pi}\int_{I\setminus F_1} \ln M\,d\theta < \delta, \qquad \frac{1}{2\pi}\int_{I\setminus F_1} M_p(\theta)\,d\theta < \delta, \tag{2.14}$$

$$\frac{1}{2\pi}\int_{I\setminus F_2} M\,d\tilde{\nu}(\theta) < \delta. \tag{2.15}$$

We now construct a continuous 2π-periodic function $k_\varepsilon(\theta)$ such that $k_\varepsilon(\theta) = k(\theta)$ if $\theta \in F_1$; $k_\varepsilon(\theta) = 1/M$ if $\theta \in F_2$; and $1/M \le k_\varepsilon(\theta) \le M$ for all θ.

By (2.12), (2.14), and the construction of k_ε, we have

$$|\mathscr{T}_1(k_\varepsilon) - \mathscr{T}_1(k)| \le \frac{2}{2\pi}\int_{I\setminus F_1} \ln M\,d\theta < 2\delta, \tag{2.16}$$

$$|\mathscr{T}_2(k_\varepsilon) - \mathscr{T}_2(k)| \le \frac{2}{2\pi}\int_{I\setminus F_1} Mp(\theta)\,d\theta < 2\delta. \tag{2.17}$$

By (2.13), (2.15), and the construction of k_ε, we have

$$\frac{1}{2\pi}\int_0^{2\pi} k_\varepsilon(\theta)\,d\tilde{\nu}(\theta) \le \frac{1}{2\pi}\int_{F_2} \frac{1}{M}d\tilde{\nu}(\theta) + \frac{1}{2\pi}\int_{I\setminus F_2} M\,d\tilde{\nu}(\theta) < 2\delta. \tag{2.18}$$

Consequently, by (2.11) and (2.16)–(2.18),

$$\mathscr{W}(k_\varepsilon) \le \exp\{\mathscr{T}_1(k) + 2\delta\}(\mathscr{T}_2(k) + 2\delta + 2\delta) \le \mathscr{V}(k) + \varepsilon.$$

Thus we have proved that for all $k \in \mathscr{K}^+$ and $\varepsilon > 0$ there is a function $k_\varepsilon \in \mathscr{K}^+$ such that $\mathscr{W}(k_\varepsilon) \le \mathscr{V}(k) + \varepsilon$. From this, (2.10) evidently follows.

Comparing (2.9) and (2.10), we obtain

$$v = w. \tag{2.19}$$

Let us show that

$$v = \mathfrak{G}(p), \tag{2.20}$$

from which the conclusion of the lemma follows by using (2.19). We define the sequence of functions

$$p_n(\theta) = \begin{cases} n & \text{if } p(\theta) > n, \\ p(\theta) & \text{if } 1/n \le p(\theta) \le n, \\ 1/n & \text{if } p(\theta) < 1/n, \end{cases}$$

where $n = 2, 3, \ldots$. Evidently $p_n(\theta) \to p(\theta)$ as $n \to \infty$ for all θ. In addition,

$$|\ln p_n(\theta)| \le |\ln p(\theta)|$$

and

$$|p(\theta)/p_n(\theta)| \le \max\{1, p(\theta)\}.$$

Let the function p satisfy the Szegő condition; then it follows from Lebesgue's theorem on taking limits under the integral sign that

$$\begin{aligned} \lim_{n\to\infty} \mathscr{V}\left(\frac{1}{p_n}\right) &= \lim_{n\to\infty} \exp\left\{\frac{1}{2\pi}\int_0^{2\pi} \ln p_n(\theta)\, d\theta\right\} \frac{1}{2\pi}\int_0^{2\pi} \frac{p(\theta)}{p_n(\theta)}\, d\theta \\ &= \exp\left\{\frac{1}{2\pi}\int_0^{2\pi} \ln p(\theta)\, d\theta\right\} = \mathfrak{G}(p). \end{aligned} \tag{2.21}$$

If p does not satisfy the Szegő condition, (2.21) is equally valid by the very definition of the Lebesgue integral.

By (2.21), for every $\varepsilon > 0$ we can find $\delta > 0$ and n such that

$$\exp\left\{\mathscr{T}_1\left(\frac{1}{p_n}\right) + \delta\right\}\left(\mathscr{T}_2\left(\frac{1}{p_n}\right) + \delta\right) < \mathfrak{G}(p) + \varepsilon. \tag{2.22}$$

By Luzin's theorem, for every $\eta > 0$ there is a continuous 2π-periodic function l_η such that

$$\frac{1}{n} \le l_\eta \le n$$

and

$$\operatorname{meas}\{\theta \in I : l_\eta(\theta) \ne p_n(\theta)\} < \eta.$$

We can choose η so that

$$\mathscr{T}_1\left(\frac{1}{l_\eta}\right) < \mathscr{T}_1\left(\frac{1}{p_n}\right) + \delta, \qquad \mathscr{T}_2\left(\frac{1}{l_\eta}\right) < \mathscr{T}_2\left(\frac{1}{p_n}\right) + \delta.$$

Then (2.22) implies the inequality

$$\mathscr{V}\left(\frac{1}{l_\eta}\right) < \mathfrak{G}(p) + \varepsilon.$$

This implies the inequality $v \leq \mathfrak{G}(p)$. The opposite inequality $v \geq \mathfrak{G}(p)$ was proved in Lemma 2.1; therefore (2.20) is established. ▷

Theorem 2.1 follows from Lemma 2.2.

5. Theorem 2.1 initiated many investigations on the theory of orthogonal polynomials and the theory of boundary values of analytic functions. Here is an application of this theorem.

THEOREM 2.2. *Let σ be a measure of the form* (2.5). *The system $\{z^n\}_0^\infty$ is complete in L^2_σ if and only if the weight function p fails to satisfy the Szegő condition.*

PROOF. If p satisfies the Szegő condition, Theorem 2.1 implies

$$\min_R \frac{1}{2\pi}\int_{\boldsymbol{T}}\left|\frac{1}{z} + R(z)\right|^2 d\sigma(z) = \mathfrak{G}(p) > 0,$$

where the minimum is taken over all polynomials. Consequently the function $1/z \in L^2_\sigma$ cannot be approximated arbitrarily closely by polynomials (in the L^2_σ norm).

If p does not satisfy the Szegő condition, we have

$$\min_R \frac{1}{2\pi}\int_{\boldsymbol{T}}\left|\frac{1}{z^n} + \frac{1}{z^{n-1}}R(z)\right|^2 d\sigma(z) = \mathfrak{G}(p) = 0,$$

$n = 1, 2, \ldots$. Consequently the function $1/z^n$ belongs to the closure of the set of algebraic polynomials in L^2_σ. Since the system $\{z^n\}_{-\infty}^\infty$ is complete in L^2_σ, the system $\{z^n\}_0^\infty$ is also complete in L^2_σ. ▷

PROBLEMS.

1. Prove the inequality (2.3) between the arithmetic and geometric means.
2. Prove the Riesz-Fejér theorem (see subsection 3).

§3. Boundary values of analytic functions

We present a brief survey of the theory of boundary values of functions that are analytic in the unit disk.

1. Let σ be a complex measure on $\boldsymbol{T}$, and

$$\sum_{n=-\infty}^{+\infty} A_n e^{in\theta} \tag{3.1}$$

its Fourier series, i.e.,

$$A_n = \frac{1}{2\pi}\int_0^{2\pi} e^{-in\theta}\, d\tilde{\sigma}(\theta), \qquad n \in \mathbf{Z} \tag{3.2}$$

($\tilde{\sigma}$ is the distribution function of σ). The *Poisson-Abel mean* of the series (3.1) is the function

$$u(re^{i\theta}) = \sum_{n=-\infty}^{+\infty} A_n r^{|n|} e^{in\theta}, \qquad 0 \le r < 1, \theta \in \mathbf{R}, \tag{3.3}$$

harmonic in the unit disk $U = \{z \in \mathbf{C} \colon |z| < 1\}$. (Here we are using an extended interpretation of the term "harmonic function." By this we mean a complex-valued function whose real and imaginary parts are harmonic.) If we substitute (3.2) into (3.3) and integrate term by term, we obtain the *Poisson integral*

$$u(re^{i\theta}) = \frac{1}{2\pi} \int_0^{2\pi} P_r(\theta - t)\, d\tilde{\sigma}(t),$$

where

$$P_r(\varphi) = \frac{1 - r^2}{1 - 2r\cos\varphi + r^2}, \qquad 0 \le r < 1, \varphi \in \mathbf{R},$$

is the *Poisson kernel.*

We shall frequently use *Fatou's theorem*, the proof of which can be found, for example, in [43].

THEOREM (Fatou). *At every point θ_0 for which the limit*

$$\lim_{h\to 0} \frac{1}{2h} \int_{\theta_0 - h}^{\theta_0 + h} d\tilde{\sigma}(\theta) = p(\theta_0) \tag{3.4}$$

exists, we have the limit

$$\lim_{z \to e^{i\theta_0}} u(z) = p(\theta_0),$$

where $z \in U$ tends nontangentially to $e^{i\theta_0}$, i.e., remaining inside an arbitrary (but fixed) sector with vertex $e^{i\theta_0}$ which is symmetric with respect to the radius through this point and has opening less than π.

We note that condition (3.4) is satisfied for almost all points $\theta_0 \in [0, 2\pi)$ (with respect to Lebesgue measure).

2. Let us introduce the *Hardy classes* H^δ, $0 < \delta \le \infty$. A function f, analytic in the disk U, belongs to class H^δ if

$$\|f\|_\delta = \sup_{0 \le r < 1} \int_0^{2\pi} |f(re^{i\theta})|^\delta \, d\theta < +\infty$$

for $0 < \delta < 1$;

$$\|f\|_\delta = \sup_{0 \le r < 1} \left(\int_0^{2\pi} |f(re^{i\theta})|^\delta \, d\theta \right)^{1/\delta} < +\infty$$

for $1 \le \delta < \infty$; and

$$\|f\|_\infty = \sup_{z \in U} |f(z)| < +\infty$$

for $\delta = \infty$.

It is clear that if $1 \le \delta \le \infty$ the function $\|\cdot\|_\delta$ is a norm in the linear space H^δ. However, if $0 < \delta < 1$, $\|\cdot\|_\delta$ is not positively homogeneous, but at least satisfies the triangle inequality $\|f+g\|_\delta \le \|f\|_\delta + \|g\|_\delta$, so that $\|f-g\|_\delta$ provides H^δ with the structure of a metric space.

We discuss the special case $\delta = 2$ in more detail. A function

$$f(z) = \sum_{n=0}^{\infty} A_n z^n \tag{3.5}$$

that is analytic in U belongs to class H^2 if and only if

$$\sum_{n=0}^{\infty} |A_n|^2 < +\infty. \tag{3.6}$$

Let us prove this. We have the Parseval equality

$$\frac{1}{2\pi}\int_0^{2\pi} |f(re^{i\theta})|^2\,d\theta = \sum_{n=0}^{\infty} |A_n|^2 r^{2n}, \qquad 0 \le r < 1. \tag{3.7}$$

If inequality (3.6) is satisfied, it is evident that the numbers (3.7) are collectively bounded. Conversely, let

$$\sum_{n=0}^{\infty} |A_n|^2 r^{2n} \le C < \infty, \qquad 0 \le r < 1.$$

Then for all $N = 0, 1, 2, \ldots$ we have

$$\sum_{n=0}^{N} |A_n|^2 = \lim_{r\to 1-} \sum_{n=0}^{N} |A_n|^2 r^{2n} \le C. \quad \triangleright$$

Thus the function $f(e^{i\theta}) = \sum_{n=0}^{\infty} A_n e^{in\theta} \in L^2(0, 2\pi)$ is defined (the series converges in the $L^2(0, 2\pi)$ norm). Then (see subsection 1)

$$f(re^{i\theta}) = \frac{1}{2\pi}\int_0^{2\pi} P_r(\theta - t) f(e^{it})\,dt, \qquad 0 \le r < 1,\ \theta \in \mathbf{R}.$$

And then by Fatou's theorem

$$\lim_{z\to e^{i\theta}} f(z) = f(e^{i\theta}) \quad \text{a.e. } \theta \in [0, 2\pi).$$

It also follows from the preceding discussion that

$$\lim_{r\to 1-} \int_0^{2\pi} |f(re^{i\theta}) - f(e^{i\theta})|^2\,d\theta = 0,$$

in particular,

$$\|f\|_2 = \left(\int_0^{2\pi} |f(e^{i\theta})|^2\,d\theta\right)^{1/2}. \tag{3.8}$$

We also observe that if a sequence of functions $f_n \in H^2$ converges to $f \in H^2$ in this space as $n \to \infty$, the convergence will be uniform on compact subsets of U. In fact, for every function f of the form (3.5) in class H^2 we have, by the Cauchy-Bunyakovskiĭ inequality and (3.7),

$$|f(z)|^2 = \left|\sum_{n=0}^{\infty} A_n z^n\right|^2 \le \sum_{n=0}^{\infty} |A_n|^2 \sum_{n=0}^{\infty} |z|^{2n} = \frac{\|f\|_2^2}{1-|z|^2}.$$

Besides the classes H^δ we introduce the wider class (N), the *Nevanlinna class*. We say that $f \in$ (N) if f is analytic in the disk U and the number

$$N(f) = \sup_{0\le r<1} \frac{1}{2\pi}\int_0^{2\pi} \ln^+|f(re^{i\theta})|\,d\theta < +\infty$$

is finite. The class (N) is the widest class for which it is natural to study boundary properties of analytic functions.

3. We shall establish a number of properties of functions of class (N). Before starting, we recall the *Poisson-Jensen formula* and introduce the *Blaschke product*.

If f is analytic in the disk U and has no zeros in the disk $|z| \le r < 1$, then by the mean value theorem for harmonic functions we have

$$\frac{1}{2\pi}\int_0^{2\pi} \ln|f(re^{i\theta})|\,d\theta = \ln|f(0)|.$$

If f is analytic in U, has zeros at the points $\{a_k\}$ in the domain $0 < |z| < r$ (counted according to multiplicity), has a zero of order m at $z = 0$: $f(z) = cz^m + \cdots$, and has no other zeros in the disk $|z| \le r$, we can apply the preceding formula to the function

$$F(z) = f(z)\left(\frac{z}{r}\right)^{-m}\left(\prod_k \frac{(z-a_k)r}{r^2-\bar{a}_k z}\right)^{-1},$$

for which

$$\left|\left(\frac{z}{r}\right)^m \prod_k \frac{(z-a_k)r}{r^2-\bar{a}_k z}\right| = 1$$

for $|z| = r$, obtaining the Poisson-Jensen formula

$$\frac{1}{2\pi}\int_0^{2\pi} \ln|f(re^{i\theta})|\,d\theta = \ln|c| + m\ln r + \sum_k \ln\frac{r}{|a_k|}. \tag{3.9}$$

In particular, from (3.9) we obtain *Jensen's inequality*

$$\frac{1}{2\pi}\int_0^{2\pi} \ln|f(re^{i\theta})|\,d\theta \ge \ln|f(0)|.$$

Let there be given a sequence $\{a_k\}$ of complex numbers such that $0 < |a_k| < 1$, $k = 1, 2, \ldots$. We construct an infinite product of fractional linear automorphisms of the unit disk:

$$B(z) = \prod_{k=1}^{\infty} \frac{|a_k|}{a_k} \frac{a_k - z}{1 - \overline{a}_k z}, \tag{3.10}$$

which (if convergent) is called a Blaschke product, and consider its convergence.

If $z = 0$, the product

$$B(0) = \prod_{k=1}^{\infty} |a_k| \tag{3.11}$$

is known to converge if and only if the series

$$\sum_{k=1}^{\infty} (1 - |a_k|) < +\infty \tag{3.12}$$

converges. Conversely, let (3.12) be satisfied, and, in particular, let $|a_k| \to 1$ as $k \to \infty$. We have

$$1 - \frac{|a_k|}{a_k} \frac{a_k - z}{1 - \overline{a}_k z} = (1 - |a_k|) \frac{a_k + z|a_k|}{a_k(1 - \overline{a}_k z)}. \tag{3.13}$$

For every compact set $K \subset U$, let there exist an index $k_0 = k_0(K)$ and a number $C = C(K)$ such that

$$\left| \frac{a_k + z|a_k|}{a_k(1 - \overline{a}_k z)} \right| \le C, \qquad k = 1, 2, \ldots, \quad z \in K,$$

and the factors in (3.10) have no zeros in K when $k \ge k_0$. Consequently, the series of terms (3.13) converges absolutely and uniformly on K, and the product (3.10) with $k \ge k_0$ converges uniformly on K. In this case, we simply say that the product (3.10) converges uniformly inside U.

Thus we have proved that the product (3.10) converges uniformly inside U if and only if (3.12) is satisfied. In this case, $B(z)$ is analytic in U by the Weierstrass theorem. Evidently $B(a_k) = 0$ for $k = 1, 2, \ldots$ (zeros counted according to multiplicity), and $|B(z)| < 1$ for $z \in U$.

We may consider finite Blaschke products, for which there is no question of convergence. If a finite number of the a_k are zero, the corresponding factor is to be replaced by z.

LEMMA 3.1. *If $f \in (\mathrm{N})$ and $\{a_k\}_1^{\infty}$ are its zeros (counted according to multiplicity) for which $a_k \neq 0$, the series*

$$\sum_{k=1}^{\infty} (1 - |a_k|) < \infty \tag{3.14}$$

converges.

PROOF. If $f \in (\mathrm{N})$, then

$$\frac{1}{2\pi}\int_0^{2\pi} \ln|f(re^{i\theta})|\,d\theta \le \frac{1}{2\pi}\int_0^{2\pi} \ln^+|f(re^{i\theta})|\,d\theta \le N(f)$$

for $0 \le r < 1$. Then it follows by the Poisson-Jensen formula that for $0 < r < 1$, $r \ne |a_k|$,

$$\sum_{|a_k|<r} \ln\frac{r}{|a_k|} = \frac{1}{2\pi}\int_0^{2\pi} \ln|f(re^{i\theta})|\,d\theta - \ln|c| - m\ln r$$
$$\le N(f) - \ln|c| - m\ln r, \tag{3.15}$$

where $f(z) = cz^m + \cdots$. But the numbers on the right of (3.15) are collectively bounded. Hence the series

$$\sum_{k=1}^{\infty} \ln\frac{1}{|a_k|}$$

is convergent, and consequently so is the series (3.14). ▷

LEMMA 3.2. *If $f \in (\mathrm{N})$ we have*

$$f(z) = B(z)h(z),$$

where B is a function analytic in U, such that $|B(z)| \le 1$ for $z \in U$, $h \in (\mathrm{N})$, and h has no zeros in U.

PROOF. It follows from Lemma 3.1 that we can construct a Blaschke product $B(z)$ with the zeros $\{a_k\}$ of a function f in class (N). Then the function

$$h(z) = f(z)/B(z)$$

is analytic in U and has no zeros in this disk. Let us show that $h \in (\mathrm{N})$. For $0 < r < 1$ and $r \ne |a_k|$, we have

$$\frac{1}{2\pi}\int_0^{2\pi} \ln|B(re^{i\theta})|\,d\theta = \ln c + m\ln r + \sum_{0<|a_k|<r} \ln\frac{r}{|a_k|} \ge \ln c + m\ln r,$$

where m is the order of the zero of f at $z = 0$, and

$$c = \prod_{a_k \ne 0} |a_k|.$$

Consequently,

$$\frac{1}{2\pi}\int_0^{2\pi} \ln^+\left|\frac{f(re^{i\theta})}{B(re^{i\theta})}\right| d\theta$$
$$\le \frac{1}{2\pi}\int_0^{2\pi} \ln^+|f(re^{i\theta})|\,d\theta - \frac{1}{2\pi}\int_0^{2\pi} \ln|B(re^{i\theta})|\,d\theta$$
$$\le N(f) - \ln c - m\ln r. \quad ▷$$

THEOREM 3.1. *A function f that is analytic in the disk U belongs to class* (N) *if and only if*

$$f(z) = \frac{a(z)}{b(z)}, \tag{3.16}$$

where a and b are functions that are analytic in the unit disk and satisfy the conditions $|a(z)| \leq 1$, $|b(z)| \leq 1$ for $z \in U$.

PROOF. According to Lemma 3.2, in order to establish the necessity of (3.16) it is enough to consider a function $f \in (\mathrm{N})$ with no zeros in U. Without loss of generality, let $f(0) > 0$.

The function $\ln f(z)$ is analytic in U (we take the branch for which $\operatorname{Im} \ln f(0) = 0$). For $|\zeta| < r < 1$, we have

$$\ln f(\zeta) = \frac{1}{2\pi} \int_0^{2\pi} \ln |f(re^{i\theta})| \frac{re^{i\theta} + \zeta}{re^{i\theta} - \zeta}\, d\theta.$$

This is the *Poisson-Schwarz formula* (applied to $\ln f(z)$), which recovers a function, analytic in the disk, from its real part on the boundary of the disk (compare the Poisson formula in subsection 1). Hence

$$f(\zeta) = \frac{a_r(\zeta)}{b_r(\zeta)}, \qquad |\zeta| < r, \tag{3.17}$$

where

$$a_r(\zeta) = \exp\left\{ -\frac{1}{2\pi} \int_0^{2\pi} \ln^+ \left| \frac{1}{f(re^{i\theta})} \right| \frac{re^{i\theta} + \zeta}{re^{i\theta} - \zeta}\, d\theta \right\},$$

$$b_r(\zeta) = \exp\left\{ -\frac{1}{2\pi} \int_0^{2\pi} \ln^+ |f(re^{i\theta})| \frac{re^{i\theta} + \zeta}{re^{i\theta} - \zeta}\, d\theta \right\}.$$

Clearly $a_r(\zeta)$ and $b_r(\zeta)$ are functions that are analytic in the disk $|\zeta| < r$ and satisfy

$$|a_r(\zeta)| \leq 1, \qquad |b_r(\zeta)| \leq 1 \tag{3.18}$$

for $|\zeta| < r$. Now, by the Montel compactness principle, we can construct a sequence $r_1 < r_2 < \cdots$, $r_k \to 1$ as $k \to \infty$, such that

$$\lim_{k\to\infty} a_{r_k}(\zeta) = a(\zeta), \qquad \lim_{k\to\infty} b_{r_k}(\zeta) = b(\zeta)$$

for $|\zeta| < 1$, where a and b are functions that are analytic in U. We notice that $b \not\equiv 0$, since when $0 \leq r < 1$

$$b_r(0) = \exp\left\{ -\frac{1}{2\pi} \int_0^{2\pi} \ln^+ |f(re^{i\theta})|\, d\theta \right\} \geq \exp\{-N(f)\} > 0.$$

It remains only to take the limit as $k \to \infty$ in (3.17) and (3.18), where $r = r_k$.

Let $f(z) = a(z)/b(z)$ be the representation from the conclusion of the theorem. Let us show that $f \in (\mathrm{N})$. Without loss of generality, we may suppose that $b(0) \neq 0$. Using Jensen's inequality, we obtain

$$\frac{1}{2\pi}\int_0^{2\pi} \ln^+ \left|\frac{a(re^{i\theta})}{b(re^{i\theta})}\right| d\theta \leq \frac{1}{2\pi}\int_0^{2\pi} \ln \frac{1}{|b(re^{i\theta})|} d\theta \leq \ln \frac{1}{|b(0)|}. \quad \triangleright$$

COROLLARY. *If $f \in (\mathrm{N})$, the nontangential boundary value*

$$\lim_{z\to e^{i\theta}} f(z) = f(e^{i\theta}) \tag{3.19}$$

exists, and

$$\int_0^{2\pi} \ln^+ |f(e^{i\theta})|\, d\theta < \infty. \tag{3.20}$$

PROOF. We use the representation (3.16), where, without loss of generality, we assume $b(0) \neq 0$. Since $a, b \in H^\infty \subset H^2$, by what was proved in subsection 2, the nontangential boundary values $a(e^{i\theta})$ and $b(e^{i\theta})$ of these functions exist for a.e. θ. To prove (3.19), it is enough to prove that $b(e^{i\theta}) \neq 0$ on a set of full measure. Suppose the contrary, that is, let $b(e^{i\theta}) = 0$ for $\theta \in E \subset I = [0, 2\pi)$, where the Lebesgue measure of E is positive: $|E| > 0$. It is enough to consider the case $|E| < 2\pi$. We set

$$\chi(\theta) = \begin{cases} 1/|E| & \text{if } \theta \in E, \\ -1/|I\backslash E| & \text{if } \theta \in I\backslash E, \end{cases}$$

$$\varphi(z) = \exp\left\{\frac{1}{2\pi}\int_0^{2\pi} \frac{e^{i\theta}+z}{e^{i\theta}-z}\chi(\theta)\, d\theta\right\}.$$

Then φ is analytic and bounded in U, with $\varphi(0) = 1$. For every k, we have (see the discussion in subsection 2)

$$\varphi(0)^k b(0) = \frac{1}{2\pi}\int_0^{2\pi} \varphi(e^{i\theta})^k b(e^{i\theta})\, d\theta = \frac{1}{2\pi}\int_{I\backslash E} \varphi(e^{i\theta})^k b(e^{i\theta})\, d\theta.$$

Hence

$$|b(0)| \leq \exp\{-k/|I\backslash E|\}, \qquad k = 1, 2, \ldots, \tag{3.21}$$

since by Fatou's theorem

$$|\varphi(e^{i\theta})| = \exp\{-1/|I\backslash E|\}$$

for a.e. $\theta \in I\backslash E$. Since k is arbitrary, it follows from (3.21) that $b(0) = 0$; but this contradicts the choice of b.

Thus (3.19) is proved. Then (3.20) follows by Fatou's lemma. $\triangleright$

PROPOSITION 3.1. *If $f \in (\mathrm{N})$, we have*

$$\int_0^{2\pi} \Big|\ln |f(e^{i\theta})|\Big|\, d\theta < \infty. \tag{3.22}$$

PROOF. Without loss of generality, we may assume that $f(0) \neq 0$. Then it follows from the definition of (N) and Jensen's inequality that

$$\frac{1}{2\pi}\int_0^{2\pi}\left|\ln|f(re^{i\theta})|\right|d\theta \leq 2N(f) - \ln|f(0)|, \qquad 0 \leq r < 1.$$

Then (3.22) follows by Fatou's lemma. ▷

PROPOSITION 3.2. *Let* $B(z)$ *be the Blaschke product* (3.10). *Then*

$$|B(e^{i\theta})| = 1$$

for a.e. $\theta \in I = [0, 2\pi)$.

PROOF. Since $|B(z)| \leq 1$ for $z \in U$, we have

$$\ln|B(e^{i\theta})| \leq 0 \tag{3.23}$$

for a.e. $\theta \in I$. Without loss of generality, let $B(0) \neq 0$. Let us write the Poisson-Jensen formula:

$$\ln|B(0)| = \sum_{|a_n|<r} \ln\frac{|a_n|}{r} + \frac{1}{2\pi}\int_0^{2\pi} \ln|B(re^{i\theta})|\,d\theta, \tag{3.24}$$

where $0 < r < 1$, $r \neq a_n$. Since $|B(0)|$ is given by (3.11), we may write (3.24) in the form

$$\frac{1}{2\pi}\int_0^{2\pi} \ln|B(re^{i\theta})|\,d\theta = \sum_{|a_n|<r} \ln\frac{r}{|a_n|} - \sum_{n=1}^{\infty} \ln\frac{1}{|a_n|}. \tag{3.25}$$

If we take the limit in (3.25) as $r \to 1-$, we obtain

$$\overline{\lim_{r\to 1-}}\, \frac{1}{2\pi}\int_0^{2\pi} \ln|B(re^{i\theta})|\,d\theta \geq 0. \tag{3.26}$$

It follows from (3.23), (3.26), and Fatou's lemma that

$$\ln|B(e^{i\theta})| = 0$$

for a.e. $\theta \in I$. ▷

4. Now we are in a position to prove the factorization theorem. For this purpose, we introduce the concept of a *singular function*, namely

$$S(z) = \exp\left\{-\frac{1}{2\pi}\int_0^{2\pi} \frac{e^{i\theta}+z}{e^{i\theta}-z}\,d\mu(\theta)\right\},$$

where the nonnegative measure μ is singular, i.e., $\mu' = 0$ a.e.

THEOREM 3.2. *Every function* $f \in$ (N), $f \not\equiv 0$, *can be represented as a product*

$$f(z) = e^{ic}B(z)\frac{S_1(z)}{S_2(z)} \exp\left\{\frac{1}{2\pi}\int_0^{2\pi} \ln|f(e^{i\theta})|\frac{e^{i\theta}+z}{e^{i\theta}-z}\,d\theta\right\}, \tag{3.27}$$

where $c \in \mathbf{R}$, $B(z)$ *is the Blaschke product formed with the zeros of* f, *and* S_1 *and* S_2 *are singular functions.*

PROOF. By Lemma 3.2 and Proposition 3.2, we may suppose that $f(z) \neq 0$ in U. In addition, let $f(0) > 0$ (by the choice of c).

The family of charges

$$d\nu_r(\theta) = \ln|f(re^{i\theta})|\, d\theta, \qquad 0 \le r < 1, \tag{3.28}$$

is uniformly bounded on $[0, 2\pi]$ (see Proposition 3.1). Consequently we can construct a sequence $r_1 < r_2 < \cdots$ with $r_k \to 1$ as $k \to \infty$, such that ν_{r_k} converges weakly to a charge

$$d\nu(\theta) = \tilde{k}(\theta)\, d\theta + d\mu(\theta),$$

where $\tilde{k} \in L^1(0, 2\pi)$, and μ is the difference of two singular measures: $\mu = \mu_1 - \mu_2$.

By the Poisson-Schwarz formula, we have

$$f(r_k z) = \exp\left\{\frac{1}{2\pi}\int_0^{2\pi} \ln|f(r_k e^{i\theta})| \frac{e^{i\theta} + z}{e^{i\theta} - z}\, d\theta\right\}.$$

If we take the limit as $k \to \infty$, we obtain

$$f(z) = \exp\left\{\frac{1}{2\pi}\int_0^{2\pi} \tilde{k}(\theta)\frac{e^{i\theta} + z}{e^{i\theta} - z}\, d\theta + \frac{1}{2\pi}\int_0^{2\pi} \frac{e^{i\theta} + z}{e^{i\theta} - z}\, d\mu(\theta)\right\}.$$

By Fatou's theorem,

$$\lim_{z \to e^{i\theta}} \ln|f(z)| = \tilde{k}(\theta)$$

for a.e. $\theta \in [0, 2\pi)$. Therefore $\tilde{k}(\theta) = \ln|f(e^{i\theta})|$. To complete the proof, we have only to set

$$S_{1,2}(z) = \exp\left\{-\frac{1}{2\pi}\int_0^{2\pi} \frac{e^{i\theta} + z}{e^{i\theta} - z}\, d\mu_{2,1}(\theta)\right\}. \quad \triangleright$$

LEMMA 3.3. *Let* k_n *be a nonnegative measurable function on the interval* $I = [0, 2\pi]$, *with the property that*

$$\int_I k_n(\theta)^{1+\delta}\, d\theta \le c < \infty, \qquad n \in \mathbf{N},$$

where $\delta > 0$, *and the measures* $k_n(\theta)\, d\theta$ *converge weakly to a measure* μ *as* $n \to \infty$. *Then* μ *is absolutely continuous with respect to Lebesgue measure.*

PROOF. Let the closed set $E \subset I$ have Lebesgue measure zero: $|E| = 0$. Choose any $\eta > 0$. We find a number $K > 0$ for which $cK^{-\delta} < \eta/2$, and then an $\varepsilon > 0$ such that $K\varepsilon < \eta/2$. Let $G \supset E$ be an open subset of I and $|G| < \varepsilon$. We denote by ψ a continuous function on I for which $0 \le \psi(\theta) \le 1$, $\psi(\theta) = 1$ for $\theta \in E$, and $\psi(\theta) = 0$ for $\theta \in I \setminus G$. Then

$$\mu(E) \le \int_G \psi(\theta)\, d\mu(\theta) = \lim_{n \to \infty} \int_G \psi(\theta) k_n(\theta)\, d\theta \le \overline{\lim_{n \to \infty}} \int_G k_n(\theta)\, d\theta.$$

We set

$$G_n = \{\theta \in I \colon k_n(\theta) \le K\}, \qquad n \in \mathbf{N}.$$

Then

$$\begin{aligned}\int_G k_n(\theta)\,d\theta &= \left(\int_{G_n} + \int_{G\setminus G_n}\right) k_n(\theta)\,d\theta \\ &\le K|G| + K^{-\delta}\int_{G\setminus G_n} k_n(\theta)^{1+\delta}\,d\theta \le K\varepsilon + cK^{-\delta} < \eta.\end{aligned}$$

Since $\eta > 0$ was arbitrary, we obtain $\mu(E) = 0$. ▷

The preceding lemma lets us obtain the following supplement to Theorem 3.2.

THEOREM 3.3. *If $f \in H^p$, $p > 0$, $f \not\equiv 0$, we can take $S_2(z) \equiv 1$ in* (3.27).

PROOF. The positive parts of the charges (3.8) are

$$d\nu_r^+(\theta) = \ln^+|f(re^{i\theta})|\,d\theta, \qquad 0 \le r < 1.$$

Since $f \in H^p$, we evidently have

$$\sup_{0\le r<1}\int_0^{2\pi}(\ln^+|f(re^{i\theta})|)^{1+\delta}\,d\theta < \infty$$

for every $\delta > 0$. Then, by Lemma 3.3, the measure $d\nu^+$ is absolutely continuous, i.e., $\mu_1 \equiv 0$. ▷

PROBLEMS

1. If $0 < \delta_1 < \delta_2$, we have $H^{\delta_1} \subset H^{\delta_2}$, and also $H^\delta \subset (\mathrm{N})$ for $\delta > 0$.

2. When $\delta \ge 1$, H^δ is a Banach space.

3. If f is continuous on $\boldsymbol{T}$, and $u(z)$ is its Poisson integral for the unit disk, we have $u(z) \to f(e^{i\theta})$ as $z \to e^{i\theta}$, uniformly in $\theta \in \mathbf{R}$.

Hint. The Poisson kernel $P_r(\theta)$ forms a family of periodic functions of δ-type as $r \to 1-$, i.e.,

a) $P_r(\theta + 2\pi) = P_r(\theta)$;
b) $P_r(\theta) > 0$;
c) $\frac{1}{2\pi}\int_0^{2\pi} P_r(\theta)\,d\theta = 1$;
d) $\max\{P_r(\theta) \colon \varepsilon \le \theta \le 2\pi - \varepsilon\} \to 0$, $r \to 1-$, for every $\varepsilon > 0$.

4. If $f(e^{i\theta}) \in L^p(0, 2\pi)$, $1 \le p < \infty$, and $u(z)$ is the Poisson integral of this function, we have the following convergence in L^p:

$$\frac{1}{2\pi}\int_0^{2\pi}|u(re^{i\theta}) - f(e^{i\theta})|^p\,d\theta \to 0, \qquad r \to 1-.$$

Hint. Use the integral analog of the triangle inequality.

§4. Outer and inner functions

1. We showed in §3 that every function $f \not\equiv 0$ of class H^δ $(\delta > 0)$ is representable in the form

$$f(z) = e^{ic}B(z)S(z)f_0(z),$$

where $c \in \mathbf{R}$, $B(z)$ is the Blaschke product over the zeros of $f(z)$, $S(z)$ is a singular function, and

$$f_0(z) = \exp\left\{\frac{1}{2\pi}\int_0^{2\pi} \ln|f(e^{i\theta})|\frac{e^{i\theta}+z}{e^{i\theta}-z}\,d\theta\right\}.$$

DEFINITION. A function of the form

$$\varphi(z) = e^{ic}B(z)S(z),$$

where $c \in \mathbf{R}$, B is a Blaschke product, and S is a singular function, is called an *inner function*.

It follows from Proposition 3.2 and Fatou's theorem that

$$\lim_{z\to e^{i\theta}} |\varphi(z)| = 1$$

for a.e. θ. In addition, it is obvious that $\varphi \in H^\infty$ and $\|\varphi\|_\infty = 1$. It follows from these considerations that the following definition is equivalent to the one given above: φ is an inner function if $\varphi \in H^\infty$ and $|\varphi(e^{i\theta})| = 1$ for a.e. θ.

DEFINITION. A function of the form

$$F(z) = \exp\left\{\frac{1}{2\pi}\int_0^{2\pi} k(\theta)\frac{e^{i\theta}+z}{e^{i\theta}-z}\,d\theta\right\}, \tag{4.1}$$

where $k \in L^1(0, 2\pi)$ and k is a real function, is called an *outer function*.

It follows from Fatou's theorem that

$$|F(e^{i\theta})| = \exp\{k(\theta)\}$$

for a.e. $\theta \in [0, 2\pi)$.

2. Let us consider the class H^2 in more detail. For $f \in H^2$ we set

$$k(\theta) = \ln|f(e^{i\theta})|;$$

then

$$k \in L^1, \qquad \exp\{2k\} \in L^1. \tag{4.2}$$

Conversely, if k has the properties (4.2), the function (4.1) belongs to H^2. In fact, by (2.3) we have

$$\begin{aligned}|f(re^{i\theta})|^2 &= \exp\left\{\frac{1}{2\pi}\int_0^{2\pi} 2k(t)P_r(\theta-t)\,dt\right\}\\ &\le \frac{1}{2\pi}\int_0^{2\pi}\exp\{2k(t)\}P_r(\theta-t)\,dt.\end{aligned}$$

Consequently,

$$\begin{aligned}\int_0^{2\pi}|f(re^{i\theta})|^2\,d\theta &\le \int_0^{2\pi}\left(\frac{1}{2\pi}\int_0^{2\pi}P_r(\theta-t)\,d\theta\right)\exp\{2k(t)\}\,dt\\ &= \int_0^{2\pi}\exp\{2k(t)\}\,dt. \quad \triangleright\end{aligned}$$

In particular, if p is a weight function that satisfies the Szegő condition, there is a unique outer function $D \in H^2$ for which

$$|D(e^{i\theta})|^2 = p(\theta)$$

for a.e. θ. In fact, let us take $k(\theta) = \frac{1}{2} \ln p(\theta)$ in the preceding discussion. This establishes the existence of the outer function D. Its uniqueness follows from Fatou's theorem. Therefore the Szegő condition selects for us the squared moduli of the functions of class H^2.

3. Inner functions arise naturally in many problems in analysis. As an example, we now introduce the *Beurling-Helson theorem.*

Let A be the operator of multiplication by z in the Hilbert space $L^2(\boldsymbol{T})$. The Beurling-Helson theorem describes the *invariant subspaces* of A, i.e., the closed subspaces $Q \subset L^2(\boldsymbol{T})$ with the property that $Ag \in Q$ for every $g \in Q$.

THEOREM 4.1 (Beurling-Helson). *Let Q be an invariant subspace of the operator A. Then either $Q = \chi_E L^2(\boldsymbol{T})$, where χ_E is the characteristic function of a measurable set E, or $Q = \varphi H^2$, where $|\varphi| = 1$ a.e. on $\boldsymbol{T}$.*

REMARK.

Here H^2 is the closed subspace of $L^2(\boldsymbol{T})$ generated by the functions z^n, $n \in \mathbf{Z}_+$. Later, $\overline{H}^2$ will be the closed subspace generated by the functions $\overline{z}^n$, $n \in \mathbf{Z}_+$.

PROOF. First let

$$AQ = Q; \tag{4.3}$$

then

$$A^{-1}Q = Q. \tag{4.4}$$

Let f denote the element of Q closest to 1. It follows from (4.3) and (4.4) that $z^n f(z) \in Q$, $n \in \mathbf{Z}$. Hence $1 - f(z) \perp z^n f(z)$, $n \in \mathbf{Z}$, i.e.,

$$\int_{\boldsymbol{T}} z^n (f(z) - |f(z)|^2)\, d\theta = 0, \qquad n \in \mathbf{Z},$$

so that $f = |f|^2$ a.e.; therefore $f = \chi_E$ for some measurable set E.

Thus $\chi_E L^2(\boldsymbol{T}) \subset Q$. Let $g \in Q$ be orthogonal to $\chi_E L^2(\boldsymbol{T})$. Then

$$\int_{\boldsymbol{T}} \chi_E(z) g(z) z^n\, d\theta = 0, \qquad n \in \mathbf{Z},$$

and consequently $g\chi_E = 0$ a.e.; therefore $g = 0$ a.e. on E. On the other hand, since $z^n g \in Q$ for $n \in \mathbf{Z}$, and $1 - \chi_E \perp Q$, we have

$$\int_{\boldsymbol{T}} (1 - \chi_E(z)) g(z) z^n\, d\theta = 0, \qquad n \in \mathbf{Z},$$

and consequently $g = 0$ almost everywhere outside E. Thus $g = 0$, i.e., $\chi_E L^2(\boldsymbol{T}) = Q$.

Now let $AQ \neq Q$. The subspace AQ is closed; consequently there is a $\varphi \in Q$ with $\varphi \neq 0$ and $\varphi \perp AQ$. Then $\varphi \perp z^n\varphi$, $n \in \mathbf{N}$, i.e.,

$$\int_{\boldsymbol{T}} |\varphi(z)|^2 z^n \, d\theta = 0, \qquad n \in \mathbf{N}. \tag{4.5}$$

If we take the complex conjugate in (4.5), we obtain

$$\int_{\boldsymbol{T}} |\varphi(z)|^2 \overline{z}^n \, d\theta = 0, \qquad n \in \mathbf{N}. \tag{4.6}$$

It follows from (4.5) and (4.6) that $|\varphi| = \text{const}$ a.e. Without loss of generality, we may suppose that $|\varphi| = 1$ a.e.

Thus

$$\varphi H^2 \subset Q. \tag{4.7}$$

It is clear that

$$L^2(T) \ominus \varphi H^2 = \overline{z}\varphi \overline{H}^2 \tag{4.8}$$

($\ominus$ denotes taking the orthogonal complement). If $f \in Q$ and $g \in H^2$, we have $fg \in Q$ and $\varphi \perp zfg$; that is,

$$Q \perp \overline{z}\varphi\overline{H}^2,$$

and consequently, by (4.8),

$$Q \subset \varphi H^2. \tag{4.9}$$

It follows from (4.7) and (4.9) that $Q = \varphi H^2$. ▷

The following result is a corollary of Theorem 4.1.

THEOREM 4.2 (Beurling). *Let Q be a nontrivial invariant subspace in H^2 for the operator A. Then $Q = \varphi H^2$, where φ is an inner function.*

REMARKS

1. The exposition in §§3, 4 follows the book [57] by P. Koosis. That book contains an extensive bibliography on boundary properties of analytic functions, the theory of Hardy spaces, and other related problems. See also [29], [43], [13], and [80].
2. The theorem of the brothers Riesz (1917) is important for these questions.

THEOREM. *If μ is a complex measure on the circle $\boldsymbol{T}$ such that*

$$\int_{\boldsymbol{T}} z^n \, d\mu(z) = 0, \qquad n = 1, 2, \ldots,$$

then μ is absolutely continuous with respect to Lebesgue measure.

§5. Szegő's asymptotic formula for $|z| > 1$

1. Let

$$d\sigma(\theta) = p(\theta)\, d\theta + d\nu(\theta), \qquad \theta \in [0, 2\pi),$$

define a positive measure on $\boldsymbol{T}$. (Here $p \in L^1(0, 2\pi)$, and ν is a singular measure.) We denote by H^2_σ the closure in $L^2_\sigma(\boldsymbol{T})$ of the set of polynomials in z. Then H^2_σ is a Hilbert subspace of $L^2_\sigma(\boldsymbol{T})$. By Theorem 2.2, $H^2_\sigma \neq L^2_\sigma(\boldsymbol{T})$ if and only if the measure σ satisfies the Szegő condition. We shall suppose that this condition is satisfied. Then the functions of class H^2_σ will be the boundary values of functions that are analytic in the unit disk. Let us prove this.

We construct the *Szegő function* corresponding to the weight function p:

$$D(z) = \exp\left\{\frac{1}{4\pi}\int_0^{2\pi} \ln p(\theta)\frac{e^{i\theta}+z}{e^{i\theta}-z}\,d\theta\right\}.$$

This is an outer function (in particular, $D(0) > 0$), and

$$|D(e^{i\theta})|^2 = p(\theta)$$

for a.e. $\theta \in [0, 2\pi)$ (so that $D \in H^2$). We note that these conditions determine D uniquely.

Let $f(e^{i\theta}) \in H^2_\sigma$. (We are going to interpret f in two ways: 1) as an element of H^2_σ; and 2) as a function representing this element, where any two representatives differ only on a set of σ-measure zero, hence of Lebesgue measure zero.) There is a sequence of polynomials $T_n \to f$ in L^2_σ. This implies that $T_n \to f$ in L^2_p, i.e.,

$$\lim_{n\to\infty}\frac{1}{2\pi}\int_0^{2\pi}|f(e^{i\theta})D(e^{i\theta}) - T_n(e^{i\theta})D(e^{i\theta})|^2\,d\theta = 0. \tag{5.1}$$

Consequently, $\{T_n(z)D(z)\}$ is a fundamental sequence in H^2 and therefore converges in this space to a function Φ, where by (5.1)

$$\Phi(e^{i\theta}) = f(e^{i\theta})D(e^{i\theta})$$

for a.e. θ. Then $f(e^{i\theta})$ is the boundary function of the analytic function

$$F(z) = \Phi(z)/D(z).$$

It is clear that $F(z)$ is independent of the auxiliary sequence $\{T_n\}$.

Notice that $F(e^{i\theta})$ might not agree with $f(e^{i\theta})$ as an element of H^2_σ (these functions are equal only a.e. with respect to Lebesgue measure!). Nevertheless, we consider $F(z)$ as a continuation of f outside the disk, and we write $F(z) = f(z)$.

2. The Szegő-Kolmogorov-Kreĭn theorem from §2 can be restated in the following way.

THEOREM 5.1. *Let the measure σ satisfy Szegő's condition. Then*
(a) *we have*

$$\inf_{\substack{f\in H^2_\sigma\\ f(0)=1}}\left\{\frac{1}{2\pi}\int_0^{2\pi}|f(e^{i\theta})|^2\,d\sigma(\theta)\right\} = D(0)^2, \tag{5.2}$$

where

$$D(0) = \exp\left\{\frac{1}{4\pi}\int_0^{2\pi}\log p(\theta)\,d\theta\right\};$$

(b) *there is a unique extremal function* f *for the problem* (5.2),

$$f(z) = \frac{D(0)}{D(z)}, \qquad |z| < 1; \tag{5.3}$$

(c) *for every minimizing sequence* $\{f_n\}$ *for the problem* (5.2), *we have*

$$f_n(z) \to f(z) \qquad (n \to \infty)$$

uniformly inside U.

PROOF. Since the set of polynomials is dense in H^2_σ, conclusion (a) follows from Theorem 2.1.

The existence and uniqueness of an *extremal function* follows from general properties of the geometry of Hilbert spaces. In fact, let $\{f_n\}$ be any minimizing sequence for the problem (5.2), i.e., $f_n \in H^2_\sigma$, $f_n(0) = 1$, and

$$\|f_n\|^2 \to D(0)^2 \qquad (n \to \infty) \tag{5.4}$$

($\|\cdot\|$ is the norm in H^2_σ). We write the parallelogram equation in H^2_σ:

$$\left\|\frac{f_n - f_m}{2}\right\|^2 + \left\|\frac{f_n + f_m}{2}\right\|^2 = \frac{1}{2}(\|f_n\|^2 + \|f_m\|^2). \tag{5.5}$$

By (5.2), we have

$$\left\|\frac{f_n + f_m}{2}\right\|^2 \geq D(0)^2. \tag{5.6}$$

It follows from (5.4)–(5.6) that

$$\lim_{n,m\to\infty} \|f_n - f_m\|^2 = 0,$$

i.e., the sequence $\{f_n\}$ is fundamental in H^2_σ and therefore converges in this space to an extremal element f. The uniqueness of the extremal element is established by using the parallelogram equation.

Since (5.2) remains valid, with the same right-hand side, for minimization in H^2_p, it follows that f is also minimized for the problem in H^2_p (more precisely, any representatives of extremal elements of the problems in H^2_σ and in H^2_p agree a.e. with respect to Lebesgue measure). We can verify directly that (5.3) is an extremal function for the problem in H^2_p. Hence (b) is established.

If $\{f_n\}$ is a minimizing sequence in H^2_σ, it is also a minimizing sequence in H^2_p. Then $f_n \to D(0)/D$ in H^2_p or, equivalently, $f_n D \to D(0)$ in H^2 and, therefore, converges uniformly inside U; this proves (c). ▷

3. The *Szegő asymptotic formula* is a consequence of the preceding theorem.

THEOREM 5.2. *Let $\{\varphi_n\}$ be the polynomials that are orthogonal with respect to σ and satisfy the Szegő condition. Then*

$$\varphi_n(z) = \frac{z^n}{\overline{D}(1/z)}(1 + o(1)) \qquad (n \to \infty) \tag{5.7}$$

uniformly on compact subsets of the domain $|z| > 1$.

PROOF. As in §1, we set $\Phi_n = \varphi_n/\alpha_n$ and $\Phi_n^* = z^n\overline{\Phi}_n(1/z)$. The polynomials $\{\Phi_n^*\}$ form a minimizing sequence for the problem (5.2). In fact, $\Phi_n^*(0) = 1$ and

$$\begin{aligned}\frac{1}{2\pi}\int_0^{2\pi} &|\Phi_n^*(e^{i\theta})|^2\,d\sigma(\theta)\\ &= \frac{1}{\alpha_n^2}\frac{1}{2\pi}\int_0^{2\pi}|\varphi_n(e^{i\theta})|^2\,d\sigma(\theta) = \frac{1}{\alpha_n^2} \to D(0)^2.\end{aligned}$$

Consequently,

$$\Phi_n^*(z) \to \frac{D(0)}{D(z)} \qquad (n \to \infty) \tag{5.8}$$

uniformly inside U. But (5.8) is equivalent to (5.7). ▷

4. As a corollary of Theorem 5.2, we obtain an asymptotic formula for the polynomials that are orthogonal on an interval Δ of the real line. Without loss of generality, we take $\Delta = [-1, 1]$.

Let μ be a measure on Δ. We define a corresponding measure σ on $\boldsymbol{T}$: $\sigma\{\pm 1\} = \pi\mu\{\pm 1\}$; if $E = (e^{i\theta_1}, e^{i\theta_2})$ is an arc of the circle, with $0 < \theta_1 < \theta_2 < \pi$, we take

$$\sigma(E) = \pi\mu(\cos\theta_2, \cos\theta_1);$$

if $\pi < \theta_1 < \theta_2 < 2\pi$, we take

$$\sigma(E) = \pi\mu(\cos\theta_1, \cos\theta_2).$$

We extend σ to a Borel σ-algebra on $\boldsymbol{T}$ in the usual way.

We denote by $\varphi_n(z) = \alpha_n z^n + \cdots$ the polynomials orthogonal with respect to σ; and by $Q_n(x) = \beta_n x^n + \cdots$ the polynomials orthogonal with respect to μ.

THEOREM 5.3. *The polynomials $\{\varphi_n\}$ and $\{Q_n\}$ are connected by the relations*

$$Q_n(x) = \frac{1}{\sqrt{2}}(z^{-n}\varphi_{2n}(z) + z^n\varphi_{2n}(z^{-1})), \qquad n \geq 1, \tag{5.9}$$

where

$$x = \frac{1}{2}\left(z + \frac{1}{z}\right). \tag{5.10}$$

PROOF. Formula (5.9) defines a polynomial in x of degree n with leading coefficient

$$\beta_n = \frac{1}{\sqrt{2}}\alpha_{2n}2^n.$$

Let us show that this is the nth orthogonal polynomial with respect to μ.

It follows from the orthogonality relations for φ_{2n} that

$$\int_T \varphi_{2n}(z)z^{-n}\frac{1}{2^m}(z+z^{-1})^m\,d\sigma(z)=0, \qquad m=0,\dots,n-1. \tag{5.11}$$

By construction, the measure σ is invariant under complex conjugation, so that φ_{2n} has real coefficients. Since $(z+z^{-1})$ is real for $z\in \boldsymbol{T}$, if we take the real part in (5.11) we obtain

$$\int_T(\varphi_{2n}(z)z^{-n}+\varphi_{2n}(z^{-1})z^n)x^m\,d\sigma(z)=0, \qquad m=0,\dots,n-1. \tag{5.12}$$

If we use (5.9) and the definition of σ, we see that (5.12) is equivalent to

$$\int_\Delta Q_n(x)x^m\,d\mu(x)=0, \qquad m=0,\dots,n-1.$$

This establishes the orthogonality relations. Let us verify the normality. Arguing in a similar way, we obtain

$$\begin{aligned}\int_\Delta Q_n(x)^2\,d\mu(x)&=\beta_n\int_\Delta Q_n(x)x^n\,d\mu(x)\\ &=\beta_n\frac{1}{\sqrt{2}}\int_\Delta(\varphi_{2n}(z)z^{-n}+\varphi_{2n}(z^{-1})z^n)x^n\,d\mu(x)\\ &=\beta_n\frac{1}{\sqrt{2}}\frac{1}{2\pi}\int_T(\varphi_{2n}(z)z^{-n}+\varphi_{2n}(z^{-1})z^n)\frac{1}{2^n}(z+z^{-1})^n\,d\sigma(z)\\ &=\beta_n\frac{1}{\sqrt{2}}\frac{1}{2^n}2\operatorname{Re}\frac{1}{2\pi}\int_T\varphi_{2n}(z)z^{-n}(z+z^{-1})^n\,d\sigma(z)\\ &=\alpha_{2n}\operatorname{Re}\frac{1}{2\pi}\int_T\varphi_{2n}(z)z^{-2n}\,d\sigma(z)=1. \quad \triangleright\end{aligned}$$

Let $w(x)\,dx$ be absolutely continuous with respect to μ. We say that μ satisfies the *Szegő condition on the interval* Δ if

$$\int_{-1}^{1}\frac{\ln w(x)\,dx}{\sqrt{1-x^2}}>-\infty.$$

Let $p(\theta)\,d\theta$ be the absolutely continuous part of the measure σ constructed from μ; then

$$p(\theta)=\pi w(\cos\theta)|\sin\theta|.$$

Consequently μ satisfies the Szegő condition on Δ if and only if σ satisfies the Szegő condition on the circle. Therefore, the next theorem follows from Theorems 5.2 and 5.3.

THEOREM 5.4. *Let $\{Q_n\}_0^\infty$ be the polynomials orthogonal on $\Delta=[-1,1]$ with respect to a measure μ that satisfies the Szegő condition; then, uniformly inside the domain $\mathbf{C}\backslash\Delta$,*

$$Q_n(x)=A(x)(x+\sqrt{x^2-1})^n(1+o(1)), \qquad n\to\infty, \tag{5.13}$$

where

$$A(x) = (D(x - \sqrt{x^2 - 1}))^{-1},$$

D is the Szegő function constructed from the measure σ obtained from μ, and $\sqrt{x^2 - 1} > 0$ for $x > 1$.

PROOF. Let x and z be connected by (5.10), with $|z| > 1$. By (5.7),

$$\varphi_{2n}(z) z^{-n} = \frac{z^n}{\overline{D}(1/z)}(1 + o(1)), \qquad n \to \infty. \tag{5.14}$$

By (5.8),

$$\varphi_{2n}(1/z) z^n = O(z^{-n}), \qquad n \to \infty. \tag{5.15}$$

Then (5.9), (5.14), and (5.15) imply (5.13). ▷

REMARK

We note a result of E. A. Rakhmanov [83].

THEOREM. *Let μ be a measure on $\boldsymbol{T}$, and $\{\varphi_n\}$ the corresponding polynomials. If $\tilde{\mu}'(\theta) > 0$ for a.e. θ ($\tilde{\mu}$ is the distribution function), the limit of the quotient*

$$\lim_{n \to \infty} \frac{\varphi_{n+1}(z)}{\varphi_n(z)} = z$$

exists uniformly inside the domain $|z| > 1$.

§6. Asymptotic formula on the circle

1. Let σ be a measure on $\boldsymbol{T}$ with an infinite spectrum, and $\{\varphi_n\}$ the polynomials orthogonal with respect to σ. We form the *Szegő kernel*

$$G_n(z, \zeta) = \sum_{\nu=0}^{n} \varphi_\nu(z) \overline{\varphi_\nu(\zeta)}.$$

We have the following lemma.

LEMMA 6.1. *The unique extremal polynomial in the following problem: determine*

$$\inf \frac{1}{2\pi} \int_{\boldsymbol{T}} |T_n(z)|^2 \, d\sigma(z); \qquad T_n(\zeta) = 1,$$

where T_n is a polynomial of degree at most n and ζ is a given complex number, is the polynomial

$$T_n(z) = \frac{G_n(z, \zeta)}{G_n(\zeta, \zeta)}.$$

PROOF. Let

$$T_n(z) = \sum_{\nu=0}^{n} c_\nu \varphi_\nu(z), \qquad T_n(\zeta) = 1.$$

Then

$$\frac{1}{2\pi} \int_{\boldsymbol{T}} |T_n(z)|^2 \, d\sigma(z) = \sum_{\nu=0}^{n} |c_\nu|^2 \geq \frac{|\sum_{\nu=0}^{n} c_\nu \varphi_\nu(\zeta)|^2}{\sum_{\nu=0}^{n} |\varphi_\nu(\zeta)|^2} = \frac{1}{G_n(\zeta, \zeta)}. \tag{6.1}$$

In the Cauchy-Bunyakovskiĭ inequality (6.1), equality is attained only for $c_\nu = \overline{\varphi_\nu(\zeta)}/G_n(\zeta, \zeta)$. ▷

The formulas

$$\alpha_n \varphi_n^*(z) = G_n(z, 0) \tag{6.2}$$

and

$$\alpha_n^2 = \sum_{\nu=0}^{n} |\varphi_\nu(0)|^2 \tag{6.3}$$

that were proved in §1 also follow directly from Lemma 6.1.

2. Let us consider (6.2) in more detail:

$$\alpha_n \varphi_n^*(z) = 1 - \sum_{\nu=1}^{n} \alpha_\nu \alpha_{\nu-1} \varphi_\nu(z). \tag{6.4}$$

We recall (see §1) that

$$a_\nu = -\overline{\Phi_{\nu+1}(0)} = -\frac{\overline{\varphi_{\nu+1}(0)}}{\alpha_{n+1}}$$

are the circular parameters. From (6.3) we obtain $\alpha_{n+1}^2 - \alpha_n^2 = |a_n|^2 \alpha_{n+1}^2$, from which

$$\frac{1}{\alpha_n^2} = \frac{1}{\alpha_0^2} \prod_{\nu=0}^{n-1} (1 - |a_\nu|^2).$$

Then $|a_n| < 1$, and by the Szegő-Kolmogorov-Kreĭn theorem, σ satisfies the Szegő condition if and only if

$$\sum_{n=0}^{\infty} |a_n|^2 < \infty.$$

Therefore, by (6.4), if the Szegő condition is satisfied, the problem of the asymptotics of $\varphi_n(e^{i\theta})$, or equivalently of the convergence of the sequence $\varphi_n^*(e^{i\theta})$, is equivalent to the problem of the convergence of the orthogonal series

$$\sum_{\nu=1}^{\infty} \alpha_\nu a_{\nu-1} \varphi_\nu(e^{i\theta}) \tag{6.5}$$

with coefficients in l_2.

By the Men′shov-Rademacher theorem, it is sufficient for the convergence almost everywhere of this series that the series

$$\sum_{\nu=1}^{\infty} |a_\nu|^2 \log^2 \nu < \infty$$

converges. Note that

$$\sum_{\nu=n+1}^{\infty} |\varphi_\nu(0)|^2 = \alpha^2 - \alpha_n^2,$$

where $\alpha = \lim_{n\to\infty} \alpha_n$. Therefore, if we want to have information about the rate of the convergence of α_n to α, we have to obtain information about the speed with which a_n approaches zero.

LEMMA 6.2. *If* $\alpha - \alpha_n = o(\log^{-3-\varepsilon} n)$, $\varepsilon > 0$, *we have the asymptotic formula*

$$\varphi_n(e^{i\theta}) = \frac{e^{in\theta}}{\overline{D}(e^{i\theta})}(1+o(1)), \qquad n \to \infty,$$

almost everywhere with respect to Lebesgue measure on $\boldsymbol{T}$.

PROOF. We have

$$\sum_{\nu=2^k+1}^{2^{k+1}} |\varphi_\nu(0)|^2 \le \frac{C}{k^{3+\varepsilon}},$$

where C is a constant. Hence

$$\sum_{\nu=2^k+1}^{2^{k+1}} |\varphi_\nu(0)|^2 \log^2 \nu \le \frac{C_1}{k^{1+\varepsilon}},$$

and the series

$$\sum_{\nu=1}^{\infty} |\varphi_\nu(0)|^2 \log^2 \nu$$

converges. ▷

Lemma 6.2 lets us obtain sufficient conditions for the convergence of the series (6.5); for example, in terms of the smoothness of the weight function. However, we shall not go into this. We note only that establishing an asymptotic formula on the circle when the Szegő condition is satisfied (and without any other restrictions) remains an open problem.

CHAPTER 4

Simultaneous Padé approximants

§1. The general Padé table

1. Let r be a positive integer, and $f = (f_1, \dots, f_r)$ a set of formal power series

$$f_j(z) = \sum_{k=0}^{\infty} f_{j,k} z^k, \qquad j = 1, \dots, r, \tag{1.1}$$

with complex coefficients; let $n, m \in \mathbf{Z}_+^r$.

PROBLEM A. *Find a polynomial* $q = q_{n,m} \not\equiv 0$, $\deg q \le |n|$, *and polynomials* $p_1 = p_{n,m,1}, \dots, p_r = p_{n,m,r}$ *of degrees at most* $m_1, \dots, m_r$, *respectively, such that for each* j, $j = 1, \dots, r$,

$$R_j = qf_j - p_j = C_j z^{n_j+m_j+1} + \cdots. \tag{1.2}$$

Problem A has a solution, since it reduces to a system of $|n|$ homogeneous linear equations in the $|n| + 1$ unknown coefficients of the polynomial q:

$$(qf_j)_k = 0, \qquad k = m_j + 1, \dots, m_j + n_j, \ j = 0, \dots, r,$$

where $(\varphi)_k$ is the coefficient of z^k in the series φ. The polynomials p_j are determined by the formulas

$$p_j(z) = \sum_{k=0}^{m_j} (qf_j)_k z^k, \qquad j = 1, \dots, r.$$

DEFINITION. Let (q, p), $p = (p_1, \dots, p_r)$, be a solution of Problem A. The rational functions

$$\pi_{n,m} = (\pi^{(1)}, \dots, \pi^{(r)}), \qquad \pi^{(j)} = p_j/q, \tag{1.3}$$

are called *simultaneous Padé approximants of type* (n, m) for the set f of series.

When $r = 1$, we obtain the ordinary Padé approximants of the power series, the special case $(n = m)$ of which was discussed in Chapter 2, §1.

When $r \ge 2$, the vector $\pi_{n,m}$ is, in general, not uniquely determined. This is already shown by the trivial example in which $f_1 = \cdots = f_r$. We therefore introduce the set $\Pi_{n,m} = \{\pi_{n,m}\}$, where $\pi_{n,m}$ is defined in (1.3), and (q, p) runs through the solutions of Problem A.

DEFINITION. The $2r$-dimensional table

$$\Pi = \{\Pi_{n,m}\colon (n, m) \in \mathbf{Z}_+^r \times \mathbf{Z}_+^r\}$$

is called the *general Padé table* for the set f.

DEFINITION. The indices (n, m): $n_1 + m_1 = \cdots = n_r + m_r \geq |n|$ are the *Mahler indices*; the approximants corresponding to them, and the $(r+1)$-dimensional table of these, are called the *Mahler-Padé approximants* and *table*.

DEFINITION. The indices (n, m): $m_1 = \cdots = m_r = |n|$, and the corresponding approximants, are *diagonal*, and their r-dimensional table is the *diagonal* general Padé table.

If we add to Problem A the requirement that the polynomial q has the lowest possible degree, the solution of the problem will evidently be unique (up to a numerical multiplier), and the vector $\pi_{n,m}$ will be uniquely determined.

In the general case, the following proposition serves as an analog of the uniqueness of $\pi_{n,m}$.

PROPOSITION 1.1. *If $(q^{(l)}, p^{(l)})$, $l = 0, 1, \ldots, r$, is a solution of Problem A, we have*

$$\Delta = \begin{vmatrix} q^{(0)} & p_1^{(0)} & \cdots & p_r^{(0)} \\ q^{(1)} & p_1^{(1)} & \cdots & p_r^{(1)} \\ \cdots & \cdots & \cdots & \cdots \\ q^{(r)} & p_1^{(r)} & \cdots & p_r^{(r)} \end{vmatrix} \equiv 0.$$

PROOF. The determinant Δ is a polynomial. It follows from its complete expansion that

$$\deg \Delta \leq |n| + |m|. \tag{1.4}$$

If we subtract the column $(q^{(l)})f_j$, $j = 1, \ldots, r$, from the column $(p_j^{(l)})$, we obtain

$$\Delta = (-1)^r \begin{vmatrix} q^{(0)} & R_1^{(0)} & \cdots & R_r^{(0)} \\ q^{(1)} & R_1^{(1)} & \cdots & R_r^{(1)} \\ \cdots & \cdots & \cdots & \cdots \\ q^{(r)} & R_1^{(r)} & \cdots & R_r^{(r)} \end{vmatrix}, \tag{1.5}$$

where $R_j^{(l)} = q^{(l)} f_j - p_j^{(l)}$. Let us denote by $\tau(\varphi)$ the smallest exponent in the power of z that appears in the power series φ with a nonzero coefficient $(\tau(0) = +\infty)$. From the formula for the complete expansion of the determinant (1.5), together with (1.2), we obtain

$$\tau(\Delta) \geq |n| + |m| + r. \tag{1.6}$$

It follows from (1.4) and (1.6) that $\deg \Delta \leq \tau(\Delta) - r$, i.e., $\Delta \equiv 0$. ▷

For $r = 1$, the formula

$$\Delta = \begin{vmatrix} q^{(0)} & p^{(0)} \\ q^{(1)} & p^{(1)} \end{vmatrix} = q^{(0)}p^{(1)} - p^{(0)}q^{(1)} \equiv 0$$

implies the uniqueness of $\pi_{n,m}$.

2. Let f be the set of power series (1.1), $n \in \mathbf{Z}_+^r$, $n \neq 0$.

PROBLEM B. *Find a set of polynomials*

$$a_n = (a_{n,1}, \dots, a_{n,r}) \not\equiv 0, \qquad \deg a_{n,j} \le n_j - 1, \quad j = 1, \dots, r,$$

such that the linear form

$$L_n = a_{n,1}f_1 + \cdots + a_{n,r}f_r$$

has the representation

$$L_n(z) = Cz^{|n|-1} + \cdots.$$

This problem has a solution.

We shall also be interested in Problem B formulated for the expanded set of series $\overline{f} = (1, f_1, \dots, f_r)$ and the $(r+1)$-dimensional multi-index $n = (n_0, n_1, \dots, n_r)$.

There is an analog of Proposition 1.1.

PROPOSITION 1.2. *If* $a^{(l)}$, $l = 0, 1, \dots, r$, *solve Problem* B *for the data* $n \in \mathbf{Z}_+^{r+1}$ *and* $\overline{f}$, *then*

$$\Delta = \begin{vmatrix} a_0^{(0)} & a_1^{(0)} & \cdots & a_r^{(0)} \\ a_0^{(1)} & a_1^{(1)} & \cdots & a_r^{(1)} \\ \cdots & \cdots & \cdots & \cdots \\ a_0^{(r)} & a_1^{(r)} & \cdots & a_r^{(r)} \end{vmatrix} \equiv 0.$$

3. DEFINITION. An index $n \in \mathbf{Z}_+^r$, $n \neq 0$, is said to be *normal for* f *with respect to Problem* B if, for each solution $a = (a_1, \dots, a_r)$ of the problem with index n, we have

$$\deg a_j = n_j - 1, \qquad j = 1, \dots, r.$$

It is clear that if an index is normal, the solution of Problem B is unique.

DEFINITION. A system f is said to be *complete* if all indices $n \in \mathbf{Z}_+^r$, $n \neq 0$, are normal for f with respect to B.

Completeness of the extended system $\overline{f}$ can be defined in terms of Problem A.

DEFINITION. Indices n and m are said to be *normal for* f *with respect to Problem* A if we have, for every solution (q, p) of the problem with indices (n, m),

$$\deg q = |n|, \qquad \deg p_j = m_j, \quad j = 1, \dots, r.$$

Once again, normality implies uniqueness of the solution.

PROPOSITION 1.3. *The system* $\overline{f}$ *is complete if all Mahler indices are normal for* f *with respect to* A.

PROOF. We need to write out explicitly the system of linear equations to which Problem B reduces for the system $\overline{f}$ and that to which Problem A reduces for the system f and the Mahler indices. The condition that the indices are normal means that the corresponding determinants are different from zero. But the determinants in the two cases differ from each other only by transposition (for an appropriate choice of indices). ▷

REMARKS

For the classical Padé table, see the remarks on Chapter 2, §1. The formal theory connected with Problems A and B was developed in [148], [131], and [142]. The approximation of p_j/q was first discussed by de Bruin [125]. The general Padé table is fully presented in [45]. A construction that generalizes Problems A and B is presented in [92].

PROBLEMS

1. A system f is complete if, for every solution a of Problem B with arbitrary multi-indices $n \in \mathbf{Z}_+^r$, $n \neq 0$, we have

$$\tau(a_1 f_1 + \cdots + a_r f_r) = |n| - 1.$$

2. Let $\mathscr{R}_{n,m}$ be the set of rational functions for which the degree of the numerator is at most m, and that of the denominator is at most n $(n, m \in \mathbf{Z}_+)$. Prove that $\tau(f - r)$, $r \in \mathscr{R}_{n,m}$, attains its maximum only for $\pi_{n,m}$.

3. Let $(n, m) \in \mathbf{Z}_+^2$ be a normal index with respect to Problem A. Determine $l = 0, 1, \ldots, +\infty$ from $\tau(f - \pi_{n,m}) = n + m + 1 + l$, and introduce the square block indices

$$B_{n,m}(l) = \{(n+i, m+j) \colon i, j = 0, \ldots, l\}.$$

Prove the *block theorem* $(f(0) \neq 0)$:

$$\{(\nu, \mu) \colon \pi_{\nu,\mu} = \pi_{n,m}\} = B_{n,m}(l).$$

4. For every solution (q, p) of Problem A with indices $(\nu, \mu) \in B_{n,m}(l)$ we have 1) $\deg q = \nu \Leftrightarrow \nu = n$ ("upper bound" block); 2) $\deg p = \mu \Leftrightarrow \mu = m$ ("left boundary" block); 3) the pair (q, p) is unique (up to a multiplier) $\Leftrightarrow \nu = n$, or $\nu = n + l$, or $\mu = m$, or $\mu = m + l$ ("boundary" block).

5. Introduce the *Hadamard determinants*

$$\Delta_{m,n} = \begin{vmatrix} f_{m-n+1} & \cdots & f_m \\ f_m & \cdots & f_{m+n-1} \end{vmatrix}, \qquad n \in \mathbf{N},$$

$\Delta_{m,0} = 1$, $m \in \mathbf{Z}_+$; here $f(z) = f_0 + f_1 z + \cdots + f_n z^n + \cdots$, $f_k = 0$ for $k < 0$. Then 1) (n, m) belongs to the "upper boundary" of some block $\Leftrightarrow \Delta_{n,m}\Delta_{n,m+1} \neq 0$; 2) (n, m) belongs to the "left boundary" of some block $\Leftrightarrow \Delta_{n,m}\Delta_{n+1,m} \neq 0$; 3) (n, m) is a normal index $\Leftrightarrow \Delta_{n,m}\Delta_{n+1,m}\Delta_{n,m+1} \neq 0$.

§2. Hermite integrals

1. The first example of a complete system, for $r \geq 3$, was essentially given by Charles Hermite.

By the completeness of a system of entire functions we will mean the completeness of their Taylor series with center at zero.

THEOREM 2.1. *Let $\lambda_1, \ldots, \lambda_r$ be distinct complex numbers; then*

$$e^{\lambda_1 z}, \ldots, e^{\lambda_r z}$$

is a complete system.

The proof is by induction. For $r = 1$ the theorem is trivial. Let $r \geq 2$, $n \in \mathbf{Z}_+^r$, $n \neq 0$. We are to find a nonzero set of polynomials $a_1, \ldots, a_r$ such that

$$\deg a_j \leq n_j - 1, \qquad j = 1, \ldots, r,$$

$$L(z) = a_1(z)e^{\lambda_1 z} + \cdots + a_r(z)e^{\lambda_r z} = C \cdot z^{|n|-1} + \cdots. \tag{2.1}$$

If we multiply (2.1) by $e^{-\lambda_r z}$, we obtain

$$\begin{aligned} L_1(z) &= L(z)e^{-\lambda_r z} \\ &= a_1(z)e^{\tilde{\lambda}_1 z} + \cdots + a_{r-1}(z)e^{\tilde{\lambda}_{r-1} z} + a_r(z) = C \cdot z^{|n|-1} + \cdots, \end{aligned} \tag{2.2}$$

where $\tilde{\lambda}_j = \lambda_j - \lambda_r$, $j = 1, \ldots, r-1$. We differentiate (2.2) n_r times (see Chapter 2, §2):

$$\begin{aligned} \widetilde{L}(z) &= D^{n_r} L_1(z) \\ &= \tilde{a}_1(z)e^{\tilde{\lambda}_1 z} + \cdots + \tilde{a}_{r-1}(z)e^{\tilde{\lambda}_{r-1} z} = \widetilde{C} \cdot z^{|n|-n_r-1} + \cdots, \end{aligned} \tag{2.3}$$

where

$$\tilde{a}_j = (D + \tilde{\lambda}_j)^{n_r} a_j, \qquad j = 1, \ldots, r-1. \tag{2.4}$$

Since $\tilde{\lambda}_j \neq 0$, it follows from (2.3), (2.4), and the inductive assumption that

$$\deg a_j = \deg \tilde{a}_j = n_j - 1, \qquad j = 1, \ldots, r-1.$$

If we change the numbering of the exponents, we obtain $\deg a_r = n_r - 1$. ▷

2. Let $\lambda_1, \ldots, \lambda_r$ be different nonzero complex numbers. By Theorem 2.1 and Proposition 1.3, the Mahler indices are normal for the system $\{e^{\lambda_j z}\}_1^r$. For these indices, Problem A takes the following form. Given $n \in \mathbf{Z}_+^r$, $m \in \mathbf{Z}_+$; find a polynomial $Q \not\equiv 0$, $\deg Q \leq |n|$, and polynomials $P_1, \ldots, P_r$, of degree not exceeding $|n| + m - n_1, \ldots, |n| + m - n_r$, respectively, such that for $j = 1, \ldots, r$,

$$R_j(z) = Q(z)e^{\lambda_j z} - P_j(z) = C_j z^{|n|+m+1} + \cdots.$$

THEOREM 2.2. *With the normalization $Q(0) = 1$, we have the following Hermite formulas $(j = 1, \ldots, r)$:*

$$Q(z) = \frac{z^{|n|+m+1}}{(|n|+m)!} \int_0^\infty T(x)e^{-zx}\,dx, \tag{2.5}$$

$$P_j(z) = \frac{e^{\lambda_j z} z^{|n|+m+1}}{(|n|+m)!} \int_{\lambda_j}^\infty T(x)e^{-zx}\,dx, \tag{2.6}$$

$$R_j(z) = \frac{e^{\lambda_j z} z^{|n|+m+1}}{(|n|+m)!} \int_0^{\lambda_j} T(x)e^{-zx}\,dx, \tag{2.7}$$

where

$$T(x) = x^m \prod_{j=1}^{r} (x - \lambda_j)^{n_j}; \tag{2.8}$$

in (2.5) *and* (2.6), *the integration is over a contour going to* $+\infty$, *and* $\operatorname{Re} z > 0$.

PROOF. Let P, Q_j, and R_j be defined by (2.5), (2.6), and (2.7); then

$$R_j(z) + P_j(z) = e^{\lambda_j z} Q(z), \qquad j = 1, \dots, r;$$

$R_j(z)$ is an entire function that has a zero of order $|n| + m + 1$ at $z = 0$.

Performing the multiplication in (2.8) we obtain

$$T(x) = x^{|n|+m} + \cdots + b x^m, \qquad b \neq 0.$$

Since

$$\int_0^\infty x^k e^{-zx}\, dx = \frac{k!}{z^{k+1}}, \qquad k \in \mathbf{Z}_+,$$

we have that

$$Q(z) = 1 + \cdots + b \frac{m!}{(|n| + m)!} z^{|n|}$$

is a polynomial of degree $|n|$, with $Q(0) = 1$.

In (2.6), we replace x by $x - \lambda_j$; then

$$P_j(z) = \frac{z^{|n|+m+1}}{(|n| + m)!} \int_0^\infty \left((x + \lambda_j)^m \prod_{k \neq j} (x + \lambda_j - \lambda_k)^{n_k} \right) x^{n_j} e^{-zx}\, dx.$$

Hence, as before, it follows that P_j is a polynomial of degree $|n| + m - n_j$ with $P_j(0) = 1$. $\triangleright$

3. Integrals of the forms (2.5)–(2.8) were introduced by C. Hermite in 1874 for proving the *transcendence of the number e*. The following proof was given by David Hilbert.

Let us suppose that

$$a_0 + a_1 e + \cdots + a_n e^n = 0 \tag{2.9}$$

for some positive integer n, where a_j are integers, and a_0, $a_n \neq 0$. Let p be a prime with $p > n$, $p > a_0$. We set (for $\nu = 1, \dots, n$)

$$M = \int_0^\infty t(x) e^{-x}\, dx, \qquad M_\nu = e^\nu \int_\nu^\infty t(x) e^{-x}\, dx,$$

$$\varepsilon_\nu = e^\nu \int_0^\nu t(x) e^{-x}\, dx,$$

where

$$t(x) = \frac{1}{(p-1)!} x^{p-1} ((x - 1) \cdots (x - n))^p.$$

Then $Me^{\nu} = M_{\nu} + \varepsilon_{\nu}$, and it follows from (2.9) that

$$(a_0 M + a_1 M_1 + \cdots + a_n M_n) + (a_1 \varepsilon_1 + \cdots + a_n \varepsilon_n) = 0. \tag{2.10}$$

We have

$$t(x) = \frac{x^{p-1}}{(p-1)!}\left(\sum_{\rho=1}^{np} c_{\rho} x^{\rho} + (-1)^{np}(n!)^p\right),$$

where c_{ρ} is an integer. Consequently

$$M = (-1)^{np}(n!)^p + \sum_{\rho=1}^{np} c_{\rho}\frac{(p+\rho-1)!}{(p-1)!}$$

is an integer and not divisible by p. In fact, for $\rho \geq 1$ the number $(p+\rho-1)!/(p-1)!$ is an integer divisible by p, and $n!$ is not divisible by p if $p > n$. We make the change of variable $\xi = x - \nu$ in the integral M_{ν}:

$$M_{\nu} - \frac{1}{(p-1)!}\int_0^{\infty} (\xi+\nu)^{p-1}((\xi+\nu-1)\cdots(\xi+\nu-n))^p e^{-\xi} d\xi. \tag{2.11}$$

If we expand the product in (2.11), as above, we see that M_{ν} is an integral multiple of p. Thus the first parenthesis in (2.10) is an integer that is not divisible by p, and therefore, in particular, is not zero. At the same time, $\varepsilon_{\nu} \to 0$ as $p \to \infty$. ▷

REMARKS

1. Linear forms for exponentials were calculated by C. Hermite [138] in 1893; simultaneous approximations [137], in 1874.

2. The convergence of Padé approximants for exponentials were investigated in [154] and [157]. The convergence of simultaneous Padé approximants for a set of exponentials was investigated in [127].

Let $Q = Q_{n,m}$ and $P_j = P_{n,m,j}$ be as defined in (2.5) and (2.6); $\pi_{n,m,j} = P_{n,m,j}/Q_{n,m}$, $j = 1, \ldots, r$. The following theorem was proved in [2]:

THEOREM. *The approximants $\pi_{n,m,j}(z)$ converge to $e^{\lambda_j z}$, $j = 1, \ldots, r$, uniformly on compact subsets of $\mathbf{C}$ when $|n| + m \to \infty$.*

A similar theorem was proved in [3] for a set of confluent hypergeometric functions (see also [161], [126], and [12]). For the normality of general Padé tables for a set of exponentials, see [146]. For properties of the zeros and poles of Padé approximants for confluent hypergeometric functions, see [163]–[165] and [74].

3. The proof of the transcendence of e is presented in the book [49]. In 1882, F. Lindemann [145], using Hermite's method, obtained more general results. We quote one of these.

THEOREM. *Let the algebraic numbers $\alpha_1, \ldots, \alpha_m$ be linearly independent (over the field $\mathbf{Q}$). Then the numbers $e^{\alpha_1}, \ldots, e^{\alpha_m}$ are algebraically independent.*

COROLLARY. *The number π is transcendental.*

PROOF. $e^{\pi i} = -1$. ▷

Hermite integrals are discussed in the book [149]. A further development of Hermite's method was obtained by C. L. Siegel [166]. A significant improvement of this method was given by A. B. Shidlovskiĭ (see, for example, [115] and [116]). See also [105], [24], [27], and [28].

Results on the arithmetic properties of values of functions that are analytic in the unit disk (not entire), obtained by the use of linear forms and simultaneous approximants, are given in, for example, [107], [25], and [26], and also in [65], [68], [70], and [94].

PROBLEM

1. Using the proof of Theorem 2.1, find an explicit form for the coefficients of the polynomials $a_j(z)$.

§3. Polyorthogonal polynomials

1. Let $f = (f_1, \dots, f_r)$ be a set of formal power series

$$f_j(z) = \sum_{k=0}^{\infty} \frac{s_{j,k}}{z^{k+1}}, \qquad j = 1, \dots, r, \tag{3.1}$$

with complex coefficients, $n \in \mathbf{Z}_+^r$. We denote by $\tau(\varphi)$ the smallest exponent of a power of $1/z$ that appears in the series φ with a nonzero coefficient.

PROBLEM C. *Find the polynomial* $Q = Q_n \not\equiv 0$, $\deg Q \le |n|$, *that satisfies the inequalities* $\tau(R_j) \ge n_j + 1$ *for certain polynomials* $P_1 = P_{n,1}, \dots, P_r = P_{n,r}$, *where* $R_j = Qf_j - P_j$, $j = 1, \dots, r$.

A solution of Problem C evidently exists.

DEFINITION. If (Q, P), where $P = (P_1, \dots, P_r)$, is a solution of Problem C with index $n \in \mathbf{Z}_+^r$, the rational function

$$\pi_n = \left(\frac{P_1}{Q}, \dots, \frac{P_r}{Q}\right)$$

is said to be the *nth simultaneous diagonal Padé approximant* for the set (3.1) of formal power series.

Generally speaking, the vector π_n is not uniquely defined. A sufficient condition for uniqueness is that the index n is normal.

DEFINITION. An index $n \in \mathbf{Z}_+^r$ is said to be *normal for Problem* C if every solution (Q, P) with this index has $\deg Q = |n|$.

DEFINITION. If all $n \in \mathbf{Z}_+^r$ are normal for C, we call f a *complete system.*

The following problem is a dual of Problem C.

PROBLEM D. *Find polynomials* $A_1 = A_{n,1}, \dots, A_r = A_{n,r}$, *not simultaneously identically zero* $(n \neq 0)$, *whose degrees do not exceed* $n_1 - 1, \dots, n_r - 1$, *respectively, with the property that* $\tau(L) \ge |n|$ *for some polynomial* $B = B_n$, *where* $L = A_1 f_1 + \dots + A_r f_r - B$.

A solution of Problem D exists.

DEFINITION. An index $n \in \mathbf{Z}_+^r$, $n \neq 0$, is said to be *normal with respect to Problem* D if we have $\deg A_j = n_j - 1$, $j = 1, \dots, r$, for every solution of the problem with index n.

If n is normal, Problem D has a unique solution.

2. The solutions of Problems C and D are closely connected. Let us introduce the notation

$$n^{(s)} = (n_1^{(s)}, \dots, n_r^{(s)}), \qquad n_j^{(s)} = n_j + \delta_{s,j},$$

where $n \in \mathbf{Z}_+^r$, $\delta_{s,j}$ is the Kronecker symbol, and $j, s = 1, \dots, r$.

PROPOSITION 3.1. *Let the index $n \in \mathbf{Z}_+^r$ be normal with respect to* C, *and* $(A_1^{(s)}, \dots, A_r^{(s)}, B^{(s)})$ *a solution of Problem* D *with indices* $n^{(s)}$, $s = 1, \dots, r$. *If*

$$Q = \det \begin{pmatrix} A_1^{(1)} & \cdots & A_r^{(1)} \\ \dots & \dots & \dots \\ A_1^{(r)} & \cdots & A_r^{(r)} \end{pmatrix},$$

and P_j is the determinant obtained from Q by replacing the jth column by the column $(B^{(1)}, \dots, B^{(r)})^{\mathrm{T}}$, $j = 1, \dots, r$, *then* $(Q, P_1, \dots, P_r)$ *is a solution of Problem* C *with index n.*

PROOF. Since n is normal, $\deg A_j^{(j)} = n_j$, $j = 1, \dots, r$. The determinants Q and P_j are polynomials; $\deg Q = |n|$. We have

$$R_j = Qf_j - P_j = \det \begin{pmatrix} A_1^{(1)} & \cdots & A_j^{(1)} f_j - B^{(1)} & \cdots & A_r^{(1)} \\ \dots & \dots & \dots & \dots & \dots \\ A_1^{(r)} & \cdots & A_j^{(r)} f_j - B^{(r)} & \cdots & A_r^{(r)} \end{pmatrix}.$$

If we add to the jth column of this determinant all the other columns, having first multiplied them by the corresponding series f_k, we obtain

$$R_j = \det \begin{pmatrix} A_1^{(1)} & \cdots & L^{(1)} & \cdots & A_r^{(1)} \\ \dots & \dots & \dots & \dots & \dots \\ A_1^{(r)} & \cdots & L^{(r)} & \cdots & A_r^{(r)} \end{pmatrix}, \tag{3.2}$$

where

$$L^{(s)} = A_1^{(s)} f_1 + \cdots + A_r^{(s)} f_r - B^{(s)}, \qquad s = 1, \dots, r.$$

We expand the determinant (3.2) in terms of the jth column:

$$R_j = L^{(1)} \Delta_{1,j} + \cdots + L^{(r)} \Delta_{r,j}, \tag{3.3}$$

where $\Delta_{s,j}$ are the corresponding algebraic complements. Since

$$\deg \Delta_{s,j} \le |n| - n_j, \qquad j = 1, \dots, r,$$
$$\tau(L^{(s)}) \ge |n| + 1, \qquad s = 1, \dots, r,$$

it follows from (3.3) that

$$\tau(R_j) \ge n_j + 1, \qquad j = 1, \dots, r. \quad \triangleright$$

REMARK. Proposition 3.1 remains valid if instead of the indices $n^{(s)}$ we take

$$(n_1 + 1, \dots, n_s + 1, n_{s+1}, \dots, n_r), \qquad s = 1, \dots, r.$$

3. DEFINITION. *Proper indices* are defined to be

$$n = (\underbrace{t+1, \dots, t+1}_{s}, \underbrace{t, \dots, t}_{r-s}),$$

where $t \in \mathbf{Z}_+$, $s = 0, \dots, r-1$. We denote a proper index of order $m \in \mathbf{Z}_+$ by ν_m.

DEFINITION. Solutions of Problems C and D corresponding to proper indices are said to be *purely diagonal*.

DEFINITION. A system f is a *weakly complete system if* all proper indices are normal for it with respect to C.

The next proposition follows from the proof of Proposition 3.1 and the remark following it.

PROPOSITION 3.2. *If a proper index* ν_m *is normal with respect to* C, *and* $(A_1^{(s)}, \dots, A_r^{(s)}, B^{(s)})$ *is a solution of Problem* D *with proper indices* ν_{m+s}, $s = 1, \dots, r$, *then the vector* (Q, P) *defined in Proposition* 3.1 *is a solution of Problem* C *with index* ν_m.

4. We present an example of a complete system that first appeared in 1919 in a paper by Angelesco [120] (also see [66]).

Let μ_j be a measure on the real line with a compact infinite spectrum S_{μ_j}, let Δ_j be the shortest interval containing S_{μ_j}, i.e., the support of the measure, and let

$$f_j(z) = \int_{\Delta_j} \frac{d\mu_j(x)}{z - x}, \qquad z \in \overline{\mathbf{C}} \backslash \Delta_j, \tag{3.4}$$

be its Markov function, $j = 1, \dots, r$. By Problems C and D for the functions (3.4) we mean these problems for the Laurent expansions

$$f_j(z) = \sum_{k=0}^{\infty} \frac{1}{z^{k+1}} \int_{\Delta_j} x^k \, d\mu_j(x), \qquad j = 1, \dots, r.$$

Just as in the case $r = 1$ (see Chapter 2, §5) we can show that (Q, P) is a solution of Problem C with index n, if

$$\deg Q \le |n|, \qquad Q \not\equiv 0, \tag{3.5}$$

and the orthogonality relations

$$\int_{\Delta_j} Q(x) x^k \, d\mu_j(x) = 0, \qquad k = 0, \dots, n_j - 1; \, j = 1, \dots, r, \tag{3.6}$$

are satisfied; the polynomials of the second kind are defined by the formulas

$$P_j(z) = \int_{\Delta_j} \frac{Q(z) - Q(x)}{z - x} \, d\mu_j(x), \qquad j = 1, \dots, r.$$

DEFINITION. A polynomial Q that satisfies conditions (3.5) and (3.6) is called an n*th polyorthogonal* (more precisely, *r-orthogonal*) *polynomial* (*of type* II).

Similarly, (A, B) is a solution of Problem D with index $n \neq 0$ when

$$\deg A_j \le n_j - 1, \qquad j = 1, \dots, r, \ \exists j_0 \colon A_{j_0} \not\equiv 0, \tag{3.7}$$

and the corresponding orthogonality conditions

$$\sum_{j=1}^{r}\int_{\Delta_j} A_j(x)x^k d\mu_j(x) = 0, \qquad k = 0, \dots, |n|-2, \tag{3.8}$$

are satisfied; the polynomial of the second kind is defined by the formula

$$B(z) = \sum_{j=1}^{r}\int_{\Delta_j} \frac{A_j(z) - A_j(x)}{z-x} d\mu_j(x).$$

DEFINITION. The polynomials $A_1, \dots, A_r$ that satisfy (3.7) and (3.8) are called *nth polyorthogonal* (more precisely, *r-orthogonal*) *polynomials* (*of type* I).

We say that intervals are disjoint if they have no common interior points.

PROPOSITION 3.3. *If the intervals* $\{\Delta_j\}$ *are pairwise disjoint, then for every* $n \in \mathbf{Z}_+^r$ *the nth r-orthogonal polynomial* Q *has* n_j *simple zeros inside* Δ_j, $j = 1, \dots, r$.

PROOF. Since (3.6) reduces to a system of homogeneous linear equations with real coefficients, we may suppose without loss of generality that Q has real coefficients.

Let us suppose that, for some j, the polynomial Q changes sign on Δ_j only at the points $x_1, \dots, x_m$, $0 \le m \le n_j - 1$. Let T be a polynomial of degree m with zeros $\{x_j\}$. We may then suppose that $Q(x)T(x) \ge 0$, $x \in \Delta_j$. The measure μ_j has infinitely many points of increase; consequently

$$\int_{\Delta_j} Q(x)T(x)d\mu_j(x) > 0,$$

contradicting (3.6). ▷

COROLLARY. *If the* $\{\Delta_j\}$ *are pairwise disjoint, the system* (3.4) *is complete.*

The following proposition is proved similarly.

PROPOSITION 3.4. *If the intervals* $\{\Delta_j\}$ *are pairwise disjoint, and* $A_1, \dots,$ A_r *are r-orthogonal polynomials of type* I *corresponding to the index* $n \in \mathbf{Z}_+^r$, $n \neq 0$, *then, for every* $j = 1, \dots, r$, *the polynomial* A_j *has* $n_j - 1$ *simple zeros inside* Δ_j.

5. As in the case $r = 1$, there are quadrature formulas connected with polyorthogonal polynomials. We shall suppose that the intervals $\{\Delta_j\}$ are pairwise disjoint.

We have

$$f_k(z) - \frac{P_{n,k}(z)}{Q_n(z)} - \frac{C_k}{z^{|n|+n_k+1}} + \cdots, \qquad k = 1, \dots, r. \tag{3.9}$$

Let $\lambda_{j,1}, \dots, \lambda_{j,n_j}$ be the zeros of Q_n on Δ_j, $j = 1, \dots, r$. Then

$$\frac{P_{n,k}(z)}{Q_n(z)} = \sum_{j=1}^{r} \sum_{l=1}^{n_j} \frac{\mu_{k,j,l}}{z - \lambda_{j,l}}, \tag{3.10}$$

where

$$\mu_{k,j,l} = \operatorname*{res}_{z=\lambda_{j,l}} \frac{P_{n,k}(z)}{Q_n(z)}. \tag{3.11}$$

As in the classical case, we call the numbers (3.11) the *Christoffel coefficients.*

Let T be a polynomial with $\deg T \le |n| + n_k - 1$. If we multiply both sides of (3.9) by $T(z)$ and integrate over a contour Γ that surrounds all the intervals $\{\Delta_j\}$, we obtain

$$\frac{1}{2\pi i} \oint_\Gamma T(z) f_k(z)\, dz - \frac{1}{2\pi i} \oint_\Gamma T(z) \frac{P_{n,k}(z)}{Q_n(z)} dz = 0. \tag{3.12}$$

By using (3.4), (3.10), Fubini's theorem, and the Cauchy integral formula, we can rewrite (3.12) in the form

$$\int_{\Delta_k} T(x) d\mu_k(x) = \sum_{j=1}^{r} \sum_{l=1}^{n_j} T(\lambda_{j,l}) \mu_{k,j,l} \qquad (\deg T \le |n| + n_k - 1). \tag{3.13}$$

There naturally arises the question of transferring the proof of the Markov theorem (Chapter 2, §6) to the system (3.4). The central feature of that theorem is the positivity of the Christoffel coefficients. However, this is not valid for $r \ge 2$.

PROPOSITION 3.5. *The Christoffel coefficients have the following properties*:

1) $\mu_{k,k,l} > 0$;

2) *if* $j \ne k$, *the coefficients* $\mu_{k,j,l}$ *corresponding to successive zeros* $\lambda_{j,l}$ *have opposite signs, and the coefficients corresponding to the zeros closest to* Δ_k *are positive.*

PROOF. Set

$$Q_{n,j}(x) = \prod_{l=1}^{n_j} (x - \lambda_{j,l}), \qquad j = 1, \dots, r;$$

then $Q_n = Q_{n,1} \cdots Q_{n,r}$. Let

$$T(x) = \frac{Q_n(x)}{(x - \lambda_{k,l}) Q_n'(\lambda_{k,l})} \cdot \frac{Q_{n,k}(x)}{(x - \lambda_{k,l}) Q_{n,k}'(\lambda_{k,l})}. \tag{3.14}$$

Then $\deg T = |n| + n_k - 2$, $T(\lambda_{k,l}) = 1$, T is zero at the other zeros of Q_n, and $T(x) \ge 0$ for $x \in \Delta_k$. If we apply the quadrature formula (3.13) to the polynomial (3.14), we obtain

$$0 < \int_{\Delta_k} T(x) d\mu_k(x) = \mu_{k,k,l}.$$

If $j \neq k$, we set

$$T(x) = \frac{Q_n(x)}{(x-\lambda_{j,l})Q_n'(\lambda_{j,l})} \cdot \frac{Q_{n,k}(x)}{Q_{n,k}(\lambda_{j,l})}. \tag{3.15}$$

Then $\deg T = |n| + n_k - 1$, $T(\lambda_{j,l}) = 1$, T is zero at the other zeros of Q_n, has constant sign on Δ_k, and the sign is determined by the rule stated in property 2). If we apply (3.13) to (3.15), we obtain the required property. ▷

§4. Chebyshev systems

1. Chebyshev systems have numerous applications in approximation theory, mathematical statistics, and other fields. In terms of these systems, we can describe a wide class of complete and weakly complete systems.

DEFINITION. A set of real functions

$$u_0(t), u_1(t), \dots, u_n(t), \tag{4.1}$$

continuous on the interval $[a, b]$, is called a *Chebyshev system* (or *T-system*) of order $n \in \mathbf{Z}_+$, if every polynomial

$$P(t) = \alpha_0 u_0(t) + \alpha_1 u_1(t) + \dots + \alpha_n u_n(t),$$
$$\alpha_0, \alpha_1, \dots, \alpha_n \in \mathbf{R}, \qquad \alpha_0^2 + \alpha_1^2 + \dots + \alpha_\eta^2 > 0,$$

in this system vanishes at most at n points of $[a, b]$.

This definition is equivalent to the statement that the determinant

$$U(t_0, t_1, \dots, t_n) = \det \begin{pmatrix} u_0(t_0) & u_1(t_0) & \cdots & u_n(t_0) \\ u_0(t_1) & u_1(t_1) & \cdots & u_n(t_1) \\ \dots & \dots & \dots & \dots \\ u_0(t_n) & u_1(t_n) & \cdots & u_n(t_n) \end{pmatrix}$$

of the system

$$\sum_{k=0}^{n} \alpha_k u_k(t_j) = 0, \qquad j = 0, 1, \dots, n,$$

of homogeneous linear equations is different from zero for arbitrary distinct points $t_0, t_1, \dots, t_n$ of $[a, b]$. Since U is a continuous function, and $U \neq 0$ on the set $a \le t_0 < t_1 < \dots < t_n \le b$, it has a constant sign there. If $U > 0$, (4.1) is called a *T_+-system*. The following elementary properties of *T-systems* are evident.

1) For any n distinct points $t_1, \dots, t_n$ of $[a, b]$, there is a nontrivial polynomial $P(t)$ in a given T-system of order n with zeros at these points. This polynomial is determined up to a constant factor: $P(t) = CU(t, t_1, \dots, t_n)$, $C \neq 0$.

2) There is exactly one polynomial $P(t)$ in a given T-system of order n that takes preassigned values at $n+1$ given points of $[a, b]$.

Property 1) can be strengthened (see [55]).

DEFINITION. A root of the polynomial $P(t)$ (in a given T-system) is a *change-knot* if it is either an endpoint of the interval, or an interior point of an interval at which $P(t)$ changes sign.

DEFINITION. A root of $P(t)$ is a *permanence-knot* if it is an interior point at which $P(t)$ does not change sign.*

THEOREM 4.1. *For arbitrary k points $t_1, \dots, t_k$ inside $[a, b]$ and l points $t_{k+1}, \dots, t_{k+l}$ on $[a, b]$, there is a polynomial $P(t)$ in a given T-system of order n that has $t_1, \dots, t_k$ as permanence-knots and $t_{k+1}, \dots, t_{k+l}$ as change-knots, and no other knots on $[a, b]$, if and only if $2k + l \leq n$.*

Let us give some examples of T-systems.

1) $1, t, \dots, t^n$ is a T_+-system of order $n \in \mathbf{Z}_+$ on any interval $[a, b] \subset \mathbf{R}$;

$$U(t_0, \dots, t_n) = \det(t_i^j) = \prod_{0 \leq i < j \leq n} (t_j - t_i)$$

is a Vandermonde determinant.

2) If $\alpha_0, \alpha_1, \dots, \alpha_n$ are distinct real numbers,

$$\frac{1}{t + \alpha_0}, \frac{1}{t + \alpha_1}, \dots, \frac{1}{t + \alpha_n}$$

is a T-system of order n on any interval $[a, b]$ that does not contain the points $\{\alpha_j\}$;

$$\begin{aligned} U(t_0, \dots, t_n) &= \det\left(\frac{1}{t_i + \alpha_j}\right) \\ &= \left(\prod_{0 \leq i < j \leq n} (t_j - t_i)(\alpha_j - \alpha_i)\right) \Big/ \left(\prod_{i, j=0}^{n} (t_i + \alpha_j)\right) \end{aligned}$$

is a Cauchy determinant.

3) Let $\beta_1, \dots, \beta_r$ be distinct real numbers, $n \in \mathbf{Z}_+^r$, $n \neq 0$. Then

$$e^{\beta_1 t}, \dots, t^{n_1 - 1} e^{\beta_1 t}, \dots, e^{\beta_r t}, \dots, t^{n_r - 1} e^{\beta_r t}$$

is a T-system of order $|n| - 1$ on any interval $[a, b] \subset \mathbf{R}$. (The proof is by induction on r: $P(t) \mapsto D^{n_r} e^{-\beta_r t} P(t)$, using Rolle's theorem.)

4) Let $\alpha_0, \alpha_1, \dots, \alpha_n$ be distinct real numbers; then $t^{\alpha_0}, t^{\alpha_1}, \dots, t^{\alpha_n}$ is a T-system of order n on any interval $[a, b] \subset (0, \infty)$. In fact, the change of variable $x = \ln t$ yields $\{e^{\alpha_j x}\}_0^n$.

2. DEFINITION The system (4.1) is a *Markov system* or *M-system* (*M_+-system*) of order n on $[a, b]$ if, for any $m = 0, \dots, n$ the system $u_0(t), \dots, u_m(t)$ is a T-system (T_+-system) of order m on $[a, b]$.

---

* *Translator's note.* There seems to be no standard terminology in English for the two types of knots (or nodes). Literal translations might be "root-knot" and "root-bundle."

If the functions (4.1) are n times continuously differentiable on $[a, b]$, we can form their Wronskians $(m = 0, \dots, n)$,

$$W_m(t) = W_m(u_0, \dots, u_m; t) = \begin{vmatrix} u_0(t) & u_1(t) & \cdots & u_m(t) \\ u_0'(t) & u_1'(t) & \cdots & u_m'(t) \\ \dots & \dots & \dots & \dots \\ u_0^{(m)}(t) & u_1^{(m)}(t) & \cdots & u_m^{(m)}(t) \end{vmatrix}. \tag{4.2}$$

A. A. Markov proved the following theorem.

THEOREM 4.2. *A necessary condition for a system* (4.1) *of n times differentiable functions to be an M-system on $[a, b]$ is that they do not change sign on $[a, b]$, and a sufficient condition is that none of the Wronskians* (4.2) *has zeros on $[a, b]$.*

PROOF. We choose $m = 0, \dots, n$ arbitrarily. Since $u_0, \dots, u_m$ is a T-system, the determinant $U(t_0, \dots, t_m)$ does not change sign in the domain $a < t_0 < t_1 < \cdots < t_m < b$; for example, let it be positive. Set $t_j = t + jh$, $j = 0, \dots, m$, where $a < t < b$, and h is a sufficiently small positive number. If we subtract the rows of U from each other, we can reduce it to the form

$$\begin{vmatrix} u_0(t) & u_1(t) & \cdots & u_m(t) \\ \Delta u_0(t) & \Delta u_1(t) & \cdots & \Delta u_m(t) \\ \dots & \dots & \dots & \dots \\ \Delta^m u_0(t) & \Delta^m u_1(t) & \cdots & \Delta^m u_m(t) \end{vmatrix} > 0, \tag{4.3}$$

where Δ is the finite difference operator with step h: $(\Delta\varphi)(t) = \varphi(t+h) - \varphi(t)$, and $\Delta^k \varphi = \Delta(\Delta^{k-1}\varphi)$ are powers of this operator. If we divide both sides of (4.3) by $h^{m(m+1)/2}$ and take the limit as $h \to 0+$, we obtain $W_m(t) \geq 0$.

We may suppose without loss of generality that $W_m(t) > 0$, $a \leq t \leq b$, and $m = 0, \dots, n$. We introduce the functions

$$\rho_0(t) = W_0(t), \qquad \rho_1(t) = \frac{W_1(t)}{W_0(t)^2},$$

$$\rho_k(t) = \frac{W_{k-2}(t) W_k(t)}{W_{k-1}(t)^2}, \qquad k = 2, \dots, n.$$

Then ρ_b has $n - k$ continuous derivatives and $\rho_k(t) > 0$, $a \leq t \leq b$, $k = 0, \dots, n$. We introduce the differential operators

$$D_k = \frac{d}{dt} \frac{1}{\rho_k(t)}, \qquad k = 0, \dots, n.$$

Each of the following formulas is easily established by induction:

$$u_0 = \rho_0, \quad D_0 u_1 = \rho_1, \quad \dots, \quad D_{n-1} \cdots D_0 u_n = \rho_n. \tag{4.4}$$

Then

$$D_0 u_0 = 0, \quad D_1 D_0 u_1 = 0, \quad \dots, \quad D_n \cdots D_1 D_0 u_n = 0. \tag{4.5}$$

Let m be the smallest index for which $u_0, \dots, u_m$ is not a T-system, i.e., there are real numbers $\alpha_0, \dots, \alpha_m$, not all zero, such that the function

$$P(t) = \alpha_0 u_0(t) + \cdots + \alpha_m u_m(t)$$

has at least $m+1$ zeros on $[a, b]$. Since m is minimal, the coefficient α_m is not zero. By Rolle's theorem, $P_1(t) = D_{m-1} \cdots D_0 P(t)$ has a zero on $[a, b]$. On the other hand, by using (4.4) and (4.5) we obtain

$$P_1(t) = \alpha_m D_{m-1} \cdots D_0 u_m(t) = \alpha_m \rho_m(t) \neq 0,$$

a contradiction. ▷

3. DEFINITION. A set of continuous real functions

$$u_1(t), \dots, u_r(t), \tag{4.6}$$

defined on $[a, b]$, is called an *AT-system for the index* $n \in \mathbf{Z}_+^r$, $n \neq 0$, if

$$u_1(t), \dots, t^{n_1-1} u_1(t), \dots, u_r(t), \dots, t^{n_r-1} u_r(t)$$

is a T-system of order $|n| - 1$ on $[a, b]$.

DEFINITION. A system (4.6) is called an *AT-system* if it is an AT-system for every $n \in \mathbf{Z}_+^r$, $n \neq 0$.

THEOREM 4.3. *If* (4.6) *is an AT-system for the index* $n \in \mathbf{Z}_+^r$, $n \neq 0$, *on the interval* $[a, b]$, *and* σ *is a measure on this interval, with an infinite number of points of increase, the index n is normal with respect to Problem* C *for the system of functions*

$$f_j(z) = \int_a^b \frac{u_j(t) d\sigma(t)}{z - t}, \qquad j = 1, \dots, r. \tag{4.7}$$

PROOF. Let Q be the nth polyorthogonal polynomial corresponding to the functions (4.7):

$$Q \not\equiv 0, \qquad \deg Q \leq |n|,$$

$$\int_a^b Q(t) t^k u_j(t) d\sigma(t) = 0, \qquad k = 0, \dots, n_j - 1; \, j = 1, \dots, r. \tag{4.8}$$

Let $t_1, \dots, t_s$ be the points of (a, b) at which Q changes sign. Suppose that $s < |n|$. Then, by Theorem 4.1, there is a real polynomial

$$P(t) = A_1(t) u_1(t) + \cdots + A_r(t) u_r(t), \qquad \deg A_j \leq n_j - 1,$$

for which $t_1, \dots, t_s$ are change-knots, and there are no other zeros of the polynomial on $[a, b]$. The function $P(t)Q(t)$ does not change sign on $[a, b]$ and has finitely many zeros on this interval. This contradicts the relation

$$\int_a^b P(t) Q(t) d\sigma(t) = 0$$

that follows from (4.8). Consequently $s = |n|$ and $\deg Q = n$. ▷

COROLLARY. *If σ is a measure on $[a, b]$ with an infinite number of points of increase, and* (4.6) *is an AT-system on $[a, b]$, it follows that* (4.7) *is a complete system of functions. All the zeros of all the polynomials $Q_n(z)$ are real and simple, and lie in the interval (a, b).*

Therefore, we can obtain examples of complete systems if we know examples of AT-systems. Examples 3) and 4) of subsection 1 lead to the following corollaries.

3′) If $\beta_1, \dots, \beta_r$ are distinct real numbers, the system $e^{\beta_1 t}, \dots, e^{\beta_r t}$ is an AT-system on every interval $[a, b] \subset \mathbf{R}$.

4′) If the real numbers $\alpha_1, \dots, \alpha_r$ are pairwise incongruent modulo 1, i.e., $\alpha_i - \alpha_j \notin \mathbf{Z}$, $i \neq j$, the system $t^{\alpha_1}, \dots, t^{\alpha_r}$ is an AT-system on every interval $[a, b] \subset (0, \infty)$.

4. DEFINITION. We say that the set (4.6) of real functions that are continuous on $[a, b]$ is a *weak AT-system* if the sequence

$$u_1(t), \dots, u_r(t), \dots, t^m u_1(t), \dots, t^m u_r(t), \dots \tag{4.9}$$

is a Markov system on $[a, b]$; in other words, (4.6) is an AT-system for every proper index $\nu_m \neq 0$.

The following corollary is a consequence of Theorem 4.3.

COROLLARY. *If* (4.6) *is a weak AT-system on $[a, b]$, the set* (4.7) *is a weakly complete system.*

The following theorem provides a method for constructing weakly complete AT-systems.

THEOREM 4.4. *On an interval $[c, d]$, disjoint from $[a, b]$, let there be given a weak AT-system*

$$\chi_1(t), \dots, \chi_r(t) \tag{4.10}$$

and a measure σ with infinitely many points of increase. Then

$$1, u_1(x), \dots, u_r(x), \tag{4.11}$$

where

$$u_j(x) = \int_c^d \frac{\chi_j(t)}{t-x} d\sigma(t), \qquad x \in [a, b],\ j = 1, \dots, r,$$

is a weak AT-system on $[a, b]$.

PROOF. We are to show that a sequence of the form (4.9) generated by (4.11) is a Markov system. We apply Theorem 4.2. We note that the condition $W_{m-1}(x) \neq 0$, $x \in [a, b]$, is equivalent to the condition that no nontrivial linear combination of the first m functions of the system can have a zero of order m at any point of the interval. Let us suppose that for some proper $(r+1)$-dimensional multi-index $n = (n_0, \dots, n_r)$ and point $x_0 \in [a, b]$

$$\begin{aligned} F(x) &= A_0(x) + A_1(x)u_1(x) + \cdots + A_r(x)u_r(x) \\ &= O(x - x_0)^{|n|}, \qquad x \to x_0, \end{aligned} \tag{4.12}$$

where $\deg A_j \le n_j - 1$, $j = 0, 1, \dots, r$. We have

$$\begin{aligned} A_j(x)u_j(x) &= \int_c^d \frac{A_j(x) - A_j(t)}{t-x}\chi_j(t)d\sigma(t) + \int_c^d \frac{A_j(t)\chi_j(t)}{t-x}d\sigma(t) \\ &= S_j(x) + R_j(x), \qquad j = 1, \dots, r, \end{aligned} \tag{4.13}$$

where $S_j(x)$ (the first integral in (4.13)) is a polynomial with $\deg S_j \le n_0 - 2$. Consequently

$$F(x) = G(x) + \sum_{j=1}^{r} \int_c^d \frac{A_j(t)\chi_j(t)}{t-x} d\sigma(t),$$

where $\deg G \le n_0 - 1$. Then

$$F^{(n_0)}(x) = n_0! \sum_{j=1}^{r} \int_c^d \frac{A_j(t)\chi_j(t)}{(t-x)^{n_0+1}} d\sigma(t).$$

We introduce the notation

$$R(t) = \sum_{j=1}^{r} A_j(t)\chi_j(t).$$

By (4.12),

$$\int_c^d \frac{R(t)d\sigma(t)}{(t-x_0)^{n_0+1+k}} = 0, \qquad k = 0, \dots, |n| - n_0 - 1. \tag{4.14}$$

Let $t_1, \dots, t_s$ be the points of (c, d) at which $R(t)$ changes sign. Since (4.10) is a weak AT-system and $\deg A_j \le n_j - 1$, $j = 1, \dots, r$, we have $s \le |n| - n_0 - 1$. The function

$$P(t) = \prod_{k=1}^{s} \left(\frac{1}{t - x_0} - \frac{1}{t_k - x_0} \right)$$

changes sign at the points $t_1, \dots, t_s$ and has no other zeros on $[c, d]$. We may suppose that $R(t) \cdot P(t) \ge 0$ on $[c, d]$. Consequently

$$\int_c^d \frac{R(t)P(t)d\sigma(t)}{(t-x_0)^{n_0+1}} > 0,$$

which contradicts (4.14). ▷

DEFINITION. An *MT-system of order* r is a system of functions

$$\begin{aligned} f_1(z) &= \int_{\Delta_1} \frac{d\mu_1(x_1)}{z - x_1}, \\ f_2(z) &= \int_{\Delta_1} \frac{d\mu_1(x_1)}{z - x_1} \int_{\Delta_2} \frac{d\mu_1(x_2)}{x_1 - x_2}, \\ &\dots\dots\dots\dots\dots\dots \\ f_r(z) &= \int_{\Delta_1} \frac{d\mu_1(x_1)}{z - x_1} \cdots \int_{\Delta_r} \frac{d\mu_r(x_r)}{x_{r-1} - x_r}, \end{aligned} \tag{4.15}$$

defined and holomorphic in the domain $\overline{\mathbf{C}}\setminus\Delta_1$, where $\Delta_1, \dots, \Delta_r$ are intervals on the real line with $\Delta_j \cap \Delta_{j+1} = \varnothing$, $j = 1, \dots, r-1$, and $\mu_1, \dots, \mu_r$ are measures on these intervals, with infinitely many points of increase.

The following corollary is a consequence of Theorem 4.4.

COROLLARY. *Every MT-system* (4.15) *is weakly complete, and*

$$1, f_1(z), \dots, f_r(z) \tag{4.16}$$

is a weak AT-system on every interval Δ_0 disjoint from Δ_1. Moreover, for (4.15) *the indices $n \in \mathbf{Z}_+^r$ of the form $n_1 \geq n_2 \geq \cdots \geq n_r$ are normal (with respect to Problem* C), *and* (4.16) *is an AT-system for indices $n \in \mathbf{Z}_+^{r+1}$: $n_0 \geq n_1 \geq \cdots \geq n_r$.*

REMARKS

1. On T-systems, see [47], [155], [42]. The exposition in subsections 1 and 2 closely follows [55].

The following generalization of the Markov theorem was obtained in [69].

THEOREM. *Let $[a, b]$ and $[c, d]$ be disjoint intervals on the real line; $d\mu(x)$ and $d\sigma(t)$, measures on $[a, b]$ and $[c, d]$ with infinitely many points of increase, and*

$$f_1(z) = \int_a^b \frac{d\mu(x)}{z - x}, \qquad f_2(z) = \int_a^b \frac{u(x)d\mu(x)}{z - x},$$

where

$$u(x) = \int_c^d \frac{d\sigma(t)}{x - t}.$$

Then the simultaneous Padé approximants $\pi_{n,j}$ to the functions f_1 and f_2 of index $(n, n) \in \mathbf{Z}_+^2$ converge to these functions (as $n \to \infty$), uniformly inside the domain $\overline{\mathbf{C}}\setminus[a, b]$.

Here the Christoffel coefficients of the quadrature formulas are positive.

PROBLEMS

1. Prove Theorem 4.1.

2. Let $W_n = W_n(u_0, \dots, u_n; t)$, $\widetilde{W}_n = (u_0, \dots, u_{n-1}, u_{n+1}; t)$ be Wronskians. Prove the identity

$$\frac{d}{dt}\widetilde{W}_n \cdot W_n - \widetilde{W}_n \cdot \frac{d}{dt}W_n = W_{n-1} \cdot W_{n+1}.$$

3. Let $\{u_i\}_1^n$ and $\{v_j\}_1^n$ be T-systems on the interval $[a, b] \subset \mathbf{R}$, and let μ be a measure on this interval with infinite spectrum. Prove that

$$\det\left(\int_a^b u_i(t)v_j(t)\,d\mu(t)\right)_{i,j=1,\dots,n} \neq 0.$$

§5. The Jacobi-Perron algorithm

1. Let $f = (f_1, \dots, f_r)$ be the set (3.1) of formal power series.

DEFINITION. If $f_r \neq 0$, we call the vector

$$Tf = \frac{1}{f} = \left(\frac{1}{f_r}, \frac{f_1}{f_r}, \dots, \frac{f_{r-1}}{f_r}\right)$$

unity divided by f.

It follows from the definition that

$$T^{-1}f = \left(\frac{f_2}{f_1}, \dots, \frac{f_r}{f_1}, \frac{1}{f_1}\right),$$

if $f_1 \neq 0$; $T^{r+1}f = f$, if all $f_i \neq 0$.

Using the operation T, we describe an algorithm analogous to the Euclidean algorithm (see Chapter 2, §4). If $f_1 \neq 0$, we have $f = 1/g_1$, where $g_1 = T^{-1}f$. Let $p_1 = (p_{1,1}, \dots, p_{1,r})$, where $p_{1,j}$ is the polynomial part of the series $g_{1,j}$. Let $\varphi_1 = g_1 - p_1$ be the set of series of the form (3.1). Then

$$f = \frac{1}{p_1 + \varphi_1}.$$

If $\varphi_{1,1} \neq 0$, we have $\varphi_1 = 1/g_2$, where $g_2 = T^{-1}\varphi_1$. As above, we introduce vectors p_1 and φ_2; then

$$f = \cfrac{1}{p_1 + \cfrac{1}{p_1 + \varphi_2}}.$$

Continuing in this way, we obtain the formulas:

$$f = \cfrac{1}{p_1 + \cfrac{1}{p_2 + \ddots + \cfrac{1}{p_n + \varphi_n}}} \tag{5.1}$$

There are two possible cases: 1) $\varphi_{N,1} = 0$ for some $N \in \mathbf{Z}_+$; then n runs from 0 to N in (5.1); and 2) $\varphi_{n,1} \neq 0$ for all $n \in \mathbf{Z}_+$; then n in (5.1) is any nonnegative integer. We shall consider case 1) in more detail in subsection 4. In case 2), with f we formally associate the *continued fraction*

$$\cfrac{1}{p_1 + \cfrac{1}{p_2 + \ddots + \cfrac{1}{p_n + \ddots}}}$$

where $p_n = (p_{n,1}, \dots, p_{n,r})$, and $p_{n,j}$ are polynomials with $\deg p_{n,r} \geq \deg p_{n,j} + 1$, $j = 1, \dots, r-1$. We may study such continued fractions independently of the algorithm described above. The vector

$$\alpha_n = \cfrac{1}{p_1 + \cfrac{1}{p_2 + \ddots + \cfrac{1}{p_n}}} \tag{5.3}$$

is called the *nth convergent* of the continued fraction (5.2), $n \in \mathbf{Z}_+$.

2. In studying the continued fraction (5.2), it is convenient to pass to a projective space. Let $\mathbf{K} = \mathbf{C}((1/z))$ be the field of formal power series (see Chapter 2, §3), and $\mathbf{K}^{r+1}$ an $(r+1)$-dimensional linear space over this field, for example the space of columns

$$\overline{f} = (f_0, f_1, \dots, f_r)^{\mathrm{T}}.$$

Vectors $\overline{f}, \overline{g} \in \mathbf{K}^{r+1} \backslash \{\overline{0}\}$ are said to be equivalent ($\overline{f} \sim \overline{g}$) if there is an $h \in \mathbf{K}$, $h \neq 0$, for which $\overline{g} = h\overline{f}$. Then $\mathbf{K}^{r+1} \backslash \{\overline{0}\}$ is divided into equivalence classes, the set of which we denote by $\mathbf{PK}^r$ and call the *r-dimensional projective space over* $\mathbf{K}$. We denote the equivalence class of $\overline{f}$ by $\hat{\overline{f}}$. The *natural embedding* of $\mathbf{K}^r$ in $\mathbf{PK}^r$ is the mapping

$$\omega \colon (f_1, \dots, f_r) \mapsto (1, f_1, \dots, f_r)^{\wedge}.$$

The inverse mapping, which we call the *natural projection*, is defined on the classes $\hat{\overline{f}}$ with $f_0 \neq 0$:

$$(f_0, f_1, \dots, f_r)^{\wedge} \mapsto (f_1 \colon f_0, \dots, f_r \colon f_0).$$

The square matrix A of order $r+1$ with elements $a_{ij} \in \mathbf{K}$ generates a linear operator on $\mathbf{K}^{r+1}$: $\overline{f} \mapsto A\overline{f}$, which induces a mapping on the projective space: $A\hat{\overline{f}} = (A\overline{f})^{\wedge}$, and a mapping on $\mathbf{K}^r$ (not defined on all elements) for which the diagram

$$\begin{array}{ccc} \mathbf{K}^r & \xrightarrow{A} & \mathbf{K}^r \\ \omega\big\downarrow & & \big\downarrow\omega \\ \mathbf{PK}^r & \xrightarrow[A]{} & \mathbf{PK}^r \end{array}$$

commutes. Here a product of matrices corresponds to a composition of mappings.

We rewrite the relation

$$g_n = p_n + \frac{1}{g_{n+1}}$$

in the form $g_n = P_n g_{n+1}$, where

$$P_n = \begin{pmatrix} 0 & 0 & 0 & \cdots & 0 & 1 \\ 1 & 0 & 0 & \cdots & 0 & p_{n,1} \\ 0 & 1 & 0 & \cdots & 0 & p_{n,2} \\ \multicolumn{6}{c}{\dotfill} \\ 0 & 0 & 0 & \cdots & 1 & p_{n,r} \end{pmatrix},$$

$n \in \mathbf{Z}_+$, $g_0 = f$, $p_0 = 0$. Here $f = M_n g_{n+1}$, where $M_n = P_0 P_1 \cdots P_n$. It follows from the preceding formula that the matrix M_n has the form

$$M_n = \begin{pmatrix} A_{n-r,0} & \cdots & A_{n-1,0} & A_{n,0} \\ A_{n-r,1} & \cdots & A_{n-1,1} & A_{n,1} \\ \multicolumn{4}{c}{\dotfill} \\ A_{n-r,r} & \cdots & A_{n-1,r} & A_{n,r} \end{pmatrix},$$

where $A_{n,i}$ are polynomials that satisfy the recurrent relations

$$A_{n,i} = \sum_{l=0}^{i} p_{n,r-l} A_{n-l-1,i}, \tag{5.4}$$

and the initial conditions

$$A_{j,i} = \delta_{i-j,r+1}, \qquad j = -1, \ldots, -r-1,$$

where $p_{n,0} = 1$ for $i = 0, 1, \ldots, r$, and $\delta_{i,j}$ is the Kronecker symbol.

From (5.3) we obtain

$$\alpha_n = M_n e, \qquad e = (0, \ldots, 0, 1)^\frown \in \mathbf{PK}^r,$$

from which

$$\alpha_{n,j} = \frac{A_{n,j}}{A_{n,0}}, \qquad j = 1, \ldots, r. \tag{5.5}$$

Since $\det P_n = (-1)^r \neq 0$, the polynomials $A_{n,0}, \ldots, A_{n,r}$ are relatively prime. If $s_n = \deg p_{n,r}$, $S_n = s_1 + \cdots + s_n$, and $S_0 = 0$, it follows from (5.4) by induction that $\deg A_{n,0} = S_n$.

3. We can now establish the convergence of the continued fraction (5.2) to the series from which it was obtained (in the norm of the field $\mathbf{C}((1/z))$ for each component).

THEOREM 5.1. *For the remainders of the continued fraction, we have the expansion*

$$f_i(z) - \alpha_{n,i}(z) = C_{n,i} \cdot (1/z)^{S_n + 1 - [(i-(n+1))/r]} + \cdots. \tag{5.6}$$

PROOF. We use the notation

$$D_{k,l,i} = A_{k,i} A_{l,0} - A_{l,i} A_{k,0}, \tag{5.7}$$

where $k, l \geq -r-1$ and $i = 1, \dots, r$. We have

$$D_{k,l,i} = -D_{l,k,i}, \qquad D_{k,0,i} = A_{k,i},$$
$$D_{k,l,i} = \sum_{j=0}^{r} p_{k,r-j} D_{k-j-1,l,i}, \qquad k \geq 0. \tag{5.8}$$

LEMMA 5.1. *For* $0 \leq l < k$,

$$\deg D_{k,l,i} \leq S_k + \left[\frac{i-(l+1)}{r}\right] - 1.$$

The lemma is established by double induction: first on k for a given l, then on l. Here we use (5.8). The proof is a purely technical (although complicated) computation, and we omit it (see [75]).

We now turn to the proof of the theorem. According to (5.1), $f = M_{n-1}(p_n + \varphi_n)$ or, explicitly,

$$f_i = \frac{\sum_{l=0}^{r} A_{n-1-l,i}(p_{n,r-l} + \varphi_{n,r-l})}{\sum_{l=0}^{r} A_{n-1-l,0}(p_{n,r-l} + \varphi_{n,r-l})},$$

where $\varphi_{n,0} = 0$. Hence, using (5.4), (5.5), and (5.7), we obtain

$$f_i - \alpha_{n,i} = \frac{-\sum_{l=0}^{r-1} D_{n,n-1-l,i}\varphi_{n,r-l}}{A_{n,0}(A_{n,0} + \sum_{l=0}^{r-1} A_{n-1-l,0}\varphi_{n,r-l})}. \tag{5.9}$$

The denominator of (5.9) contains a polynomial of degree $2S_n$; according to Lemma 5.1,

$$\deg D_{n,n-1-l,i} \leq S_n + \left[\frac{i+r-1-n}{r}\right] - 1,$$

and finally $\varphi_{n,i} = c_{n,i}(1/z) + \cdots$. Then (5.6) follows from everything that has been said. ▷

REMARK. It follows from Lemma 5.1 that $\|\alpha_{n,i} - \alpha_{n+1,i}\| \to 0$ as $n \to \infty$; consequently, every continued fraction (5.2) converges to some vector-series.

4. Let us prove the following generalization of the Kronecker theorem.

THEOREM 5.2. *The Jacobi-Perron algorithm terminates if and only if the functions* $1, f_1, \dots, f_r$ *are linearly dependent over the field* $\mathbf{C}(z)$.

PROOF. For some n, let $f = M_n g_{n+1}$, $g_{n+1,1} = p$, a polynomial. Then $g_{n+1} = M_n^{-1} f$, and the elements ν_{ij} of the matrix M_n^{-1} are polynomials. In particular,

$$p = \left(\sum_{i=0}^{r} \nu_{i1} f_i\right) : \left(\sum_{i=0}^{r} \nu_{i0} f_i\right),$$

where $f_0 = 1$. Hence

$$\sum_{i=0}^{r} (p\nu_{i0} - \nu_{i1}) f_i = 0,$$

where not all the coefficients are zero, since $\det M_n^{-1} \neq 0$.

Let $\{q_i\}_0^r$ be polynomials, not all zero,

$$\sum_{i=0}^{r} q_i f_i = 0,$$

and let κ be the maximum degree of the q_i. Let us suppose that the algorithm does not terminate; then $\alpha_{n,i}$ is defined for all n. By Theorem 5.1 the series

$$R_n = \sum_{i=0}^{r} q_i(A_{n,0} f_i - A_{n,i})$$

equals

$$R_n = C_n z^{\kappa + [-(n+1/r)]} + \cdots .$$

On the other hand,

$$R_n = -\sum_{i=0}^{r} q_i A_{n,i}$$

is a polynomial. Consequently, for sufficiently large n we have $R_n = 0$ and $\det M_n = 0$, a contradiction. ▷

REMARK. If the infinite fraction (5.2) converges to a vector-series f, or f equals a finite fraction (5.1) (with $\varphi_{n,1} = 0$), then, expanding f by the Jacobi-Perron algorithm, we obtain the original fraction (uniqueness theorem). Here we are supposing that the original fraction was constructed properly, i.e., $\deg p_{n,r} \geq \deg p_{n,j} + 1$, $j = 1, \dots, r-1$.

5. In terms of the continued fraction (5.2), it is easy to state a criterion for the weak completeness of the system f (see §3).

DEFINITION. We say that an infinite continued fraction (5.2) is *regular* if $\deg p_{n,r} = 1$, $n \in \mathbf{N}$.

THEOREM 5.3. *A necessary and sufficient condition for a system f to be weakly complete is that f can be expanded in a regular continued fraction.*

PROOF. Let f be a weakly complete system. Let us recall the definition of this concept. We consider multi-indices $(m_1, \dots, m_r) \in \mathbf{Z}_+^r$ of the form $m_1 = \cdots = m_q = s+1$, $m_{q+1} = \cdots = m_r = s$, where $s \in \mathbf{Z}_+$, $q = 0, \dots, r-1$. We number them by the index $m = sr + q$. Let $Q_{m,0}$, $Q_{m,1}, \dots, Q_{m,r}$ be polynomials (simultaneous approximation of type m) of degree at most m; $Q_{m,0} \not\equiv 0$;

$$Q_{m,0} f_i - Q_{m,i} = c_{m,i} z^{-m_i - 1} + \cdots, \tag{5.10}$$

where m_i are the components of the mth index. If it follows from these hypotheses that $\deg Q_{m,0} = m$, $m \in \mathbf{Z}_+$, the system f is weakly complete. It is easily shown that the weak completeness of f is equivalent to the relations $c_{m,i} \neq 0$ for $i \equiv m+1 \pmod r$, $m \in \mathbf{Z}_+$.

We establish recurrent formulas for the $Q_{m,i}$. Set

$$Q_{m,i} = \delta_{i-m,r+1}, \qquad m = -r-1, \dots, -1. \tag{5.11}$$

Let $Q_m = (Q_{m,0}, \dots, Q_{m,r})$ and let a and b be numbers;

$$\widetilde{Q} = azQ_{n-1} + \sum_{j=0}^{r} b_j Q_{n-r-1+j}.$$

If we equate the coefficients of corresponding powers of z in the series $\widetilde{Q}_0 f_i - \widetilde{Q}_i$ to zero, we find that $\widetilde{Q}$ is the nth simultaneous approximation if $a \neq 0$, and the numbers b_j satisfy a system of $r+1$ linear equations with a triangular matrix, with the determinant

$$\prod_{j=1}^{q} c_{n-q-1+j,j} \prod_{j=q}^{r} c_{n-q-r-1+j,j} \neq 0;$$

here $q-1$ is the residue of $n-1$ modulo r. Thus if we choose the coefficient a correspondingly we find that $\{Q_n\}$ satisfies the recurrence relation

$$Q_n = \sum_{j=0}^{r} p_{n,r-j} Q_{n-j-1}, \tag{5.12}$$

where $p_{n,r}$ is a linear function; $p_{n,0}, \dots, p_{n,r-1}$ are numbers; $p_{n,0} = 1$; with initial conditions (5.11). The relations (5.11) and (5.12) mean that $\{Q_n\}$ are the denominators and numerators of the convergents of the continued fraction (5.2). By (5.10), this fraction converges to f, and consequently is obtained from f by the Jacobi-Perron algorithm.

For the proof of the sufficiency we need the following lemma.

LEMMA 5.2. *Let $Q^{(j)}$ be the simultaneous approximants to f of types $m^{(j)}$, $j = 0, \dots, r$. Then sufficient conditions for the vanishing of the determinant of the matrix Q formed from the column vectors $Q^{(j)}$ are*

1) *the positivity of*

$$r + \sum_{i=1}^{r} (m_i^{(j_i)} - m_i^{(j_0)})$$

for every permutation $(j_0, \dots, j_r)$ of the indices $0, \dots, r$;

2) *the condition $|m_i^{(j)} - m_i^{(k)}| \leq 1$, $i = 1, \dots, r$; $j, k = 0, \dots, r$, together with the condition that there exist two maximal indices $m^{(j)}$ and $m^{(k)}$ such that $j \neq k$ and $m_i^{(j)} = m_i^{(k)} \geq m_i^{(l)}$, $i = 1, \dots, r$; $l = 0, \dots, r$.*

PROOF. For each $i = 1, \dots, r$, we subtract the zeroth row of Q, multiplied by f_i, from the ith row. We obtain

$$Q = \begin{vmatrix} Q_0^{(0)} & R_1^{(0)} & \cdots & R_r^{(0)} \\ Q_0^{(1)} & R_1^{(1)} & \cdots & R_r^{(1)} \\ Q_0^{(r)} & R_1^{(r)} & \cdots & R_r^{(r)} \end{vmatrix}, \tag{5.13}$$

where

$$R_i^{(j)} = Q_i^{(i)} - Q_0^{(j)} f_i = C_i^{(j)} \cdot z^{-m_i^{(j)}-1} + \cdots .$$

If condition 1) is satisfied, in the complete expansion of the determinant (5.13) all the terms are $O(1/z)$; but Q is a polynomial, and consequently $Q \equiv 0$.

Condition 2) follows from 1). ▷

We now prove the theorem. Let the fraction (5.2) be regular. Then the vector A_n is the nth simultaneous approximation to f, and $\deg A_{n,0} = n$. For A_n, we have $c_{n,i} \neq 0$ for $i \equiv n+1 \pmod r$. In fact, in the contrary case A_n is the $(n+1)$th simultaneous approximation, and by condition 2) of Lemma 5.2, $\det M_{n+1} = 0$. This is a contradiction.

Let n be the smallest index that is not normal. Then there is an nth simultaneous approximation Q_n that is not proportional to A_n. We have $\deg Q_{n,0} = n$. If not, Q_n is an $(n-1)$th simultaneous approximation, and consequently Q_n is proportional to A_{n-1}, i.e., A_{n-1} is an nth simultaneous approximation. This contradicts what was proved above. Thus we have found a number λ such that $\deg(\lambda A_{n,0} + Q_{n,0}) < n$, and consequently, for some $\nu \neq 0$,

$$\nu A_{n-1} = \lambda A_n + Q_n. \tag{5.14}$$

We form the matrix M from the columns $A_{n-r+1}, \dots, A_{n-2}, A_n, Q_n$. By condition 2) of Lemma 5.2, $\det M = 0$. But by (5.14) $\det M = \det M_n$. We have arrived at a contradiction. ▷

REMARKS

The first generalization of the continued fraction algorithm was given by L. Euler [133]. This algorithm was studied in [140], [141], and [156]. Multidimensional C-fractions were studied in [124], [128], and [129]. The results of the present section are presented in [75] and [77].

Here we did not consider problems of the convergence of periodic and limit periodic fractions. We notice only the first classical papers [170] and [159]. The convergence of multidimensional P-fractions is investigated in [76].

§6. The Darboux method

1. Among the orthogonal polynomials, we especially single out the *classical orthogonal polynomials*:

1) the *Jacobi polynomials*, orthogonal with respect to the weight

$$(1-x)^\alpha (1+x)^\beta \tag{6.1}$$

on the interval $[-1, 1]$ $(\alpha, \beta > -1)$;

2) the *generalized Laguerre polynomials*, orthogonal with respect to the weight

$$x^\alpha e^{-x} \tag{6.2}$$

on the interval $[0, +\infty)$ $(\alpha > -1)$;

3) the *Hermite polynomials*, orthogonal with the weight

$$e^{-x^2} \tag{6.3}$$

on the whole real line.

Special cases of the Jacobi polynomials are the *Legendre polynomials* ($\alpha = \beta = 0$) and the *Chebyshev polynomials* ($\alpha = \beta = -1/2$).

The weight functions (6.1)–(6.3) satisfy *Pearson's equation*:

$$\frac{y'}{y} = \frac{A(x)}{B(x)}, \tag{6.4}$$

where A and B are polynomials of respective degrees at most 1 and 2 with real coefficients, $B \not\equiv 0$. Conversely (see [100]), every solution of (6.4) that is a weight function with finite power moments on a connected domain in $\mathbf{R}$ becomes one of the functions (6.1)–(6.3) under a linear change of variable.

Besides the general properties of orthogonal polynomials, the classical orthogonal polynomials have a number of additional properties; in particular, they are solutions of second-order differential equations, which explains their extensive application as special functions. They have explicit *Rodrigues formulas* which make it possible to investigate these polynomials in detail.

PROPOSITION 6.1. *The Jacobi polynomials can be defined by the Rodrigues formula*

$$\begin{aligned} Q_n(\alpha, \beta; x) = {} & (x-1)^{-\alpha}(x+1)^{-\beta} \\ & \times \frac{1}{2^n n!}\frac{d^n}{dx^n}((x-1)^{n+\alpha}(x+1)^{n+\beta}), \end{aligned} \tag{6.5}$$

where $n \in \mathbf{Z}_+$, and for $x > 1$ we take the principal branches of the multiple-valued functions.

PROOF. If we apply the Leibniz formula to (6.5), we see that (6.5) is a polynomial of degree n. We now verify the orthogonality relation.

Let $h(x)$ denote the weight function (6.1). If we integrate by parts m times with $m \le n-1$, we obtain

$$\begin{aligned} \int_{-1}^{1} ((x^2-1)^n h(x))^{(n)} x^m\,dx &= -\int_{-1}^{1} ((x^2-1)^n h(x))^{(n-1)} m x^{m-1}\,dx \\ &= \cdots = (-1)^m m! \int_{-1}^{1} ((x^2-1)^n h(x))^{(n-m)}\,dx \\ &= (-1)^m m!((x^2-1)^n h(x))^{(n-m-1)}\big|_{-1}^{+1} = 0. \end{aligned} \tag{6.6}$$

The integrated terms in (6.6) vanish since the function $(x^2-1)^n$ has zeros of order n at ± 1. Thus the polynomials (6.5) satisfy the orthogonality relation. ▷

PROPOSITION 6.2. *Let $x \in \mathbf{C}$, $x \neq \pm 1$; then*

$$\phi(x, w) = \sum_{n=0}^{\infty} Q_n(\alpha, \beta; x) w^n \tag{6.7}$$

is the Taylor series of a function that is holomorphic at $w = 0$:

$$\phi(x, w) = \left(\frac{t_x - 1}{x - 1}\right)^{\alpha}\left(\frac{t_x + 1}{x + 1}\right)^{\beta}\frac{1}{\sqrt{D}}, \tag{6.8}$$

where

$$t_x = \frac{1 - \sqrt{D}}{w}, \qquad D = 1 - 2wx + w^2, \tag{6.9}$$

and $\sqrt{D} = 1$ *at* $w = 0$.

PROOF. Using the Cauchy formula, we rewrite (6.5) in the form

$$Q_n(\alpha, \beta; x) = (x-1)^{-\alpha}(x+1)^{-\beta}2^{-n}\cdot\frac{1}{2\pi i}\int_{\Gamma}\frac{(t-1)^{n+\alpha}(t+1)^{n+\beta}dt}{(t-x)^{n+1}}, \tag{6.10}$$

where Γ is a circle with center at x and sufficiently small radius (such that the points ± 1 do not belong to the disk $\overline{D}$ inside Γ), surrounding Γ counterclockwise. The series

$$\sum_{n=0}^{\infty}\left(\frac{t^2 - 1}{t - x}\frac{w}{2}\right)^n = \left(1 - \frac{t^2 - 1}{t - x}\frac{w}{2}\right)^{-1} \tag{6.11}$$

converges absolutely and uniformly for $t \in \Gamma$ and $|w| < \varepsilon$ for some $\varepsilon > 0$. If we multiply (6.11) by the function

$$\frac{1}{2\pi i}\left(\frac{t-1}{x-1}\right)^{\alpha}\left(\frac{t+1}{x+1}\right)^{\beta}\frac{1}{t-x},$$

integrate the resulting series termwise, and use (6.10), we conclude that the series (6.7) converges in a neighborhood of $w = 0$, and its sum has the integral representation

$$\phi(x, w) = (x-1)^{-\alpha}(x+1)^{-\beta}\frac{1}{2\pi i}\int_{\Gamma}\frac{(t-1)^{\alpha}(t+1)^{\beta}\,dt}{t - x - (t^2 - 1)w/2}. \tag{6.12}$$

The integrand in (6.12) is holomorphic at all points of $\overline{D}$ except for the roots of the equation

$$wt^2 - 2t + 2x - w = 0,$$

at which the denominator is zero. When $w \to 0$, one root of this equation tends to infinity, and the other, t_x, tends to x. Consequently, for some $\varepsilon_1 > 0$ and all $|w| < \varepsilon_1$ the integrand has a single simple pole t_x in D. If we evaluate the integral by the residue theorem, we obtain (6.8)–(6.9). ▷

2. One of the first results on the asymptotic behavior of the classical orthogonal polynomials was found by Gaston Darboux. He obtained the asymptotics (as $n \to \infty$) of the Legendre polynomials at points of the interval $(-1, 1)$ by means of the method that he developed. This method rests on the study of the singular points on the boundary of the largest disk of holomorphy of the generating function. If there are finitely many such points, and if in the neighborhood of each one the generating function has a simple and known behavior (ordinarily, only poles and algebraic branch points are considered), we can describe the asymptotics of the Taylor coefficients.

THEOREM 6.1. *Let*

$$f(x) = \sum_{n=0}^{\infty} a_n x^n$$

be holomorphic in the unit disk, which is its maximal disk of holomorphy. Let there be k singular points of f on the unit circle: $1/\omega_j$, $j = 1, \dots, k$, *and let these all be algebraic points. In a punctured neighborhood of each point* $1/\omega_j$, *let* f *have an expansion of the form*

$$f(x) = b_0^{(j)}(1-\omega_j x)^{m_j/p_j} + b_1^{(j)}(1-\omega_j x)^{(m_j+1)/p_j} + \cdots, \tag{6.13}$$

where

$$m_j \in \mathbf{Z}, \quad p_j \in \mathbf{N}, \quad m_j/p_j \notin \mathbf{Z}_+, \quad b_0^{(j)} \neq 0,$$

and the expansion (6.13) *is carried out in powers of* $1/p_j$. *Then* (*as* $n \to \infty$)

$$a_n = \sum_{j=1}^{k} b_0^{(j)} \binom{-m_j/p_j + n - 1}{n} \omega_j^n (1 + O(n^{-1/p_j})). \tag{6.14}$$

It is sufficient to carry out the proof for $k = 1$. Let (without loss of generality) the single singular point on the unit circle be at $x = 1$, and in a neighborhood of that point let

$$f(x) = b_0(1-x)^{m/p} + b_1(1-x)^{(m+1)/p} + \cdots. \tag{6.15}$$

Take an arbitrary positive integer q and set

$$f(x) = \sum_{j=0}^{q} b_j (1-x)^{(m+j)/p} + \tilde{f}(x).$$

We denote the Taylor coefficients of $\tilde{f}$ by $\tilde{a}_n$. We know the Taylor series for each term in (6.15):

$$(1-x)^\alpha = \sum_{n=0}^{\infty} \binom{-\alpha - 1 + n}{n} x^n, \qquad |x| < 1; \tag{6.16}$$

here α is any complex number. By Stirling's formula

$$\binom{-\alpha - 1 + n}{n} = \frac{1}{\Gamma(-\alpha)} n^{-\alpha-1}(1 + O(n^{-1})), \qquad n \to \infty. \tag{6.17}$$

From (6.16) and (6.17), we obtain

$$\begin{aligned} a_n &= \sum_{j=0}^{q} b_j \binom{-(m+j)/p - 1 + n}{n} + \tilde{a}_n \\ &= b_0 \binom{-m/p - 1 + n}{n} + O(n^{-(m+1)/p-1}) + \tilde{a}_n. \end{aligned} \tag{6.18}$$

We choose a nonnegative integer $l \geq (m+1)p + 1$ and a q so that $(m+q+1)/p + 1 \geq l$. Then $\tilde{f}(x)$ is holomorphic in the open unit disk, and its lth derivative is continuous in the closed unit disk. Hence the Taylor coefficients of $\tilde{f}^{(l)}$ are bounded (and even tend to zero), and therefore

$$\tilde{a}_n = O(n^{-l}) = O(n^{-(m+1)/p-1}), \qquad n \to \infty. \tag{6.19}$$

It follows from (6.17)–(6.19) that as $n \to \infty$

$$\begin{aligned} a_n &= b_0 \binom{-m/p - 1 + n}{n} + O(n^{-(m+1)/p-1}) \\ &= b_0 \binom{-m/p - 1 + n}{n} (1 + O(n^{-1/p})). \quad \triangleright \end{aligned}$$

REMARK. A disk of arbitrary radius is carried to the unit disk by a simple change of variable. If a_n depends on a parameter, it is usually easy to establish the uniform nature of (6.14).

3. Let us apply Darboux's method to the classical Jacobi polynomials and the polyorthogonal polynomials that generalize them.

THEOREM 6.2. *Uniformly with respect to x on compact subsets of the domain $G = \mathbf{C}\backslash[-1, 1]$ (as $n \to \infty$)*

$$Q_n(\alpha, \beta; x) = \frac{1}{\sqrt{\pi n}} A(x) w(x)^n (1 + O(1/\sqrt{n})), \tag{6.20}$$

where

$$w(x) = x + \sqrt{x^2 - 1} \qquad (|w(x)| > 1), \tag{6.21}$$

$$A(x) = \left(1 + \sqrt{\frac{x+1}{x-1}}\right)^{\alpha} \left(1 + \sqrt{\frac{x-1}{x+1}}\right)^{\beta} \sqrt{\frac{x + \sqrt{x^2-1}}{2\sqrt{x^2-1}}}, \tag{6.22}$$

and the branches of the multiple-valued functions in (6.22) *are chosen so that* $A(\infty) = 2^{\alpha+\beta}$.

PROOF. The singular points of the function (6.8) are the roots of the equation

$$w^2 - 2wx + 1 = 0,$$

namely $w_{\pm}(x) = x \pm \sqrt{x^2 - 1}$, with $|w_+(x)| > 1$, $|w_-(x)| < 1$ $(w_+ = w)$. Therefore $w_-(x)$ is the singular point closest to the origin for the function (6.8). This is a second-order branch point. Let us calculate the limit

$$A(x) = \lim_{w \to w_-(x)} \phi(x, w) \sqrt{1 - w/w_-(x)}, \tag{6.23}$$

where $|w| < |w_-(x)|$, $\sqrt{1 - w/w_-(x)} = 1$ when $w = 0$. We have

$$\lim_{w \to w_-} t_x = \frac{1}{w_-} = w. \tag{6.24}$$

In addition,

$$\lim_{w\to w_-}\frac{\sqrt{1-w/w_-}}{\sqrt{1-2wx+w^2}}=\lim_{w\to w_-}\frac{1}{\sqrt{1-w/w_+}}=\sqrt{\frac{x+\sqrt{x^2-1}}{2\sqrt{x^2-1}}}. \tag{6.25}$$

If we use (6.24) and (6.25) in (6.23), we obtain (6.22). Thus, in a punctured neighborhood of $w_-(x)$,

$$\phi(x, w)=A(x)(1-w/w_-(x))^{-1/2}+\cdots.$$

Then (6.20) follows by using Theorem 6.1. ▷

By the same method, it is easy to obtain the asymptotics of $Q_n(x)$ for $x\in(-1, 1)$ and the asymptotics of the functions of the second kind.

4. Let us consider polyorthogonal polynomials that have properties close to those of the Jacobi polynomials.

Let $\Delta_0=[-1, 1]$, $\Delta_\infty=[1, \infty)$, $h(t)=|1-t|^\alpha|1+t|^\beta e^{-t}$, $t\in(-1, 1)\cup(1, +\infty)$, where $\alpha, \beta>-1$. We denote by $\Lambda_n(t)$ a polynomial of degree $2n$, $n\in\mathbf{Z}_+$, $\Lambda_n\not\equiv 0$, satisfying the orthogonality relations

$$\int_{\Delta_j}\Lambda_n(t)t^k h(t)\,dt=0,\qquad k=0, \dots, n-1;\ j=0, \infty.$$

Integrating by parts, as in Proposition 6.1, we can easily verify that Λ_n can be determined by a Rodrigues formula,

$$\Lambda_n(t)=e^t(t-1)^{-\alpha}(t+1)^{-\beta}2^{-n}(n!)^{-1}\cdot\frac{d^n}{dt^n}((t-1)^{n+\alpha}(t+1)^{n+\beta}e^{-t}), \tag{6.26}$$

where the branches of the multiple-valued functions are consistent.

We can find the generating function of the sequence $\{\Lambda_n(x)\}$ for $x\neq\pm1$:

$$\tilde\phi(x, w)=\sum_{n=0}^{\infty}\Lambda_n(x)w^n. \tag{6.27}$$

As in Proposition 6.2, we transform (6.26) by the Cauchy formula, substitute into (6.27), and change the order of operations, obtaining

$$\tilde\phi(x, w)=\frac{1}{2\pi i}\int_\Gamma\frac{((1-t)/(1-x))^\alpha((1+t)/(1+x))^\beta e^{x-t}dt}{(t-x)-(t^2-1)w/2}, \tag{6.28}$$

where Γ is a circle with center at x and sufficiently small radius. When $|w|$ is sufficiently small, the only singular point of the integrand is inside Γ, and is a simple pole, once again the point $t_x(w)$ determined in (6.9). If we calculate the integral (6.28) by the residue theorem, we obtain

$$\tilde\phi(x, w)=\exp(x-t_x(w))\phi(x, w),$$

where ϕ is the generating function of the Jacobi polynomials defined in (6.8). The function $\tilde\phi$, like ϕ, is holomorphic in the disk $|w|<|w_-(x)|$ and has a single singular point $w_-(x)$ on the boundary with $x \in G = \mathbf{C}\setminus[-1, 1]$, a branch point of second order. In a punctured neighborhood of $w_-(x)$,

$$\tilde{\phi}(x, w) = \widetilde{A}(x)(1 - w/w_-(x))^{-1/2} + \cdots,$$

where

$$\widetilde{A}(x) = A(x) \lim_{w \to w_-} \exp(x - t_x(w)) = A(x)\exp(-\sqrt{x^2 - 1}), \tag{6.29}$$

and A is defined in (6.22). We have established the following theorem.

THEOREM 6.3. *We have*

$$\Lambda_n(x) = \frac{1}{\sqrt{\pi n}} \widetilde{A}(x) w(x)^n (1 + O(1/\sqrt{n})), \qquad n \to \infty,$$

uniformly in x *on compact subsets of* G, *where* w *and* $\widetilde{A}$ *are defined in* (6.21) *and* (6.29).

5. The first result on the convergence of simultaneous Padé approximants to a set of Markov functions was obtained by V. A. Kalyagin. He considered the functions

$$f_\pm(x) = \int_{\Delta_\pm} \frac{|h(t)|dt}{t - x}, \qquad x \in G_\pm = \overline{\mathbf{C}}\setminus\Delta_\pm,$$

where $\Delta_+ = [0, 1]$, $\Delta_- = [-1, 0]$, and

$$h(t) = (1 - t)^\alpha (1 + t)^\beta t^\gamma, \qquad \alpha, \beta, \gamma > -1.$$

He studied the simultaneous Padé approximants $\pi_{n,\pm}(x)$ of index $(n, n) \in \mathbf{Z}_+^2$. The polynomial $Q_n(x)$ of degree $2n$, the common denominator of these approximants, satisfies the orthogonality relations

$$\int_{\Delta_\pm} Q_n(t) t^k |h(t)| dt = 0, \qquad k = 0, \ldots, n - 1.$$

Kalyagin's result was based on a study of the asymptotic behavior of $Q_n(x)$ and the corresponding functions of the second kind, as $n \to \infty$. As in Proposition 6.1, he established the Rodrigues formula

$$Q_n(t) = \frac{1}{n!} \frac{1}{h(t)} \frac{d^n}{dt^n}((t^2 - 1)^n t^n h(t)), \qquad n \in \mathbf{Z}_+, \tag{6.30}$$

where the branches of the multiple-valued functions are chosen consistently. We consider the generating function

$$F_*(x, w) = \sum_{n=0}^{\infty} Q_n(x) w^{2n}. \tag{6.31}$$

If we write (6.30) using the Cauchy formula and substitute it into (6.31), we obtain

$$F_*(x, w) = -\frac{1}{h(x)} \frac{1}{2\pi i} \int_\Gamma \frac{h(t)dt}{w^2 t^3 - (1 + w^2)t + x},$$

where $x \neq 0$ or ± 1, Γ is a circle with center x and sufficiently small radius, and $|w|$ is sufficiently small. When $w \to 0$, one root t_* of the equation

$$w^2t^3 - (1+w^2)t + x = 0 \tag{6.32}$$

tends to x; the two others tend to ∞. Then, by the residue theorem,

$$F_*(x, w) = -\frac{h(t_*)}{h(x)}(3w^2t_*^2 - (1+w^2))^{-1}.$$

We can find the roots of equation (6.32) by Cardano's formulas. We obtain

$$D = \frac{x^2w^2}{4} - \frac{(1+w^2)^3}{27},$$

$$g = \left(\frac{-xw}{2} + \sqrt{D}\right)^{1/3}, \qquad d = \left(\frac{-xw}{2} - \sqrt{D}\right)^{1/3},$$

when $w = 0$, $\sqrt{D} = i/\sqrt{27}$, $g = -i/\sqrt{3}$, $d = i/\sqrt{3}$. Then $t_* = (g+d)/w$. Using these formulas, we obtain

$$F_*(x, w) = -\frac{h(t_*)}{h(x)}\frac{g-d}{6\sqrt{D}}. \tag{6.33}$$

The only singular points of the function (6.33) are the zeros of the discriminant D. Consequently, we have to consider the equation

$$\frac{(u+1)^3}{27} = \frac{x^2u}{4}, \tag{6.34}$$

where $u = w^2$. The roots of the discriminant of this equation are 0 and ± 1. For $x = 0$, the equation has the root $u = -1$ of multiplicity 3; for $x = \pm 1$, it has the double root $u = 1/2$ and the simple root $u = -4$. As x tends to ∞, one root tends to zero and the others to infinity. Therefore, in the domain $G_* = \overline{\mathbf{C}}\backslash\Delta_*$, $\Delta_* = [-1, 1]$, the equation (6.4) is satisfied by three holomorphic functions, one, $u_*(x)$, with a double zero at $x = \infty$; the others, $u_\pm(x)$, with simple poles. By the symmetry of the equation under the substitution $x \mapsto -x$, we have

$$u_*(-x) = u_*(x), \quad u_-(x) = u_+(-x), \qquad x \in G_*.$$

The multiple-valued analytic function $u(x)$, the solution of equation (6.34), has second-order branch points at the points ± 1, and a third-order branch point at zero. Consequently, one function, u_+, extends to a holomorphic function in the domain G_+, and the other, u_-, in G_-. The equation (6.34) has, for $x \in \Delta_*$, real coefficients; therefore on the upper (or lower) side of the interval Δ_+ the functions u_+ and u_* have conjugate complex limits, whereas the values of u_- are real. Similarly, the limits of u_- and u_* on Δ_- are complex conjugates, and the limits of u_+ are real. Consequently, the upper side of Δ_+ transforms under u_+ to an analytic Jordan arc C_+ that joins the points -1 and $1/2$ and, as is easily seen, is in the upper halfplane. Under the same mapping, the lower side of Δ_+ is transformed to the curve C_-, symmetric to C_+ with respect to the real line. We denote by C the closed curve consisting of C_+ and C_-. The function $u_\pm$ effects a conformal mapping of the domain $G_\pm$ on the exterior of C. The function u_* effects a two-sheeted mapping of G_* on the interior of C. Moreover, $u_\mp(\Delta_\pm) = [-4, -1]$. If $x \in \Delta_\pm$ we have $|u_\mp(x)| \geq 1$. By the Vieta theorem, $u_* u_+ u_- = -1$. Consequently, for $x \in \Delta_\pm$ we have

$$|u_*(x)|^2 = |u_\mp(x)|^{-1} \leq 1.$$

The function $u_*(x)/u_\pm(x)$ is holomorphic in G_*, and its limits on Δ_* have absolute value at most 1. Consequently, by the maximum principle,

$$|u_*(x)| < |u_\pm(x)|, \qquad x \in G_*,$$

i.e., u_* is the root of equation (6.34) closest to the origin. We can obtain explicit expressions for u_* and $u_\pm$ by using Cardano's formulas.

THEOREM 6.4. *We have*

$$Q_n(x) = \frac{1}{\sqrt{2\pi n}} u_*(x)^{-n}(2A_*(x) + O(1/\sqrt{n})), \qquad n \to \infty, \tag{6.35}$$

uniformly in x *on compact subsets of* G_*, *where* A_* *is a function that is holomorphic in* G_*, *and* $A_*(x) \neq 0$.

PROOF. By (6.33), the possible singular points of F_* are $w = \pm\sqrt{u_k}$ and the points at which $-xw/2 + \sqrt{D} = 0$, i.e., the points $w = \pm i$. The points closest to the origin are $\pm w_* = \pm\sqrt{u_*}$ (we assume that $w_*(x) > 0$ for $x > 1$). It follows from the explicit forms of D, g, and h that, after a circuit around w_*, the function $\sqrt{D}$ changes sign, the new value of g agrees with the principal value of cd, where c is a cube root of unity, and the new value of d agrees with the principal value of $c^2 g$. We describe this situation by $\sqrt{D} \mapsto -\sqrt{D}$, $g \mapsto cd$, $d \mapsto c^2 g$. Let us determine the constant c. For this purpose, we continue g analytically along the curve consisting of the line-segment $[0, w_*(x)(1-\delta)]$, the circle $|w - w_*| = |w_*|\delta$ (traversed counterclockwise), and the previous line-segment (traversed in the opposite sense). Here δ is a sufficiently small positive number. Let $L^{(\delta)}$ be a curve that surrounds the point $-xw/2+\sqrt{D}$; and $L(x)$ the limit curve for $\delta \to 0$. Then $L(x)$ begins at $i/\sqrt{27}$ and ends at $-i/\sqrt{27}$. Starting from the analytic representation of the curves $L(x)$, $x \in G_*$, it is easy to deduce that they are homotopic to each other in $\mathbf{C}\backslash\{0\}$. Consequently it is sufficient to continue g analytically for any specific $x \in G_*$. Let $x > 1$; then the square roots of D are $w^2 = u_* \in (0, 1/2)$, $w^2 = u_+ > 1/2$, and $w^2 = u_- < -4$. Consequently, $L(x)$ is a Jordan arc, lying in the left halfplane, and under a circuit around this curve the argument increases by $+\pi$; the new value of g at $w = 0$ is equal to

$$-\frac{i}{\sqrt{3}} e^{\pi i/3} = \frac{i}{\sqrt{3}} \zeta^2, \qquad \zeta = \exp\left(\frac{2\pi i}{3}\right).$$

Thus $c = \zeta^2$, $\sqrt{D} \mapsto -\sqrt{D}$, $g \mapsto \zeta^2 d$, $d \mapsto \zeta g$, F_* changes its value after going around w_*, and after a second circuit returns to its original value, i.e., w_* is a second-order branch point of F_*. In addition, $\lim(g-d) \neq 0$ (as $w \to w_*$); consequently, in a punctured neighborhood of $w_*(x)$,

$$F_*(x, w) = A_*(x)(1 - w/w_*(x))^{-1/2} + \cdots, \qquad A_*(x) \neq 0.$$

Since F_* is an even function (of w), we have a similar expansion in a punctured neighborhood of $-w_*(x)$. Using this and applying Theorem 6.1, we obtain (6.35). ▷

REMARKS

1. For classical orthogonal polynomials, see [87], [100], [65], and [14]. For Darboux's method see [87], [168], and the book [58]. The classical orthogonal polynomials have also been studied by Steklov's method and the saddle-point method.

2. For the results in subsection 4, see [93]. Let us make some additions. Let

$$R_{n,k}(x) = \int_{\Delta_k} \frac{\Lambda_n(t)}{t-x} h(t)\, dt, \qquad k = 0, \infty; \ n \in \mathbf{Z}_+,$$

be functions of the second kind, $f_k = R_{0,k}$.

THEOREM. *Uniformly for x in compact subsets of G,*

$$R_{n,0}(x) = 2\sqrt{\frac{\pi}{n}} A_0(x) w(x)^{-n}(1 + O(1/\sqrt{n})), \qquad n \to \infty,$$

where $A_0(x)$ is holomorphic in G.

This was proved by Darboux's method.

Let $\pi_{n,k}(x)$ be simultaneous Padé approximants of functions $f_k(x)$, $k = 0, \infty$, with index $(n, n) \in \mathbf{Z}_+^2$.

COROLLARY. *As $n \to \infty$, we have $\pi_{n,0} \to f_0(x)$ uniformly inside G.*

THEOREM. *Uniformly for x in compact subsets of $\mathbf{C}\backslash\Delta_\infty$,*

$$R_{n,\infty}(x) = 2^{-n}\Gamma(n + \alpha + \beta)e^x(1 + O(1/n)), \qquad n \to \infty.$$

This theorem can be proved both by Darboux's method and by the saddle-point method. (The saddle-point method can be looked up in the textbooks [88], [86], and [58], as well as in [17].)

COROLLARY. *The approximants $\pi_{n,\infty}$ diverge to infinity throughout the complex plane.*

REMARK. The limit distribution (as $n \to \infty$) of the zeros of the polynomials $\Lambda_n(x)$ on the interval Δ_0 is the Chebyshev measure; the zeros in Δ_∞ tend to infinity.

3. The results in subsection 5 were proved by V. A. Kalyagin [44]. Various properties of these polynomials had already been discussed in [121], [119], [130], [139], and [118].

THEOREM. *Uniformly for x in compact subsets of $G_\pm$*

$$R_{n,\pm}(x) = \sqrt{\frac{2\pi}{n}} u_\pm(x)^{-n}(2A_\pm(x) + O(1/\sqrt{n})), \qquad n \to \infty,$$

where $A_\pm(x)$ is a function that is holomorphic in $G_\pm$.

This result was also obtained by Kalyagin by Darboux's method.

COROLLARY. *Uniformly inside G_**

$$\pi_{n,\pm}(x) \to f_\pm(x) \quad \text{as } n \to \infty.$$

The asymptotics of linear forms and of their coefficients for functions $f_\pm$ were obtained for $\alpha = \beta = 0$ in [90] and [91].

4. In this book, we have not considered the asymptotics of Laguerre polynomials and of their polyorthogonal generalizations. The first result on the asymptotic behavior of the Laguerre polynomials was obtained by L. Fejér [134]. Then O. Perron improved Fejér's method and obtained complete asymptotic expansions of these polynomials. Perron's method was described in the book [87]. In [95]–[97], the method was extended to polyorthogonal polynomials. Here we present only one result.

Let $\deg \widetilde{\Lambda}_n = n$,

$$\int_0^\infty \widetilde{\Lambda}_n(x^r)x^k e^{-x^r} x^p\,dx = 0, \qquad k = 0, \dots, n-1,$$

r a specific positive integer, $p > -1$. We set

$$\Lambda_n(x) = \widetilde{\Lambda}_n(x^r),$$

$$f(z) = \int_0^\infty \frac{e^{-x^r} x^p\,dx}{x - z},$$

$$R_n(z) = \int_0^\infty \frac{\Lambda_n(x)}{x - z} e^{-x^r} x^p\,dx, \qquad z \in \mathbf{C}\backslash\mathbf{R}_+,$$

$$\pi_n = f - R_n/\Lambda_n.$$

The polynomial Λ_n can be defined by the Rodrigues formula

$$\Lambda_n(x) = \frac{1}{n!} e^{x^r} x^{-p} \frac{d^n}{dx^n}(x^{n+p} e^{-x^r}).$$

THEOREM. *Uniformly inside* $\mathbf{C}\backslash\mathbf{R}_+$

$$\lim_{n\to\infty} n^{-r/(r+1)} \ln \widetilde{\Lambda}_n(r^r x) = (r+1)(-x/r)^{1/(r+1)},$$

$$\lim_{n\to\infty} n^{-r/(r+1)} \ln R_n(x) = -(r+1)(-x/r)^{r/(r+1)},$$

with the principal values of the logarithm and the roots.

COROLLARY. *Uniformly inside the open set* $\{x\colon x^r \in \mathbf{C}\backslash\mathbf{R}_+\}$

$$\pi_n(x) \to f(x) \quad \text{as } n \to \infty.$$

5. We present one more result. Let r be a positive integer, $\beta_1 > \cdots > \beta_r > 0$, and

$$f_j(x) = \int_0^\infty \frac{e^{-\beta_j t}\,dt}{x - t}, \qquad x \in \mathbf{C}\backslash\mathbf{R}_+,\ j = 1, \dots, r.$$

For every $n \in \mathbf{Z}_+^r$ there is one and only one polynomial G_n (up to normalization), with $\deg G_n = |n|$, such that

$$\int_0^\infty G_n(t) t^k e^{-\beta_j t}\,dt = 0, \qquad k = 0, \dots, n_j - 1,\ j = 1; \dots, r.$$

It has $|n|$ simple zeros on $(0, \infty)$. We have the formula

$$G_n(t) = \prod_{j=1}^r \left(\frac{1}{n_j!} e^{\beta_j t} \frac{d^{n_j}}{dt^{n_j}} e^{-\beta_j t}\right) t^n,$$

where the product of differential operators can be taken in any order. Let $P_{n,j}$ and $R_{n,j}$ be polynomials and functions of the second kind; $\pi_{n,j} = P_{n,j}/G_n$, $j = 1, \dots, r$.

We choose $p_j > 0$, with $p_1 + \cdots + p_r = 1$. We let

$$B = \sum_{j=1}^r \beta_j p_j, \qquad \omega(c) = \prod_{j=1}^r |1 - \beta_j c|^{p_j},$$

$$\omega_j = \max\{\omega(x)\colon 1/\beta_1 \le c \le 1/\beta_j\}, \qquad j = 1, \dots, r.$$

THEOREM. *If $n_j/|n| \to p_j$ as $|n| \to \infty$, $j = 1, \dots, r$, we have, uniformly in x on compact subsets of $\mathbf{C}\backslash\mathbf{R}_+$, where $|\arg(-x)| < \pi$,*

$$\lim_{n\to\infty} \frac{1}{\sqrt{|n|}} \ln |G_n(x)| = 2\operatorname{Re}\sqrt{-Bx}.$$

In addition, if $\omega_j < 1$,

$$\lim_{n\to\infty} \frac{1}{\sqrt{|n|}} \ln |R_{n,j}(x)| = -2\operatorname{Re}\sqrt{-Bx},$$

and if $\omega_j \geq 1$,

$$\lim_{n\to\infty} |R_{n,j}(x)|^{1/|n|} = \omega_j.$$

If $\omega_j \leq 1$, uniformly inside $\mathbf{C}\backslash\mathbf{R}_+$

$$\pi_{n,j}(x) \to f_j(x) \quad \text{as } n \to \infty;$$

if $\omega_j > 1$, we have $\pi_{n,j}(x)$ diverging to infinity.

In particular, $\pi_{n,1}$ always converges to f_1. If, for example, $r = 2$, $p_1 = p_2 = 1/2$, $\beta_1 = 1$, then $\pi_{n,2} \to f_2 \Leftrightarrow \beta_2 \geq 3 - 2\sqrt{2}$. If $\beta_1 \leq 2\beta_r$, then for all admissible p_j there is convergence for all j.

6. New general results on the asymptotics of polynomials that are orthogonal on the real line or on a half line, with exponential weight functions, have been obtained by E. A. Rakhmanov [84]. It appears to us that the results of that paper could be extended to the corresponding polyorthogonal generalizations.

PROBLEMS

1. Find all the solutions of Pearson's equation (6.4).

2. Prove the following Rodrigues formula for the generalized Laguerre polynomials (with the weight function (6.2)):

$$L_n^\alpha(x) = \frac{1}{n!} x^{-\alpha} e^x \frac{d^n}{dx^n}(x^{n+\alpha} e^{-x}),$$

find their coefficients, their L^2-norms with the weight function (6.2), and the coefficients of the recursion relations.

3. Prove the Rodrigues formula for the Hermite polynomials:

$$H_n(x) = \frac{1}{n!} e^{x^2} \frac{d^n}{dx^n}(e^{-x^2}),$$

find their coefficients, L^2-norms with the weight function (6.3), and the coefficients in the recursion relation. Express the Hermite polynomials in terms of Laguerre polynomials.

4. Find the generating functions for the Laguerre and Hermite polynomials.

5. Prove Perron's formula for the Laguerre polynomials:

$$L_n^\alpha(x) e^{-x} x^{\alpha/2} = \frac{1}{n!} \int_0^\infty t^n J_\alpha(2\sqrt{tx}) t^{\alpha/2} e^{-t} dt,$$

where the branches of the multiple-valued functions are consistent. Here $J_\alpha(z)$ is the Bessel function of the first kind and order α (see [20]).

6. Let $\deg Q_n = 3n$,

$$\int_{\Delta_k} Q_n(x) x^m e^{-x^2} dx = 0, \qquad m = 0, \dots, n-1;\ k = +, 0, -,$$

where $\Delta_- = (-\infty, -1]$, $\Delta_0 = [-1, 1]$, $\Delta_+ = [1, \infty)$. Write the Rodrigues formula for Q_n. Find the generating function. Find the asymptotics of the polynomials as $n \to \infty$. Investigate the functions of the second kind.

7. Let $r \in \mathbf{N}$, let $\alpha_1 > -1, \dots, \alpha_r > -1$ be pairwise unequal modulo 1, and $\alpha_0 > -1$. For every $n \in \mathbf{Z}_+^r$ there is a unique (up to a multiplier) polynomial B_n with $\deg B_n = |n|$, such that

$$\int_0^1 B_n(t) t^{k+\alpha_j} (1-t)^{\alpha_0} dt = 0, \qquad k = 0, \dots, n_j - 1; \ j = 1, \dots, r.$$

The polynomial B_n has $|n|$ simple zeros on $(0, 1)$ and can be defined by the formula [78]

$$B_n(t) = (1-t)^{-\alpha_0} \prod_{j=1}^{r} \left(\frac{1}{n_j!} t^{-\alpha_j} \frac{d^{n_j}}{dt^{n_j}} t^{n_j + \alpha_j} \right) (1-t)^{n+\alpha_0},$$

where the product of differential operators can be taken in any order.

CHAPTER 5

Potential theory

§1. The logarithmic potential

1. We shall use the following notation. If K is a compact subset of the complex plane $\mathbf{C}$, we denote by $\mathfrak{M}(K)$ the set of finite charges on K, i.e., real σ-additive functions defined on Borel subsets of K; by $\mathfrak{M}^+(K)$ the set of measures on K, i.e., charges that take nonnegative values; and by $\mathring{\mathfrak{M}}^+(K)$ the set of probability measures μ (for which $\mu(K) = 1$). Every charge ν can be written in just one way as the difference of two measures, $\nu = \nu_+ - \nu_-$ (Jordan decomposition). We set $|\nu| = \nu_+ + \nu_-$ and call $V(\nu) = |\nu|(K)$ the *total variation* of ν. Let us recall that $\mathfrak{M}(K)$ is a Banach space over the field of real numbers under the norm $V(\cdot)$. Furthermore, if $C(K)$ is the Banach space of continuous real functions on K with the uniform norm $\|f\| = \max|f(x)|$, $x \in K$, then by the Riesz theorem $\mathfrak{M}(K)$ is isometric to the space of continuous linear functionals which is dual to $C(K)$; the isometry is expressed by the relation

$$\nu : f \mapsto \langle \nu , f \rangle = \int_K f(x)\, d\nu(x), \qquad f \in C(K),$$

where $V(\nu) = \sup|\langle \nu , f \rangle| : \|f\| \le 1$; the positive functionals are precisely the measures μ (so that $\forall f \in C(K) : f(x) \ge 0,\ x \in K \Rightarrow \langle \mu , f \rangle \ge 0)$.

In $\mathfrak{M}(K)$ we may introduce the weak topology of the dual space, in which the open sets are generated by the following system of neighborhoods of zero:

$$\{\nu \in \mathfrak{M}(K) : \langle \nu , f \rangle < \varepsilon\}, \qquad \varepsilon > 0,\ f \in C(K).$$

The convergence of a sequence of charges ν_n to ν as $n \to \infty$ in the weak topology is called *weak convergence* and denoted by

$$\nu_n \to \nu, \qquad n \to \infty; \tag{1.1}$$

by (1.1), we mean that $\forall f \in C(K)$ $\langle \nu_n , f \rangle \to \langle \nu , f \rangle$, $n \to \infty$.

Evidently $\mathfrak{M}^+(K)$ is a convex cone, $\mathring{\mathfrak{M}}^+(K)$ is a convex set, and both sets are closed in the weak topology. Moreover, it follows from the weak compactness of balls in the dual space that $\mathring{\mathfrak{M}}^+(K)$ is compact in the weak topology; more precisely, from every sequence of probability measures on K we can select a weakly convergent subsequence.

2. Definition. Let μ be a measure on the compact set K. The *logarithmic potential of a measure* μ is a function

$$V_\mu(x) = \int_K \ln \frac{1}{|x-t|} d\mu(t), \qquad x \in \mathbf{C}. \tag{1.2}$$

The integral in (1.2) is to be thought of in the Lebesgue sense. Since, for any specified $x \in \mathbf{C}$, the function $\ln \frac{1}{|x-t|}$, $t \in K$, is bounded below, this integral is meaningful, although it may take the value $+\infty$.

Definition. The *logarithmic potential of a charge* $\nu \in \mathfrak{M}(K)$, $\nu = \nu_+ - \nu_-$, is the function

$$V_\nu(x) = \int_K \ln \frac{1}{|x-t|} d\nu(t) \overset{\text{def}}{=} V_{\nu_+}(x) - V_{\nu_-}(x), \tag{1.3}$$

where (1.3) is defined for x such that $V_{\nu_+}(x) \neq +\infty$ or $V_{\nu_-}(x) \neq +\infty$.

Theorem 1.1. *Let* μ_n, $\mu_0 \in \mathfrak{M}^+(K)$, x_n, $x_0 \in \mathbf{C}$, $n \in \mathbf{N}$, $\mu_n \to \mu_0$, *and* $x_n \to x_0$ *as* $n \to \infty$. *Then*

$$V_{\mu_0}(x_0) \leq \varliminf_{n\to\infty} V_{\mu_n}(x_n).$$

Proof. Let

$$k(x) = \begin{cases} \ln \frac{1}{|x|} & \text{if } x \in \mathbf{C}\backslash\{0\}, \\ +\infty & \text{if } x = 0. \end{cases}$$

The function $k(x)$ is called the *logarithmic kernel*. It is not continuous.

We introduce the intermediate kernels

$$k^{(\eta)}(x) = \begin{cases} k(x) & \text{if } k(x) \leq \eta, \\ \eta & \text{if } k(x) > \eta, \end{cases}$$

where $\eta > 0$. For $\mu \in \mathfrak{M}^+(K)$, we set

$$V_\mu^{(\eta)}(x) = \int_K k^{(\eta)}(x-t)\, d\mu(t), \qquad x \in \mathbf{C}.$$

We select an arbitrary $\eta > 0$. The family of functions

$$\{V_{\mu_n}^{(\eta)}(x)\}_{n=1}^\infty \tag{1.4}$$

is equicontinuous at every point x_0.

In fact, $k^{(\eta)}(x)$ is continuous in $\mathbf{C}$ and consequently, by the Cantor theorem, is uniformly continuous on any compact subset of $\mathbf{C}$; then for every $\varepsilon > 0$ there is a $\delta > 0$ such that the inequality

$$|k^{(\eta)}(x-t) - k^{(\eta)}(x_0 - t)| < \varepsilon \tag{1.5}$$

is satisfied for every $x : |x - x_0| < \delta$. Since the sequence $\{\mu_n\}$ converges weakly, we can, in particular, find a number $C > 0$ such that

$$\mu_n(K) \leq C, \qquad n \in \mathbf{N}. \tag{1.6}$$

It follows from (1.5) and (1.6) that for $|x-x_0|<\delta$ and all n

$$|V_{\mu_n}^{(\eta)}(x)-V_{\mu_n}^{(\eta)}(x_0)|\le\int_K|k^{(\eta)}(x-t)-k^{(\eta)}(x_0-t)|\,d\mu_n(t)\le C\varepsilon.$$

It follows from the continuity of $k^{(\eta)}$ and the definition of weak convergence that

$$V_{\mu_n}^{(\eta)}(x_0)\to V_{\mu_0}^{(\eta)}(x_0),\qquad n\to\infty. \tag{1.7}$$

Consequently

$$V_{\mu_0}^{(\eta)}(x_0)=\lim_{n\to\infty}V_{\mu_n}^{(\eta)}(x_n). \tag{1.8}$$

In fact,

$$|V_{\mu_0}^{(\eta)}(x_0)-V_{\mu_n}^{(\eta)}(x_n)|\le|V_{\mu_0}^{(\eta)}(x_0)-V_{\mu_n}^{(\eta)}(x_0)|+|V_{\mu_n}^{(\eta)}(x_0)-V_{\mu_n}^{(\eta)}(x_n)|. \tag{1.9}$$

By (1.7) and the equicontinuity of the family (1.4), the terms in (1.9) are less than any prescribed $\varepsilon>0$ for all sufficiently large n.

Since $k^{(\eta)}(x)\le k(x)$, $x\in\mathbf{C}$, for all $\eta>0$, it follows from (1.8) that

$$V_{\mu_0}^{(\eta)}(x_0)\le\varliminf_{n\to\infty}V_{\mu_n}(x_n). \tag{1.10}$$

It remains only to take the limit in (1.10) as $\eta\to+\infty$, and to note that

$$V_{\mu_0}(x_0)=\lim_{\eta\to+\infty}V_{\mu_0}^{(\eta)}(x_0)$$

by the monotone convergence theorem (or by the definition of the Lebesgue integral of an unbounded function). ▷

Theorem 1.1 shows that the function

$$(\mu,x)\mapsto V_\mu(x),\qquad \mu\in\mathfrak{M}^+(K),\ x\in\mathbf{C},$$

of two variables is lower semicontinuous in the weak topology on $\mathfrak{M}^+(K)$ and in the ordinary topology on $\mathbf{C}$.

COROLLARY. *The logarithmic potential of a measure is a lower semicontinuous function on* $\mathbf{C}$.

COROLLARY. *There is a reduction principle: if* μ_n, $\mu\in\mathfrak{M}^+(K)$, *and* $\mu_n\to\mu$ *as* $n\to\infty$, *we have*

$$V_\mu(x)\le\varliminf_{n\to\infty}V_{\mu_n}(x),\qquad x\in\mathbf{C}.$$

3. DEFINITION. A function $f(x)$ is *superharmonic* in a domain $G\subset\mathbf{C}$ if

1) $-\infty<f(x)\le+\infty$, $x\in G$ and $f(x)\not\equiv+\infty$;

2) f is lower semicontinuous at each point $x\in G$;

3) the inequality

$$f(x)\ge\frac{1}{2\pi}\int_0^{2\pi}f(x+re^{i\theta})\,d\theta \tag{1.11}$$

is satisfied for every $x\in G$ and every sufficiently small $r>0$.

A function f is harmonic in G only when both f and $(-f)$ are superharmonic.

THEOREM 1.2. 1) *The logarithmic potential of a measure is superharmonic in* $\mathbf{C}$.

2) *The logarithmic potential of a charge is harmonic in* $\mathbf{C}$ *outside its support.*

PROOF. The kernel $k(x)$ is superharmonic in $\mathbf{C}$ and harmonic in $\mathbf{C}\backslash\{0\}$. In fact, let $x = u + iv$ $(u, v \in \mathbf{R})$; then $k(x) = -\frac{1}{2}\ln(u^2 + v^2)$, and consequently

$$\frac{\partial^2 k}{\partial u^2} = \frac{u^2 - v^2}{(u^2+v^2)^2}, \qquad \frac{\partial^2 k}{\partial v^2} = \frac{v^2 - u^2}{(u^2+v^2)^2}.$$

Therefore $\Delta k(x) = 0$, $x \neq 0$, where

$$\Delta = \frac{\partial^2}{\partial u^2} + \frac{\partial^2}{\partial v^2}$$

is the Laplacian. This proves that the kernel is harmonic. For the proof of superharmonicity it is enough to verify the inequality (1.11) at $x = 0$. This is trivial.

Let $\mu \in \mathfrak{M}^+(K)$. We have already shown that $V_\mu(x)$ is lower semicontinuous. Let us verify (1.11). If we use the Fubini theorem and the superharmonicity of the kernel, we obtain

$$\begin{aligned}\frac{1}{2\pi r}\int_{|x-y|=r} V_\mu(y)|dy| &= \int_K \frac{1}{2\pi r}\int_{|x-y|=r} k(y-t)|dy|\,d\mu(t)\\ &\leq \int_K k(x-t)\,d\mu(t) = V_\mu(x).\end{aligned}$$

Let $\nu \in \mathfrak{M}(K)$. For $x \in \mathbf{C}\backslash S(\nu)$, where $S(\nu)$ is the support (spectrum) of the charge ν, we have the formula

$$\Delta V_\nu(x) = \int_{S(\nu)} \Delta k(x-t)\,d\mu(t) = 0,$$

since in the present case we can differentiate under the integral sign. ▷

4. We now turn our attention to two properties of the logarithmic potential.

THEOREM 1.3 (first maximum principle). *Let* μ *be a measure with compact support* $S(\mu)$. *If* $V_\mu(x) \leq M$ *for all* $x \in S(\mu)$, *this inequality is valid for all* $x \in \mathbf{C}$.

PROOF. Let G be any connected component of $\mathbf{C}\backslash S(\mu)$, and let $y_0 \in \partial G$. We choose an arbitrary $\eta > 0$. We then find a closed disk B centered at y_0 for which $\mu(B) < \eta$. (This is possible since the finiteness of $V_\mu(y_0)$ implies that $\mu(\{y_0\}) = 0$.) For an arbitrary point $x \in B \cap G$, we denote by y_1 the point of $B \cap S(\mu)$ that is closest to x. Then

$$\frac{1}{|x-y|} \leq \frac{2}{|y_1 - y|}, \qquad y \in B \cap S(\mu).$$

Consequently

$$\begin{aligned}V_\mu(x) &= \int_B \ln\frac{1}{|x-y|}\,d\mu(y) + \int_{S(\mu)\setminus B}\ln\frac{1}{|x-y|}\,d\mu(y)\\ &\le \mu(B)\ln 2 + \int_B \ln\frac{1}{|y_1-y|}\,d\mu(y)\\ &\quad + \int_{S(\mu)\setminus B}\ln\frac{1}{|y_1-y|}\,d\mu(y) + \int_{S(\mu)\setminus B}\ln\left|\frac{y_1-y}{x-y}\right|\,d\mu(y)\\ &= \mu(B)\ln 2 + V_\mu(y_1) + \int_{S(\mu)\setminus B}\ln\left|\frac{y_1-y}{x-y}\right|\,d\mu(y).\end{aligned}$$

Since $\mu(B) < \eta$, $V_\mu(y_1) \le M$,

$$\int_{S(\mu)\setminus B}\ln\left|\frac{y_1-y}{x-y}\right|\,d\mu(y) \to 0, \qquad x \to y_0,$$

and then

$$\varlimsup_{\substack{x\to y_0\\ x\in G}} V_\mu(x) \le \eta\ln 2 + M.$$

Since $\eta > 0$ was arbitrary, by the maximum principle for harmonic functions, we have $V_\mu(x) \le M$, $x \in G$. ▷

THEOREM 1.4 (principle of continuity). *Let μ be a measure with compact support $S(\mu)$. If the logarithmic potential V_μ is continuous on $S(\mu)$, it is continuous in $\mathbf{C}$.*

PROOF. Fix $x_0 \in S(\mu)$. Denote by μ_ε $(\varepsilon > 0)$ the restriction of μ to the disk $|x - x_0| < \varepsilon$, and set $\mu'_\varepsilon = \mu - \mu_\varepsilon$. Choose $\eta > 0$ arbitrarily. Find $\delta \in (0, 1/2)$ so that $V_{\mu_\delta}(x_0) < \eta$. Since V_{μ_δ} is continuous on $S(\mu)$ at x_0, there is an $\varepsilon \in (0, \delta)$ such that $V_{\mu_\delta}(x) < 2\eta$ when $x \in S(\mu)$ and $|x - x_0| < \varepsilon$. Then

$$V_{\mu_\varepsilon}(x) < V_{\mu_\delta}(x) < 2\eta, \qquad x \in S(\mu_\varepsilon).$$

By the first maximum principle, $V_{\mu_\varepsilon}(x) \le 2\eta$, $x \in \mathbf{C}$.

The function $V_{\mu'_\varepsilon}$ is continuous at x_0; hence there is an $r \in (0, 1/2)$ such that for any $x : |x - x_0| < r$, we have

$$|V_{\mu'_\varepsilon}(x_0) - V_{\mu'_\varepsilon}(x)| < \eta.$$

Consequently for all such x

$$\begin{aligned}|V_\mu(x_0) - V_\mu(x)| &\le V_{\mu_\varepsilon}(x_0) + V_{\mu_\varepsilon}(x) + |V_{\mu'_\varepsilon}(x_0) - V_{\mu'_\varepsilon}(x)|\\ &< 2\eta + 2\eta + \eta = 5\eta. \quad ▷\end{aligned}$$

PROBLEMS

1. Prove that the limit of a nondecreasing sequence of lower semicontinuous functions is a lower semicontinuous function (in any topological space).

2. Let $\Sigma(G)$ be the class of superharmonic functions in a domain G. Prove that $\Sigma(G)$ is a convex cone; the limit of a sequence $f_n \in \Sigma(G)$, uniformly convergent on compact subsets of

G, and also the limit ($\not\equiv +\infty$) of a nondecreasing sequence $f_n \in \Sigma(G)$, belong to this class; if f, $g \in \Sigma(G)$, then $\min(f(\cdot), g(\cdot)) \in \Sigma(G)$.

3. Prove the minimum principle for superharmonic functions in, for example, the following form: if f is superharmonic in the domain G, $x_0 \in G$, and $f(x) \geq f(x_0)$ for all $x \in G$, then $f(x) = f(x_0)$, $x \in G$.

Hint. $E = \{x \in G : f(x) = f(x_0)\}$ is a set that is not empty, and open and closed in G; consequently $E = G$.

§2. Energy of a charge

1. Let K be a compact subset of $\mathbf{C}$, and μ a measure on K.

DEFINITION. The *energy* of μ is the (Lebesgue) integral

$$I(\mu) = \iint_{K \times K} k(x-y)\, d\mu(x)\, d\mu(y). \tag{2.1}$$

Since $k(x-y)$ is bounded below for x, $y \in K$, the integral (2.1) is always meaningful, although it may have the value $+\infty$. Therefore (2.1) is a functional of the form

$$I(\cdot) : \mathfrak{M}^+(K) \to (-\infty, +\infty]. \tag{2.2}$$

DEFINITION. The *mutual energy* of two measures μ_1, $\mu_2 \in \mathfrak{M}^+(K)$ is the integral

$$I(\mu_1, \mu_2) = \iint_{K \times K} k(x-y)\, d\mu_1(x)\, d\mu_2(y).$$

We have

$$I(\cdot, \cdot) : \mathfrak{M}^+(K) \times \mathfrak{M}^+(K) \to (-\infty, +\infty]. \tag{2.3}$$

From the symmetry of $k(x-y)$ and the Fubini theorem, we obtain what is known as the *reciprocity law*:

$$I(\mu_1, \mu_2) = I(\mu_2, \mu_1) = \int_K V_{\mu_1}(y)\, d\mu_2(y) = \int_K V_{\mu_2}(x)\, d\mu_1(x). \tag{2.4}$$

In addition,

$$I(\mu) = I(\mu, \mu). \tag{2.5}$$

THEOREM 2.1. *The functional* (2.3) *is lower semicontinuous.*

PROOF. Let

$$\mu_n, \mu, \lambda_n, \lambda \in \mathfrak{M}^+(K), \quad \mu_n \to \mu, \quad \lambda_n \to \lambda \qquad (n \to \infty).$$

In addition, let

$$\pi_n = \mu_n \times \lambda_n, \qquad \pi = \mu \times \lambda \in \mathfrak{M}^+(K \times K),$$

be a direct product of measures. By the Stone-Weierstrass theorem, linear combinations of the functions $f(x)g(y)$, where f, $g \in C(K)$, are dense in $C(K \times K)$; consequently

$$\pi_n \to \pi \qquad (n \to \infty). \tag{2.6}$$

Let $k^{(\eta)}$ be an intermediate kernel (see §1, subsection 2). Set

$$I^{(\eta)}(\mu, \lambda) = \iint_{K\times K} k^{(\eta)}(x-y)\, d\mu(x)\, d\lambda(y).$$

Since $k^{(\eta)}(x-t)$ is continuous on $K \times K$, by (2.6) we have $I^{(\eta)}(\mu_\eta, \lambda_\eta) \to I^{(\eta)}(\mu, \lambda)$, $n \to \infty$. For every $\eta > 0$ we have $k^{(\eta)} \leq k$; consequently $I^{(\eta)}(\mu_n, \lambda_n) \leq I(\mu_n, \lambda_n)$. Therefore

$$I^{(\eta)}(\mu, \lambda) \leq \varliminf_{n\to\infty} I(\mu_n, \lambda_n). \tag{2.7}$$

If we take the limit as $\eta \to \infty$ in (2.7), we obtain

$$I(\mu, \lambda) \leq \varliminf_{n\to\infty} I(\mu_n, \lambda_n). \quad \triangleright$$

COROLLARY. *The functional* (2.2) *is lower semicontinuous.*

2. Let $\nu \in \mathfrak{M}(K)$, $\nu = \nu_+ - \nu_-$, be the Jordan decomposition, and $|\nu| = \nu_+ + \nu_-$.

DEFINITION. We say that ν is a *charge with finite energy* if $I(|\nu|) < +\infty$. We denote by $\mathscr{E}(K)$ the set of charges on K with finite energy.

The *energy of the charge* $\nu \in \mathscr{E}(K)$ is the integral

$$I(\nu) = \iint_{K\times K} k(x-y)\, d\nu(x)\, d\nu(y) \stackrel{\text{def}}{=} I(\nu_+) - 2I(\nu_+, \nu_-) + I(\nu_-).$$

The mutual energy of two charges ν_1, $\nu_2 \in \mathscr{E}(K)$ is the integral

$$\begin{aligned} I(\nu_1, \nu_2) &= \iint_{K\times K} k(x-y)\, d\nu_1(x)\, d\nu_2(y) \\ &\stackrel{\text{def}}{=} I(\nu_{1+}, \nu_{2+}) - I(\nu_{1+}, \nu_{2-}) - I(\nu_{1-}, \nu_{2+}) + I(\nu_{1-}, \nu_{2-}). \end{aligned}$$

Formulas (2.4) and (2.5) are also valid for charges with finite energy. Then $\mathscr{E}(K)$ is a linear space over the field $\mathbf{R}$; $I(\cdot, \cdot)$ is a symmetric bilinear form in this space; and $I(\cdot)$ is the associated quadratic form.

THEOREM 2.2. *If*

$$R_K \stackrel{\text{def}}{=} \max\{|x| : x \in K\} < 1,$$

we have $I(\nu) \geq 0$ *for every charge* $\nu \in \mathscr{E}(K)$, *and* $I(\nu) = 0$ *implies* $\nu = 0$.

Therefore if $R_K < 1$ the quadratic form $I(\cdot)$ is strictly positive definite. Consequently we can introduce on the linear space $\mathscr{E}(K)$ a scalar product and the corresponding norm by the formulas

$$(\nu_1, \nu_2) = I(\nu_1, \nu_2), \qquad \|\nu\| = \sqrt{I(\nu)}.$$

Then $\mathscr{E}(K)$ becomes a pre-Hilbert space.

Before proving Theorem 2.2, we establish some intermediate results.

3. Many of the facts that we shall need can be formulated in a more natural way in terms of distributions. Let us recall that the set $\mathscr{D}$ of test functions consists of infinitely differentiable (in the real sense) compactly supported functions defined on $\mathbf{C}$. (The set $\mathscr{D}(G)$ of test functions in a domain $G \subset \mathbf{C}$ is defined similarly.) A sequence $\varphi_n \in \mathscr{D}$ converges to zero in $\mathscr{D}$ if the supports of these functions belong to a compact set $K \subset \mathbf{C}$, and for all k, $l \in \mathbf{Z}_+$,

$$\frac{\partial^{k+l}}{\partial \xi^k \xi \eta^l} \varphi_n(x) \to 0, \qquad n \to \infty,$$

uniformly for $x = \xi + i\eta \in \mathbf{C}$. The set $\mathscr{D}$ is a topological linear space. The space $\mathscr{D}'$ of distributions consists of the continuous linear functionals on $\mathscr{D}$ and has a topological dual space. The action of $f \in \mathscr{D}'$ on $\varphi \in \mathscr{D}$ is denoted by $\langle f, \varphi\rangle$. A sequence $f_n \in \mathscr{D}'$ converges to zero if for any $\varphi \in \mathscr{D}$ the action $\langle f_n, \varphi_n\rangle \to 0$ as $n \to \infty$.

A function f that is locally integrable in $\mathbf{C}$ generates a regular distribution:

$$\langle f, \varphi\rangle = \int f(x)\varphi(x)\, dm(x), \qquad \varphi \in \mathscr{D}, \tag{2.8}$$

where $dm(x)$ denotes two-dimensional Lebesgue measure on $\mathbf{C}$. The integral (2.8) is often used formally to denote $\langle f, \varphi\rangle$ with any distribution f. It is known [22], [23] that f is a positive distribution ($\varphi \in \mathscr{D}$, $\varphi(x) \geq 0 \Rightarrow \langle f, \varphi\rangle \geq 0$) if

$$\langle f, \varphi\rangle = \int \varphi(x)\, d\mu(x), \qquad \varphi \in \mathscr{D},$$

where μ is a σ-finite (positive) measure on $\mathbf{C}$ (in general, with an unbounded support, and infinite).

We recall the rule for differentiating distributions:

$$\left\langle \frac{\partial^{k+l}}{\partial \xi^k \partial \eta^l} f, \varphi \right\rangle = (-1)^{k+l} \left\langle f, \frac{\partial^{k+l}}{\partial \xi^k \partial \eta^l} \varphi \right\rangle, \qquad \varphi \in \mathscr{D}.$$

The *convolution* of distributions f and g is defined by the formula

$$\langle f * g, \varphi\rangle = \int f(y) \left(\int g(x)\varphi(x+y)\, dm(x)\right) dm(y), \tag{2.9}$$

when the right-hand side of (2.9) is meaningful for all $\varphi \in \mathscr{D}$. When $\nu \in \mathfrak{M}(K)$ we may write

$$V_\nu = k * \nu. \tag{2.10}$$

By $\delta = \delta_0$ we denote the δ-function at zero:

$$\langle \delta, \varphi\rangle = \varphi(0), \qquad \varphi \in \mathscr{D},$$

and by $\delta^{(r)}$ a unit measure uniformly distributed on the circle $|x| = r$:

$$\langle \delta^{(r)}, \varphi\rangle = \frac{1}{2\pi} \int_0^{2\pi} \varphi(re^{i\theta})\, d\theta, \qquad \varphi \in \mathscr{D}.$$

The logarithmic kernel $k(x) = -\ln|x|$ is the fundamental solution of the Laplace operator:

$$-\frac{1}{2\pi}\Delta k = \delta. \tag{2.11}$$

In fact, (2.11) is another notation for the Green formula (recall the proof in §5 , subsection 1). It follows from (2.10) and (2.11) that

$$\Delta V_\nu = -2\pi\nu, \qquad \nu \in \mathfrak{M}(K).$$

In particular, we have the *uniqueness theorem.*

THEOREM 2.3. *If* $\nu \in \mathfrak{M}(K)$ *and* $V_\nu(x) = 0$ *for a.e.* $x \in \mathbf{C}$ *(with respect to Lebesgue measure), then* $\nu = 0$.

4. Besides the logarithmic kernel we introduce the *M. Riesz kernels*:

$$k_\alpha(x) = A(\alpha)|x|^{\alpha-2}, \qquad x \in \mathbf{C},$$

where

$$A(\alpha) = \pi^{\alpha-1}\frac{\Gamma(1-\alpha/2)}{\Gamma(\alpha/2)}, \qquad 0 < \alpha < 2.$$

LEMMA 2.1. *If* $0 < \alpha,\ \beta < 1$, *then*

$$k_\alpha * k_\beta = k_{\alpha+\beta}.$$

PROOF. Let us find the Fourier transform of k_α:

$$\tilde{k}_\alpha(y) = \int \exp(2\pi i \operatorname{Re}(x\overline{y}))A(\alpha)|x|^{\alpha-2}\,dm(x). \tag{2.12}$$

In (2.12), we introduce polar coordinates $r = |x|$, $\theta = \arg(x\overline{y})$:

$$\tilde{k}_\alpha(y) = \int_0^\infty \left(\int_0^{2\pi} \exp(2\pi i r|y|\cos\theta)\,d\theta\right) A(\alpha)r^{\alpha-1}\,dr.$$

Let

$$\begin{aligned} J_0(z) &= \frac{1}{2\pi}\int_0^{2\pi} \exp(iz\cos\theta)\,d\theta \\ &= \frac{1}{2\pi}\int_0^{2\pi}\left(\sum_{n=0}^\infty \frac{1}{n!}(iz\cos\theta)^n\right)d\theta = \sum_{n=0}^\infty \frac{(-1)^n}{(n!)^2}\left(\frac{z}{2}\right)^{2n} \end{aligned} \tag{2.13}$$

be the Bessel function of the first kind. Then

$$\tilde{k}_\alpha(y) = 2\pi A(\alpha)\int_0^\infty J_0(2\pi r|y|)r^{\alpha-1}\,dr = (2\pi)^{1-\alpha}A(\alpha)|y|^{-\alpha}\mathcal{J}, \tag{2.14}$$

where

$$\mathcal{J} = \int_0^\infty J_0(u)u^{\alpha-1}\,du.$$

Using (2.13), for $p > 0$ we have

$$\begin{aligned}\mathscr{J}(p) &\stackrel{\text{def}}{=} \int_0^\infty e^{-up} J_0(u) u^{\alpha-1}\,du \\ &= \frac{\Gamma(\alpha)}{2\pi}\int_0^{2\pi}(p - i\cos\theta)^{-\alpha}\,d\theta = \frac{\Gamma(\alpha)}{\pi}\int_{-1}^{1}(p-it)^{-\alpha}(1-t^2)^{-1/2}\,dt.\end{aligned}$$

In addition,

$$\begin{aligned}\mathscr{J} &= \lim_{p\to 0+}\mathscr{J}(p) = \frac{\Gamma(\alpha)}{\pi}\int_0^1 t^{-\alpha}(1-t^2)^{-1/2}\,dt\cdot(e^{\pi i\alpha} + e^{-\pi i\alpha}) \\ &= \Gamma(\alpha)\frac{\cos(\pi\alpha)}{\pi}\mathrm{B}\left(\frac{1-\alpha}{2}, \frac{1}{2}\right),\end{aligned} \tag{2.15}$$

where B is the beta function. If we substitute (2.15) in (2.14) and use the duplication and addition formulas for the gamma function, we obtain

$$\tilde{k}_\alpha(y) = |y|^{-\alpha}. \tag{2.16}$$

It is clear that (2.16) is the Fourier transform of k_α in the sense of distribution theory. Since the convolution $k_\alpha * k_\beta$ exists, we have, by the properties of the Fourier transform,

$$(k_\alpha * k_\beta)^\sim = \tilde{k}_\alpha\cdot\tilde{k}_\beta = \tilde{k}_{\alpha+\beta}. \quad \triangleright$$

If $\varphi\in\mathscr{D}$ and $\langle dm, \varphi\rangle = 0$, we use the formula

$$\lim_{\varepsilon\to 0+}\frac{|x|^{-\varepsilon}-1}{\varepsilon} = \ln\frac{1}{|x|}$$

and obtain

$$\begin{aligned}&\lim_{\alpha\to 2-}\int k_\alpha(x)\varphi(x)\,dm(x) \\ &\quad = \lim_{\varepsilon\to 0+} A(2-\varepsilon)\int(|x|^{-\varepsilon}-1)\varphi(x)\,dm(x) = 2\pi\int\ln\frac{1}{|x|}\varphi(x)\,dm(x),\end{aligned}$$

since $A(2-\varepsilon)\sim 2\pi/\varepsilon$, $\varepsilon\to 0+$. Similarly, we can verify the following lemma.

LEMMA 2.2. *If $\nu\in\mathfrak{M}(K)$ and $\nu(K) = 0$, we have $k_\alpha * \nu \to 2\pi V_\nu$ when $\alpha\to 2-$ (in the $\mathscr{D}'$ topology).*

PROOF OF THEOREM 2.2. Let $\nu\in\mathscr{E}(K)$, $\nu(K) = 0$. By Lemma 2.1,

$$\int_K (k_\alpha * \nu)(x)\,d\nu(x) = \int(k_{\alpha/2} * \nu)(x)^2\,dm(x), \qquad 0 < \alpha < 2. \tag{2.17}$$

If we take the limit as $\alpha\to 2-$ and use Lemma 2.2, we obtain

$$2\pi I(\nu) = 2\pi\int_K V_\nu(x)\,d\nu(x) = \int(k_1 * \nu)(x)^2\,dm(x) \geq 0. \tag{2.18}$$

If $I(\nu) = 0$, for every charge λ with compact support and finite energy ($\lambda(\mathbf{C}) = 0$) we have $(\nu, \lambda) = 0$ (by the Cauchy-Bunyakovskiĭ inequality, as also follows from (2.18)). Approximating continuous functions by potentials, we easily deduce that $\nu = 0$.

Let $\nu \in \mathscr{E}(K)$ be any charge, and $R_K < 1$; then we need to go over to the charge $\nu_0 = \nu - \nu(K)\delta^{(1)}$ and use what was proved above. In fact, $\nu_0(K) = 0$, and $I(\nu_0) = I(\nu)$. ▷

5. We define the kernel

$$g(x, y) = -\ln\left\{\left|\frac{x-y}{\frac{\overline{x}y}{R^2}-1}\right|\frac{1}{R}\right\} = \ln\frac{1}{|x-y|} + \ln R + \ln\left|\frac{\overline{x}y}{R^2} - 1\right|, \qquad (2.19)$$

depending on the parameter $R > 0$; x, y belong to the disk $U_R = \{z : |z| < R\}$. This differs from the logarithmic kernel only by a term that is harmonic in the disk; consequently we can generalize to (2.19) the properties of the logarithmic kernel that were established above. In addition, $g(x, y) > 0$ for $x, y \in U_R$; $g(x, y) = 0$ for $x \in U_R$ and $y \in \partial U_R$; and (2.19) is the Green function of U_R (for more about Green functions, see §5, below). For a measure μ with support in $\overline{U}_R$, we introduce the Green potential

$$G_\mu(x) = \int g(x, y)\, d\mu(y), \qquad x \in U_R, \qquad (2.20)$$

and the Green energy

$$I_g(\mu) = \iint g(x, y)\, d\mu(x)\, d\mu(y). \qquad (2.21)$$

The Green potential of the charge μ, $S(\mu) \subset \overline{U}_R$, is also defined by (2.20) for m-a.e. $x \in U_R$; the energy is defined by (2.21) provided that $I_g(|\mu|) < +\infty$. Then the set $\mathscr{E}_g$ of charges with finite Green energy becomes a pre-Hilbert space with the norm $\|\mu\|_g = \sqrt{I_g(\mu)}$. (If $S(\mu) \subset \partial U_R$, we take μ as the null element of the space.)

LEMMA 2.3. *Every function $f(x)$ that is superharmonic in the disk $U_{R'}$ admits a representation of the form*

$$f(x) = G_\mu(x) + h(x), \qquad x \in U_R, \qquad (2.22)$$

where $R < R'$, μ is a measure with support in $\overline{U}_R$, and h is a function that is harmonic in U_R.

PROOF. It is enough to establish the representation

$$f(x) = V_\mu(x) + h_1(x), \qquad x \in U_R, \qquad (2.23)$$

where h_1 is harmonic in U_R.

It follows from the definition of a superharmonic function that it is locally integrable; therefore f may be considered as a regular distribution in U_R. We set

$$\mu_1 = -\frac{1}{2\pi}\Delta f.$$

Let us prove that μ_1 is a measure in $U_{R'}$ (possibly infinite). If $\varphi \in \mathscr{D}$ is supported in $U_{R'}$, we obtain by the Taylor formula with the center at the origin

$$\lim_{r\to 0+}\left\langle \frac{1}{r^2}(\delta-\delta^{(r)}),\,\varphi\right\rangle = -\frac{1}{4}\Delta\varphi(0)\,,$$

i.e.,

$$\frac{1}{r^2}(\delta-\delta^{(r)}) \to -\frac{1}{4}\Delta\delta\,,\qquad r\to 0+\,.$$

(For the definitions of δ and $\delta^{(r)}$, see subsection 3.) Then

$$\frac{1}{r^2}(\delta-\delta^{(r)}) * f \to \frac{\pi}{2}\mu_1\,,\qquad r\to 0+\,. \tag{2.24}$$

On the left of (2.24) we have a nonnegative (since f is superharmonic) regular distribution; consequently μ_1 is a measure. Let μ be the part of μ_1 in U_R (a finite measure). Then the regular distribution $H(x) = f(x) - V_\mu(x)$ satisfies Laplace's equation $\Delta H = 0$ in U_R. There is a function h_1, harmonic in U_R, such that $H(x) = h_1(x)$ for m-a.e. x. (It is sufficient to consider the regularization of H, i.e., its convolution with a δ-like sequence of test functions ρ_n supported in the disk $|x| < 1/n$.) It follows from the definition of a superharmonic function that

$$f(x) = \lim_{r\to 0+}\frac{1}{2\pi}\int_0^{2\pi} f(x+re^{i\theta})\,d\theta\,,$$

and therefore $H(x) = h_1(x)$ for all $x\in U_R$. ▷

COROLLARY. *If f is superharmonic in $U_{R'}$, nonnegative in U_R, $R<R'$, and is majorized by the Green potential of a measure (supported in U_R), then f itself is the Green potential of a measure.*

PROOF. Under our hypotheses, we have that in the representation (2.22), $h(x) = 0$ for $x\in\partial U_R$; consequently $h(x)\equiv 0$. ▷

We now prove the second maximum principle for Green potentials.

THEOREM 2.4. *Let μ be a measure with compact support in U_R and finite Green energy. A nonnegative superharmonic function v in $\overline{U}_R$ majorizes G_μ in this disk if $G_\mu \le v$ on the support of μ.*

PROOF. We set

$$G(x) = \min(G_\mu(x),\, v(x))\,;$$

then G is superharmonic in $\overline{U}_R$, nonnegative, and majorized by the potential G_μ; consequently $G = G_\nu$, where ν is a measure in $\overline{U}_R$. We have

$$I_g(\nu) = \int G_\nu\,d\nu \le \int G_\mu\,d\nu = \int G_\nu\,d\mu \le \int G_\mu\,d\mu = I_g(\mu)\,,$$

i.e., ν has finite energy. Let

$$e = \{x : G_\mu(x) > v(x)\} = \{x : G_\mu(x) > G_\nu(x)\}\,;$$

then $\mu(e) = 0$, and on the complement of e we have $G_\mu = G_\nu$. Therefore

$$\|\mu - \nu\|_g^2 = \int_e (G_\mu - G_\nu)(-d\nu) \leq 0 \Rightarrow G_\mu = G_\nu. \quad \triangleright$$

THEOREM 2.5. *If μ and λ are probability measures with compact supports in $\mathbf{C}$, and μ has finite energy, the validity of the inequality $V_\mu(x) \leq V_\lambda(x)$ for all x in the support of μ implies its validity for all $x \in \mathbf{C}$.*

PROOF. We have

$$G_\mu(x) = V_\mu(x) + \ln R + \int \ln \left| \frac{\overline{x}y}{R^2} - 1 \right| d\mu(y), \tag{2.25}$$

and the integral in (2.25) tends to zero as $R \to \infty$, uniformly for $x \in U_R$. We can write a similar formula for λ. Now the theorem is obtained by letting $R \to \infty$ in Theorem 2.4. $\triangleright$

Theorem 2.5 is known as the *second maximum principle for the logarithmic potential.*

COROLLARY. *Under the hypotheses of Theorem 2.5, we have $V_\mu(x) = V_\lambda(x)$, $|x| > R_0$, for some $R_0 > 0$.*

PROOF. The function $V(x) = V_\mu(x) - V_\lambda(x)$, $|x| > R_0$, is harmonic and $V(\infty) = 0$. If $V(x) \not\equiv 0$, there are an even number of (curvilinear) sectors with vertex at ∞, such that on these sectors V alternately takes positive and negative values. By Theorem 2.5, $V(x) \leq 0$; therefore $V(x) \equiv 0$. $\triangleright$

PROBLEMS

Calculate the energy of the measure μ, for

1) $\mu = \delta^{(r)}$;
2) $d\mu(x) = dx$, $-1 < x < 1$;
3) $d\mu(x) = (1/\pi)(1 - x^2)^{-1/2}\,dx$, $-1 < x < 1$;
4) μ is a unit measure uniformly distributed in the annulus $0 < r \leq |x| \leq R < \infty$.

§3. Capacity of a compact set

Let us suppose that the compact set is contained in the unit disk $|x| < 1$; then we may apply Theorem 2.2. However, the restriction is only a technicality; the final result, as is easily seen, applies to any compact set in $\mathbf{C}$.

1. Let K be a compact subset of $\mathbf{C}$. We state the following extremal problem.

PROBLEM. *Find*

$$W = \inf I(\mu) : \mu \in \mathring{\mathfrak{M}}^+(K) \tag{3.1}$$

and the minimizing measure $\overline{\mu}$ for which the infimum is attained.

If $W = +\infty$, i.e., there is no measure that is different from zero on K and has finite energy, then every measure $\overline{\mu} \in \mathring{\mathfrak{M}}^+(K)$ is minimizing.

THEOREM 3.1. *If $W < +\infty$, there is a unique minimizing measure.*

PROOF. *Existence.* Let $\{\mu_n\}$ be a minimizing sequence, i.e.,

$$\mu_n \in \mathring{\mathfrak{M}}^+(K), \qquad \lim_{n\to\infty} I(\mu_n) = W.$$

By the weak compactness of $\overset{\circ}{\mathfrak{M}}{}^{+}(K)$, we can extract a weakly convergent sequence from $\{\mu_n\}$; without loss of generality, we can assume that $\mu_n \to \overline{\mu}$ for $n \to \infty$. By definition, $W \le I(\overline{\mu})$. By the corollary of Theorem 2.1

$$I(\overline{\mu}) \le \varliminf_{n\to\infty} I(\mu_n) = W.$$

Consequently $I(\overline{\mu}) = W$, i.e., $\overline{\mu}$ is a minimizing measure.

Uniqueness. Set

$$\mathscr{E}^{+}(K) = \mathscr{E}(K) \cap \mathfrak{M}^{+}(K), \qquad \overset{\circ}{\mathscr{E}}{}^{+}(K) = \mathscr{E}(K) \cap \overset{\circ}{\mathfrak{M}}{}^{+}(K).$$

For $W < +\infty$, (3.1) is equivalent to

$$W = \inf \|\mu\|^2 : \mu \in \overset{\circ}{\mathscr{E}}{}^{+}(K).$$

Let $\overline{\mu}$, $\overline{\mu}' \in \overset{\circ}{\mathscr{E}}{}^{+}(K)$, $\|\overline{\mu}\|^2 = \|\overline{\mu}'\|^2 = W$. Then from the convexity of $\overset{\circ}{\mathscr{E}}{}^{+}(K)$ and the triangle inequality we obtain

$$W \le \left\| \frac{\overline{\mu} + \overline{\mu}'}{2} \right\|^2 \le \left(\frac{1}{2}(\|\overline{\mu}\| + \|\overline{\mu}'\|) \right)^2 = W,$$

i.e., $(\overline{\mu} + \overline{\mu}')/2$ is also an extremal measure. By the parallelogram equation, we have

$$\|\overline{\mu} - \overline{\mu}'\|^2 = 2\|\overline{\mu}\|^2 + 2\|\overline{\mu}'\|^2 - 4\left\| \frac{\overline{\mu} + \overline{\mu}'}{2} \right\|^2 = 0,$$

and consequently, $\overline{\mu} = \overline{\mu}'$. ▷

2. Definition. The number $W = W(K)$ in (3.1) is the *energy*, or *Robin constant*, of K. The *logarithmic capacity* of K is the number

$$C_l(K) = \exp(-W(K)).$$

If $W(K) < +\infty$ and $\overline{\mu} = \overline{\mu}_K$ is the minimizing measure, $\frac{1}{W(K)}\overline{\mu}_K$ is called the *equilibrium measure* of K, and its logarithmic potential is the *equilibrium potential*. (The origin of these terms will be clear from what follows.)

Then $C_l(K) = 0$, or equivalently $W(K) = +\infty$, only if $\overset{\circ}{\mathscr{E}}{}^{+}(K) = \varnothing$.

Definition. A set $E \subset \mathbf{C}$ has *interior capacity zero* if the capacity of each of its subsets is zero.

We say that a property is satisfied *quasi-everywhere* (*q.e.*) on K if it is satisfied at all points of K except for a set of interior capacity zero.

A charge $\nu \in \mathfrak{M}(K)$ is said to be *C-absolutely continuous* if $C_l(E) = 0 \Rightarrow \nu(E) = 0$ for every compact set $E \subset K$.

It follows from the definition of capacity that every charge with finite energy is C-absolutely continuous. In integrating with respect to a C-absolutely continuous charge we may neglect Borel sets of zero interior capacity.

Remarks

In §§1–3, we have presented standard facts from potential theory. The fundamental exposition of this theory is given in [59]. We recommend the book [104] in which the theory of the

Newtonian potential is discussed (with the kernel $1/|x|$ in $\mathbf{R}^3$). We also mention [18], [169], [79], and [46].

§4. Extremal problems of potential theory

1. We consider a vector generalization of problem (3.1). Let r be a positive integer; $\Delta_1, \dots, \Delta_r$ compact subsets of $\mathbf{C}$; and $C = (c_{ij})$, a real symmetric nonsingular positive definite matrix of order r. We also require that $c_{ij} \geq 0$ for the pairs i, j of indices for which $\Delta_i \cap \Delta_j \neq \varnothing$. Then the family of compact sets $\Delta = (\Delta_1, \dots, \Delta_r)$ will be said to be *compatible with* C. If, for example, the compact sets $\{\Delta_i\}$ are disjoint, no additional conditions have to be imposed on C, since we have $c_{jj} \geq 0$ for a positive definite matrix C.

We introduce the direct products

$$\mathfrak{M} = \mathfrak{M}(\Delta_1) \times \cdots \times \mathfrak{M}(\Delta_r),$$
$$\mathfrak{M}^+ = \mathfrak{M}^+(\Delta_1) \times \cdots \times \mathfrak{M}^+(\Delta_r),$$

etc.

The weak and other topologies are defined for the direct products in the usual way (componentwise convergence). Let $\mu = (\mu_1, \dots, \mu_n) \in \mathfrak{M}^+$ be a vector measure (in what follows, we shall simply say "a measure"). We define the *energy* by the formula

$$J(\mu) = \sum_{i,j=1}^{r} c_{ij} I(\mu_i, \mu_j), \tag{4.1}$$

and the *mutual energy of two measures* $\mu^{(1)}$ and $\mu^{(2)}$ by

$$J(\mu^{(1)}, \mu^{(2)}) = \sum_{i,j=1}^{r} c_{ij} I(\mu_i^{(1)}, \mu_j^{(2)}). \tag{4.2}$$

PROPOSITION 4.1. *The functionals* (4.1) *and* (4.2) *are lower semicontinuous in the weak topology on* $\mathfrak{M}^+$.

PROOF. It is sufficient to consider the mutual energy, since $J(\mu) = J(\mu, \mu)$. We separate the sum (4.2) into two parts Σ_1 and Σ_2; Σ_1 contains the terms with $c_{ij} \geq 0$, and Σ_2 those with $c_{ij} < 0$. Each term in the first sum is lower semicontinuous by Theorem 2.1. Each term in the second sum is continuous on $\mathfrak{M}^+$ in the weak topology. In fact, if $c_{ij} < 0$, we have $\Delta_i \cap \Delta_j = \varnothing$ and the function $k(x-y)$ is continuous on the compact set $\Delta_i \times \Delta_j$. ▷

If now $\mu, \mu^{(1)}, \mu^{(2)} \in \mathscr{E}$ are vector charges, whose components have finite energy, the energy of the charge $J(\mu)$ and the mutual energies of the charges $J(\mu^{(1)}, \mu^{(2)})$ are defined by the formulas (4.1) and (4.2).

PROPOSITION 4.2. *The functional* $J(\cdot)$ *is a nonsingular positive definite quadratic form on the linear space* $\mathscr{E}$.

PROOF. We transform the matrix C to diagonal form Λ by using the orthogonal matrix $U: C = U^{\top}\Lambda U$, where

$$\Lambda = \begin{pmatrix} \lambda_1 & & 0 \\ & \ddots & \\ 0 & & \lambda_r \end{pmatrix}, \qquad \lambda_1, \dots, \lambda_r > 0,$$

and $U^{\top}U = E$ is the unit matrix of order r.

Let $U = (u_{ij})$; then in explicit form

$$c_{ij} = \sum_{l=1}^{r} \lambda_l u_{li} u_{lj}, \qquad i, j = 1, \dots, r. \tag{4.3}$$

If we substitute (4.3) into (4.1), we obtain

$$J(\mu) = \sum_{l=1}^{n} \lambda_l I\left(\sum_{i=1}^{n} u_{li}\mu_i, \sum_{j=1}^{n} u_{lj}\mu_j\right). \tag{4.4}$$

Since the functional $I(\cdot)$ is positive definite and the numbers λ_l in (4.4) are positive, it follows that $J(\mu) \geq 0$. Then if $J(\mu) = 0$, we have

$$\sum_{i=1}^{r} u_{li}\mu_i = 0, \qquad l = 1, \dots, r. \tag{4.5}$$

Since the matrix U is nonsingular, it follows from (4.5) that $\mu_i = 0$, $i = 1, \dots, r$. ▷

PROBLEM. *Find*

$$W = \inf J(\mu) : \mu \in \mathring{\mathfrak{M}}^{+}$$

and the extremal measure $\overline{\mu}$ *for which the infimum is attained.*

Again, if $W = +\infty$, the problem is trivial. We note that $W = +\infty$ only if the capacity of at least one of the compact sets Δ_i is zero.

PROPOSITION 4.3. *If* $W < +\infty$, *the extremal measure exists and is unique.*

The proof of this proposition is a repetition of the proof of Theorem 3.1.

2. We show that the logarithmic potentials of components of the extremal measure satisfy a system of equilibrium conditions, which in turn completely characterizes the extremal measure. The results presented below are simple generalizations of the corresponding equilibrium properties of the potential.

For an arbitrary measure $\mu \in \mathfrak{M}^{+}$ or charge $\mu \in \mathscr{E}$, in addition to the energy J we define the *partial energy* functionals

$$J_j(\mu) = \sum_{i=1}^{r} c_{ij} I(\mu_i, \mu_j), \qquad j = 1, \dots, r.$$

Then

$$J(\mu) = \sum_{j=1}^{r} J_j(\mu).$$

We introduce the notation $w_j = J_j(\overline{\mu})$, $j = 1, \dots, r$, for an extremal measure. For each j we define, on the compact set Δ_j, the following *W-potential* of a measure $\mu \in \mathfrak{M}^+$:

$$W_j^{\mu}(x) = \sum_{i=1}^{r} c_{ij} V_{\mu_i}(x), \qquad x \in \Delta_j. \tag{4.6}$$

For an extremal measure we set $W_j = W_j^{\overline{\mu}}$. We note the relation

$$J_j(\mu) = \int_{\Delta_j} W_j^{\mu}(x)\, d\mu_j(x), \qquad j = 1, \dots, r.$$

We can write similar relations for mutual energies. Let us emphasize that $W_j^{\mu}(x)$ is a lower semicontinuous function of $x \in \Delta_j$ and $\mu \in \mathfrak{M}^+$. The proof follows from Theorem 1.1 and the similar proof of Proposition 4.1. The same formula (4.6) defines the W-potential of a charge μ.

We shall frequently use the following property of convex sets.

LEMMA 4.1. *Let H be a real Hilbert space and M a convex subset of H; let there be an element $x_0 \in M$ that has the smallest norm: $\|x_0\| \le \|x\|$, $x \in M$. Then for all $x \in M$*

$$\|x_0\|^2 \le (x_0, x). \tag{4.7}$$

PROOF. Let $0 < a \le 1$. Since x_0 is extremal and M is convex, it follows that

$$\|x_0\|^2 \le \|ax + (1-a)x_0\|^2 = \|x_0\|^2 + 2a(x_0, x) - 2a\|x_0\|^2 + O(a^2).$$

Hence

$$2a((x_0, x) - \|x_0\|^2) \ge O(a^2). \tag{4.8}$$

If we divide both sides of (4.8) by a and let a tend to zero, we obtain (4.7). ▷

PROPOSITION 4.4. *For $j = 1, \dots, r$,*

$$W_j(x) \ge w_j, \quad \text{q.e. } x \in \Delta_j. \tag{4.9}$$

PROOF. Suppose the contrary, that for every j there is a compact set $K \subset \Delta_j$ with $C_l(K) > 0$ such that

$$W_j(x) < w_j, \qquad x \in K. \tag{4.10}$$

Take any measure $\nu_j \in \overset{\circ}{\mathscr{E}}{}^+(K)$. Set $\tilde{\mu} = (\tilde{\mu}_1, \dots, \tilde{\mu}_r)$, where $\tilde{\mu}_i = \overline{\mu}_i$, $i \ne j$, and $\tilde{\mu}_j = \nu_j$. Then by (4.10)

$$(\overline{\mu}, \tilde{\mu}) = \sum_{i \ne j} \int_{\Delta_i} W_i(x)\, d\overline{\mu}_i(x) + \int_{\Delta_j} W_j(x)\, d\nu_j(x) < W,$$

contradicting Lemma 4.1. ▷

PROPOSITION 4.5. *For* $j = 1, \dots, r$

$$W_j(x) \le w_j, \qquad x \in \Delta_j^* \stackrel{\text{def}}{=} S(\overline{\mu}_j).$$

PROOF. Let us suppose that for some j there is an $x_0 \in \Delta_j^*$ such that $W_j(x_0) > w_j$. From the lower semicontinuity of W_j it follows that there exists a neighborhood $\mathscr{V}$ of x_0 for which $x \in \mathscr{V} \Rightarrow W_j(x) > w_j$. Since x_0 is a point of the spectrum of $\overline{\mu}_j$, we have $\mu_j(\mathscr{V}) > 0$. Then, using Proposition 4.4, we obtain

$$\begin{aligned} w_j &= \int_{\Delta_j}^0 W_j(x)\, d\overline{\mu}_j(x) = \int_{\mathscr{V}} W_j\, d\overline{\mu}_j + \int_{\Delta_j \setminus \mathscr{V}} W_j\, d\overline{\mu}_j \\ &> w_j \overline{\mu}_j(\mathscr{V}) + w_j \overline{\mu}_j(\Delta_j \setminus \mathscr{V}) = w_j. \quad \triangleright \end{aligned}$$

3. In what follows, we shall need a proposition stronger than Proposition 4.4, in which (4.9) will be satisfied for all $x \in \Delta_j$. This is not true for an arbitrary compact set Δ_j. The stronger form holds for what are called *regular compact sets*. We shall obtain the result for compact sets that satisfy a simple sufficient condition for regularity.

DEFINITION. Let Δ be a compact set in $\mathbf{C}$, $x_0 \in \Delta$,

$$\Delta(r) = \Delta \cap \{x : |x - x_0| \le r\}, \qquad r > 0,$$

and $W(r)$ the Robin constant of $\Delta(r)$. We say that a compact set Δ has the K-*property* at x_0 if there are positive numbers K and r_0 for which

$$W(r) \le K \ln\left(\frac{1}{r}\right), \qquad r < r_0. \tag{4.11}$$

We say that a compact set Δ has the K-property ($\Delta \in (\mathrm{K})$) if it has the K-property at any of its points.

THEOREM 4.1. *Let E be any compact set in $\mathbf{C}$; μ a measure on E; Δ a compact set in $\mathbf{C}$ that has the* K*-property at some point x_0; and f a real continuous function on Δ. If*

$$V_\mu(x) + f(x) \ge 0, \quad q.e.\ x \in \Delta, \tag{4.12}$$

we have

$$V_\mu(x_0) + f(x_0) \ge 0. \tag{4.13}$$

PROOF. Let $\Delta_n = \Delta(2^{-n})$ at x_0; W_n the Robin constant of the compact set Δ_n; σ_n the corresponding measure; and V_n its logarithmic potential. Then, by (4.11), for some K and for some $n_0 \ge 1$ we have $W_n \le Kn \ln 2$, $n > n_0$. We shall show that for all $n > n_0$ and $x \in \mathbf{C}$

$$V_n(x) \le K \ln \frac{2}{|x - x_0|}. \tag{4.14}$$

First let $|x - x_0| \le 2^{1-n}$. Then by the first maximum principle we have

$$V_n(x) \le W_n \le K \ln \frac{2}{|x - x_0|}.$$

Now let $|x - x_0| \ge 2^{1-n}$. Since $|t - x_0| \le 2^{-n}$ for $t \in \Delta_n$ we have $|x - t| \ge |x - x_0|/2$, by the triangle inequality. Consequently

$$V_n(x) = \int_{\Delta_n} \ln \frac{1}{|t - x|} \, d\sigma_n(t) \le \int_{\Delta_n} \ln \frac{2}{|x - x_0|} \, d\sigma_n(t) = \ln \frac{2}{|x - x_0|}.$$

This completes the proof of (4.14).

For every n_1, the sequence $\{\sigma_n\}_{n>n_1}$ converges, in the weak topology of the space $\mathfrak{M}(\Delta_{n_1})$, to the single measure δ_{x_0} concentrated at x_0. In fact, if g is any real function that is continuous on the compact set Δ_{n_1}, the uniform continuity of g implies that

$$\left| \int_{\Delta_{n_1}} g(x) \, d\sigma_n(x) - g(x_0) \right| \le \int_{\Delta_{n_1}} |g(x) - g(x_0)| \, d\sigma_n(x) \to 0, \qquad n \to \infty.$$

In particular,

$$\int_\Delta f(x) \, d\sigma_n(x) \to f(x_0), \qquad n \to \infty, \tag{4.15}$$

and for every $x \ne x_0$

$$V_n(x) = \int_\Delta \ln \frac{1}{|x - t|} \, d\sigma_n(t) \to \ln \frac{1}{|x - x_0|}, \qquad n \to \infty. \tag{4.16}$$

If $V_\mu(x_0) = +\infty$, (4.13) is evident. Let

$$V_\mu(x_0) < +\infty. \tag{4.17}$$

By the Fubini theorem,

$$\int_\Delta V_\mu(x) \, d\sigma_n(x) = \int_E V_{\sigma_n}(t) \, d\mu(t). \tag{4.18}$$

By (4.14), (4.16), and (4.17), we may apply the Lebesgue theorem on passage to the limit to (4.18). We obtain

$$\int_\Delta V_\mu(x) \, d\sigma_n(x) \to \int_E \ln \frac{1}{|t - x_0|} \, d\mu(t) = V_\mu(x_0). \tag{4.19}$$

It follows from (4.15) and (4.19) that

$$V_\mu(x_0) + f(x_0) = \lim_{n\to\infty} \int_\Delta (V_\mu(x) + f(x)) \, d\sigma_n(x). \tag{4.20}$$

Then (4.13) follows from (4.12) and (4.20) by the C-absolute continuity of the measure σ_n. ▷

Propositions 4.4 and 4.5 and Theorem 4.1 lead to the following result.

THEOREM 4.2. *If* $\Delta_j \in (K)$, *then*

$$W_j(x)\begin{cases} = w_j, & \text{if } x \in \Delta_j^*, \\ \geq w_j, & \text{if } x \in \Delta_j \backslash \Delta_j^*, \end{cases}$$

where $j = 1, \ldots, r$.

THEOREM 4.3. *Let* E *be any compact subset of* $\mathbf{C}$; Δ *a compact subset of* $\mathbf{C}$ *with the* K*-property*; μ_n, $\mu \in \mathfrak{M}^+(E)$; f_n *and* f *continuous real functions on* Δ, *and* $n \in \mathbf{N}$. *If* $\mu_n \to \mu$ *as* $n \to \infty$, *and* $f_n(x) \rightrightarrows f(x)$, $x \in \Delta$, $n \to \infty$ *(uniform convergence), then*

$$\lim_{n\to\infty} \min_{x\in\Delta}\{V_{\mu_n}(x) + f_n(x)\} = \min_{x\in\Delta}\{V_\mu(x) + f(x)\}. \tag{4.22}$$

PROOF. Let us denote the minimum on the right-hand side of (4.22) by m, and the minimum on the left-hand side by m_n. Both minima are attained on the compact set Δ by the lower semicontinuity of the functions:

$$m = (V_\mu + f)(x_*), \qquad m_n = (V_{\mu_n} + f_n)(x_n), \qquad n \in \mathbf{N}.$$

Let $\underline{m}$ and $\overline{m}$ be the lower and upper limits of the sequence m_n.

We choose a subsequence $N \subset \mathbf{N}$ for which

$$m_n \to \underline{m}, \qquad x_n \to x_0 \qquad (n \to \infty,\ n \in N).$$

Then

$$f(x_0) = \lim_{\substack{n\to\infty \\ n\in N}} f_n(x_n), \tag{4.23}$$

and by Theorem 1.1

$$V_\mu(x_0) \leq \varliminf_{\substack{n\to\infty \\ n\in N}} V_{\mu_n}(x_n). \tag{4.24}$$

It follows from (4.23) and (4.24) that

$$m \leq \underline{m}. \tag{4.25}$$

Let us show that for every $m' > m$ the compact set

$$K = \{x \in \Delta : V_\mu(x) + f(x) \leq m'\} \tag{4.26}$$

has positive capacity

$$C_l(K) > 0. \tag{4.27}$$

Suppose the contrary: $C_l(K) = 0$. Let $\{\sigma_n\}$ be a sequence of extremal measures constructed for a point x_* and compact set Δ, as in the proof of Theorem 4.1. Then

$$\int_\Delta (V_\mu + f)\, d\sigma_n \to (V_\mu + f)(x_*) = m, \qquad n \to \infty. \tag{4.28}$$

In addition,

$$\int_\Delta (V_\mu + f)\, d\sigma_n = \int_K (V_\mu + f)\, d\sigma_n + \int_{\Delta\backslash K} (V_\mu + f)\, d\sigma_n. \tag{4.29}$$

The first integral on the left of (4.29) is zero, since the capacity of K was assumed to be zero, and σ_n is a C-absolutely continuous measure. Consequently

$$\int_\Delta (V_\mu + f)\, d\sigma_n \geq m' \cdot \sigma_n(\Delta \backslash K) = m'. \tag{4.30}$$

From (4.28) and (4.30), we obtain $m \geq m'$; but this contradicts the choice of m'. This contradiction establishes (4.27).

Let us suppose that $\overline{m} > m$. Choose m' and m'' so that $m < m' < m'' < \overline{m}$. For an infinite sequence of indices $n \in N_1 \subset \mathbf{N}$, we have

$$V_{\mu_n}(x) + f_n(x) \geq m'', \qquad x \in \Delta. \tag{4.31}$$

The compact set K is defined by (4.26). Let ν be any probability measure on K with logarithmic potential continuous in $\mathbf{C}$. (The existence of such a measure follows, for example, from the principle of continuity and the Luzin theorem applied to the logarithmic potential of any nonzero measure with finite energy.) Then

$$\int_K V_{\mu_n}\, d\nu = \int_E V_\nu\, d\mu_n \to \int_E V_\nu\, d\mu = \int_K V_\mu\, d\nu, \qquad n \to \infty.$$

Consequently

$$\int_K (V_{\mu_n} + f_n)\, d\nu \to \int_K (V_\mu + f)\, d\nu, \qquad n \to \infty. \tag{4.32}$$

By (4.31)

$$\int_K (V_{\mu_n} + f_n)\, d\nu \geq m'', \qquad n \in N_1. \tag{4.33}$$

According to (4.26)

$$\int_K (V_\mu + f)\, d\nu \leq m'. \tag{4.34}$$

It follows from (4.32), (4.33), and (4.34) that $m'' \leq m'$, contrary to the choice of these numbers. This contradiction shows that

$$\overline{m} \leq m. \tag{4.35}$$

The conclusion of the theorem follows from (4.25) and (4.35). ▹

4. Definition. Let G be a domain in $\overline{\mathbf{C}}$ that contains the point at infinity, and let $\Delta = \partial G$ be its boundary. If $\Delta \in (\mathrm{K})$, we say that Δ is *admissible.*

A real function φ defined on an admissible compact set Δ is said to be *admissible* if $\varphi = V_\lambda + f$, where λ is a measure on $\mathbf{C}$ with compact support, and f is a function that is continuous on Δ.

Let us consider the following extremal problem.

Problem A. *Let Δ be an admissible compact set, and φ an admissible function on Δ. Find a measure $\mu \in \mathring{\mathfrak{M}}^+(\Delta)$ such that*

$$w \stackrel{\text{def}}{=} \min_{x \in \Delta}(V_\mu(x) + \varphi(x)) = \max_{\nu \in \mathring{\mathfrak{M}}^+(\Delta)} \min_{x \in \Delta}(V_\nu(x) + \varphi(x)).$$

On $\mathfrak{M}^+(\Delta)$ we define the following functional with values in $(-\infty, \infty]$:

$$F(\mu) = I(\mu) + 2\langle \mu, \varphi \rangle, \tag{4.36}$$

where $I(\cdot)$ is the usual energy of the measure, and

$$\langle \mu, \varphi \rangle = \int_\Delta \varphi(x)\, d\mu(x). \tag{4.37}$$

We denote by $\widetilde{\mathscr{E}}$ the set of charges $\nu \in \mathfrak{M}(\Delta)$ for which $F(|\nu|) < +\infty$. It is clear that $\widetilde{\mathscr{E}}$ is a linear manifold in $\mathscr{E}$. Formulas (4.36) and (4.37) also define $F(\mu)$ for $\mu \in \widetilde{\mathscr{E}}$.

PROBLEM B. *Find a measure* $\mu \in \mathring{\mathfrak{M}}^+(\Delta)$ *such that*

$$W \stackrel{\text{def}}{=} F(\mu) = \min F(\nu) : \nu \in \mathring{\mathfrak{M}}^+(\Delta). \tag{4.38}$$

PROPOSITION 4.6. *The solution of Problem* B *exists and is unique.*

The proof is no different from the proof of Theorem 3.1. Under our hypotheses, $W < +\infty$. The existence of an extremal measure μ follows from the weak compactness of $\mathring{\mathfrak{M}}^+(\Delta)$ and the lower semicontinuity of the functional $F(\cdot)$. The uniqueness follows from the convexity of the set

$$\mathring{\widetilde{\mathscr{E}}}^+(\Delta) = \widetilde{\mathscr{E}}(\Delta) \cap \mathring{\mathfrak{M}}^+(\Delta) \tag{4.39}$$

and the geometry of Hilbert space. Let us explain this in more detail. It is clear that (4.38) is equivalent to

$$W \stackrel{\text{def}}{=} F(\mu) = \min F(\nu) : \nu \in \mathring{\widetilde{\mathscr{E}}}^+(\Delta).$$

Let μ_1 and μ_2 be solutions of this problem. Since (4.39) is convex and W is minimal,

$$\begin{aligned} W \le F\left(\frac{\mu_1+\mu_2}{2}\right) &= \left\|\frac{\mu_1+\mu_2}{2}\right\|^2 + 2\left\langle \frac{\mu_1+\mu_2}{2}, \varphi \right\rangle \\ &\le \frac{1}{2}(\|\mu_1\|^2 + \|\mu_2\|^2) + \langle \mu_1, \varphi\rangle + \langle \mu_2, \varphi \rangle = \frac{1}{2}(F(\mu_1) + F(\mu_2)) = W. \end{aligned}$$

Thus $(\mu_1+\mu_2)/2$ is a solution. Using the parallelogram equation, we obtain

$$\begin{aligned} \|\mu_1 - \mu_2\|^2 &= 2\|\mu_1\|^2 + 2\|\mu_2\|^2 - 4\left\|\frac{\mu_1+\mu_2}{2}\right\|^2 \\ &= 2(F(\mu_1) - 2\langle \mu_1, \varphi\rangle) + 2(F(\mu_2) - 2\langle \mu_2, \varphi \rangle) \\ &\quad - 4\left(F\left(\frac{\mu_1+\mu_2}{2}\right) - 2\left\langle \frac{\mu_1+\mu_2}{2}, \varphi\right\rangle\right) = 0. \end{aligned}$$

Consequently $\mu_1 = \mu_2$. ▷

Let us consider a generalization of Lemma 4.1.

LEMMA 4.2. *Let $\mathscr{H}$ be a real pre-Hilbert space; ϕ an additive homogeneous real functional defined on $\mathscr{H}$; and M a convex subset of $\mathscr{H}$. Suppose that the functional $\|\cdot\|^2 + 2\phi(\cdot)$ attains its minimum on M at the point $x_0 \in M$, i.e.,*

$$\|x_0\|^2 + 2\phi(x_0) \le \|x\|^2 + 2\phi(x), \qquad x \in M. \tag{4.40}$$

Then we have the inequality

$$\|x_0\|^2 + 2\phi(x_0) \le (x_0, x) + \phi(x), \qquad x \in M. \tag{4.41}$$

PROOF. Let $a \in (0, 1]$, $x \in M$. From the convexity of M and the inequality (4.40), we have

$$\begin{aligned}\|x_0\|^2 + 2\phi(x_0) &\le \|ax + (1-a)x_0\|^2 + 2\phi(ax + (1-a)x_0) \\ &= O(a^2) + 2a(x, x_0) + \|x_0\|^2 - 2a\|x_0\|^2 \\ &\quad + 2a\phi(x) + 2\phi(x_0) - 2a\phi(x_0).\end{aligned}$$

Hence

$$2a\{((x_0, x) + \phi(x)) - (\|x_0\|^2 + \phi(x_0))\} \ge O(a^2). \tag{4.42}$$

If we divide both sides of (4.42) by $2a$ and let a tend to zero, we obtain (4.41). ▷

Let us set $w = \|\mu\|^2 + \langle \mu, \varphi \rangle$, where μ is an extremal measure for F.

PROPOSITION 4.7. *For all $x \in \Delta$,*

$$V_\mu(x) + \varphi(x) \ge w. \tag{4.43}$$

PROOF. According to Theorem 4.1, it is enough to prove (4.43) for q.e. $x \in \Delta$. Let us assume the contrary, that there is a compact set $E \subset \Delta$ with $C_l(E) > 0$, such that $V_\mu(x) + \varphi(x) < w$, $x \in E$. Take any measure $\nu \in \mathring{\mathscr{E}}^+(E)$. Then

$$(\mu, \nu) + \langle \nu, \varphi \rangle = \int_E (V_\mu(x) + \varphi(x))\, d\nu(x) < w. \quad ▷$$

PROPOSITION 4.8. *For all $x \in \Delta^* = S(\mu)$,*

$$V_\mu(x) + \varphi(x) \le w.$$

The proof is an exact repetition of the proof of Proposition 4.5.

Propositions 4.7 and 4.8 imply the following theorem.

THEOREM 4.4. *The following equilibrium conditions are satisfied for an extremal measure μ for Problem* B:

$$V_\mu(x) + \varphi(x) \begin{cases} = w & \text{if } x \in \Delta^*, \\ \ge w & \text{if } x \in \Delta \setminus \Delta^*. \end{cases}$$

THEOREM 4.5. *The only solution of Problem* A *is the extremal measure for Problem* B.

PROOF. Let $\nu \in \mathring{\mathfrak{M}}^+(\Delta)$ and $V_\nu(x) + \varphi(x) \geq w$, $x \in \Delta$. Then, in particular,

$$V_\nu(x) \geq V_\mu(x) \tag{4.44}$$

for $x \in \Delta^*$. By the second maximum principle, (4.44) is satisfied for all $x \in \mathbf{C}$. Let us consider the difference $V = V_\mu - V_\nu$. Then

$$V(x) \leq 0, \qquad x \in \mathbf{C}. \tag{4.45}$$

Let G be the unbounded domain whose boundary is Δ. The function V is harmonic in G and $V(\infty) = 0$. If we use (4.45) and the maximum principle for harmonic functions, we obtain $V(x) = 0$, $x \in G$.

The function $V_\mu + \varphi$ is constant on Δ^*, and therefore continuous on Δ^*. Since φ is lower semicontinuous on Δ, we have that V_μ is upper semicontinuous on Δ^*. Then V_μ is continuous on Δ^*; consequently, by Theorem 1.4, continuous in $\mathbf{C}$, and V is upper semicontinuous in $\mathbf{C}$. Then $V(x) \geq 0$, $x \in \Delta$. It follows from this and (4.45) that $V(x) = 0$, $x \in \Delta$, and V is continuous in $\mathbf{C}$.

If now G_0 is any bounded connected component of the complement $\overline{\mathbf{C}} \backslash \Delta$, then V is harmonic in G_0, continuous in $\overline{G}_0$, and zero on ∂G_0. Consequently $V = 0$ in G_0. Thus $V_{\mu-\nu} \equiv 0$. Consequently $\mu = \nu$ by the uniqueness theorem. ▷

5. Let $\Delta = (\Delta_1, \dots, \Delta_r)$ be a set of admissible compact sets compatible with the positive definite matrix C; let the energy J and W-potential be defined by formulas (4.1) and (4.6). In what follows, an essential role will be played by the following problem.

PROBLEM C. *Find a measure* $\mu \in \mathring{\mathfrak{M}}^+$ *such that, for every* $j = 1, \dots, r$,

$$\min_{x\in\Delta_j} W_j^\mu(x) = \max_{\nu_j \in \mathring{\mathfrak{M}}^+(\Delta_j)} \min_{x\in\Delta_j} W_j^{\mu^{(j)}}(x),$$

where $\mu^{(j)}$ *is obtained from* μ *by replacing the* jth *component* μ_j *by* ν_j.

THEOREM 4.6. *The unique solution of Problem* C *is the extremal measure* $\overline{\mu}$ *of the functional* J.

PROOF. By Theorem 4.5, the measure μ is a solution of Problem C when, for each $j = 1, \dots, r$, μ_j is an extremal measure for the following problem:

$$\min\left\{ I(\nu_j) + 2\sum_{i\neq j} \frac{c_{ij}}{c_{jj}} I(\mu_i, \nu_j) \right\} : \nu_j \in \mathring{\mathfrak{M}}^+(\Delta_j). \tag{4.46}$$

In particular, μ_j is a measure with finite energy. If we multiply (4.46) by c_{jj} and add the constant

$$\sum_{k,l\neq j} c_{kl} I(\mu_k, \mu_l),$$

which is independent of ν_j, we find that Problem C is equivalent to the following problem: *find* $\mu \in \mathring{\mathfrak{M}}^+$ *such that, for each* $j = 1, \dots, r$,

$$J(\mu) = \min J(\mu^{(j)}) : \nu_j \in \mathring{\mathfrak{M}}^+(\Delta_j).$$

We call its solutions points of coordinatewise minimum of the energy functional.

Consequently, to prove Theorem 4.6 we have only to establish the following result.

LEMMA 4.3. *The energy functional* J *has no points of coordinatewise minimum except for the point* $\overline{\mu}$ *of global minimum.*

PROOF. Suppose the contrary: that there is a point $\nu \neq \overline{\mu}$ of coordinatewise minimum. We introduce a function of r real variables (a quadratic polynomial) $g(t_1, \dots, t_r) = J(\nu^{(t)})$, where

$$\nu_j^{(t)} = t_j \overline{\mu}_j + (1 - t_j)\nu_j, \qquad j = 1, \dots, r.$$

Since ν is a point of coordinatewise minimum for the functional J, we have

$$g(0) = \min_{0 \le t_j \le 1} g(0, \dots, 0, t_j, 0, \dots, 0), \qquad j = 1, \dots, r. \tag{4.47}$$

Consequently $\partial g(0)/\partial t_j \geq 0$, $j = 1, \dots, r$. We set

$$\tilde{g}(t) = g(t, \dots, t) = J(\overline{\mu} - \nu)t^2 + 2(J(\overline{\mu}, \nu) - J(\nu))t + J(\nu), \qquad t \in \mathbf{R}.$$

By (4.47),

$$\frac{d\tilde{g}(0)}{dt} = \frac{\partial g(0)}{\partial t_1} + \cdots + \frac{\partial g(0)}{\partial t_r} \geq 0. \tag{4.48}$$

Now $\overline{\mu}$ is the only point of global minimum of J; therefore $\tilde{g}(1) < \tilde{g}(0)$. In addition, $J(\overline{\mu} - \nu) > 0$. Then, using the properties of a quadratic trinomial, we obtain $d\tilde{g}(0)/dt < 0$. We have arrived at a contradiction of (4.48). ▷

REMARKS

1. The extremal problem (4.6) (for the case of disjoint intervals Δ_j and a matrix C of special form) was first investigated by A. A. Gonchar and E. A. Rakhmanov [37].

2. The concepts, introduced here, of an admissible compact set and admissible function are adequate for constructing the theory, but can be extended. For example, instead of the condition $\Delta \in (\mathrm{K})$ we can require the regularity of Δ, and we can include among the admissible functions any lower semicontinuous and approximately continuous functions (see Gonchar and Rakhmanov [38], where extremal distributions of charge are considered for the exterior field).

PROBLEMS

1. Prove that the equilibrium condition completely characterizes the extremal measure $\overline{\mu}$:

a) if $\mu \in \overset{\circ}{\mathfrak{M}}{}^{+}$, $W_j^{\mu}(x) = w_j$ for q.e. $x \in S(\mu_j)$, and $W_j(x) \le w_j$ for all $x \in S(\mu_j)$, then $\mu = \overline{\mu}$;

b) if $\mu \in \overset{\circ}{\mathscr{E}}{}^{+}$, $W_j^{\mu}(x) = a_j$ for q.e. $x \in S(\mu_j)$, and $W_j^{\mu}(x) \ge a_j$ for q.e. $x \in \Delta_j$ (here a_j is any constant), then $\mu = \overline{\mu}$.

2. If there is a sector S with nonzero opening and vertex at x, belonging to the compact set Δ, such that for some $r_0 > 0$ we have $S \cap \{y : |y - x| \le r_0\} \subset \Delta$, then Δ has the K-property at x (the converse is generally false).

§5. The Green function

1. Let D be a domain in $\mathbf{C}$, with its boundary $\Gamma = \partial D$ compact in $\mathbf{C}$.

THE DIRICHLET PROBLEM. *Given a continuous real function $u(x)$ on Γ, find a function $U(z)$ that is harmonic in D, continuous in the closure $\overline{D}$, and satisfies $U(z) = u(z)$, $z \in \Gamma$.*

If there is a solution of the Dirichlet problem, it is unique by the minimum and maximum principles for harmonic functions.

DEFINITION. If the Dirichlet problem has a solution for every $u \in C(\Gamma)$, the domain D is said to be *regular.*

In Chapter 3, §1, we solved the Dirichlet problem for the disk by using the Poisson integral. Consequently, if D is a simply-connected domain, whose boundary is a closed Jordan curve ("Jordan domain"), then the theorem on continuous correspondence of boundaries under conformal mapping shows that the Dirichlet problem is always solvable for such domains.

DEFINITION. The *Green function with singularity at a point* $z_0 \in D$, $z_0 \neq \infty$, for a regular domain D is a function $g(z, z_0)$ with the following properties.

(1) $g(z, z_0)$ is a harmonic function of z in the domain $D \backslash \{z_0\}$;

(2) the function $A(z) = g(z, z_0) - \ln \frac{1}{|z-z_0|}$ is bounded in a neighborhood of z_0;

(3) $g(z, z_0)$ is a continuous function of z on the set $\overline{D} \backslash \{z_0\}$, and $g(z, z_0) = 0$ when $z \in \Gamma$.

If $z_0 = \infty$, condition (2) in the definition of $g(z) = g(z, \infty)$ is to be replaced by the following definition:

(2′) the function $A(z) = g(z) - \ln |z|$ is bounded in a neighborhood of infinity.

If, for example, D is a simply-connected domain whose boundary consists of more than one point, then $g(z, z_0) = \ln |\varphi(z, z_0)|$, where $\varphi(z, z_0)$ is the *Riemann function*, which provides a conformal mapping of D on the exterior of the unit disk so that $\varphi(z, z_0) = \infty$ (and, by definition, $\varphi'(z_0, z) > 0$).

The determination of the Green function reduces to the solution of a Dirichlet problem. For the example of a proper domain, we show that the converse is also valid: if we know the Green function, we can describe the solution of the Dirichlet problem with arbitrary boundary conditions.

DEFINITION. A domain D is said to be *proper* if its boundary consists of a finite number of finitely many closed piecewise analytic Jordan curves.

It is well known that every proper domain is regular [32].

PROPOSITION 5.1. *Let D be a proper domain, and u a function that is harmonic in D and continuous in $\overline{D}$. Then for every point $z_0 \in D$ we have the Green formula*:

$$u(z_0) = \frac{1}{2\pi}\int_\Gamma u(\zeta)\frac{\partial}{\partial n}g(\zeta, z_0)\,ds, \tag{5.1}$$

where $\partial/\partial n$ is the derivative in the direction of the interior normal and ds is the element of arc length on Γ.

PROOF. It is enough to carry out the proof for a function u that is harmonic in a bounded domain D and continuously differentiable in $\overline{D}$. We use the notation $K_r = \{z : |z - z_0| < r\}$ and $\gamma_r = \{z : |z - z_0| = r\}$; r is a sufficiently small positive number. In addition, $D_r = D\backslash\overline{K}_r$, $\gamma_r = \partial D_r$. By the divergence theorem,

$$\frac{1}{2\pi}\int_{\Gamma_r}\left(u(\zeta)\frac{\partial}{\partial n}K(\zeta) - K(\zeta)\frac{\partial}{\partial n}u(\zeta)\right)\,ds = \frac{-1}{2\pi}\int_{D_r}(u\Delta K - K\Delta u)\,dm = 0,$$

where Δ is the Laplacian and

$$K(\zeta) = \ln\frac{1}{|\zeta - z_0|}. \tag{5.2}$$

Consequently

$$\frac{1}{2\pi}\int_\Gamma\left(u\frac{\partial K}{\partial n} - K\frac{\partial u}{\partial n}\right)\,ds = \lim_{r\to 0+}\frac{-1}{2\pi}\int_{\gamma_r}\left(u\frac{\partial K}{\partial n} - K\frac{\partial u}{\partial n}\right)\,ds.$$

By the mean value theorem, we obtain

$$\frac{1}{2\pi}\int_{\gamma_r}K\frac{\partial u}{\partial n}\,ds = r\ln\frac{1}{r}\cdot\frac{\partial u(\zeta_r)}{\partial n} \to 0, \qquad r\to 0+,$$

where $\zeta_r \in \gamma_r$. In addition, on γ_r

$$\frac{\partial K}{\partial n} = \frac{\partial}{\partial r}\ln\frac{1}{r} = \frac{-1}{r},$$

and therefore

$$\frac{1}{2\pi}\int_{\gamma_r}u\frac{\partial K}{\partial n}\,ds = -u(\zeta_r') \to -u(z_0), \qquad r\to 0+\,.$$

Thus

$$u(z_0) = \frac{1}{2\pi}\int_\Gamma\left(u\frac{\partial K}{\partial n} - K\frac{\partial u}{\partial n}\right)\,ds. \tag{5.3}$$

Again applying the divergence theorem, in (5.3) we may replace the kernel (5.2) by

$$K(\zeta) = \ln\frac{1}{|\zeta - z_0|} + A(\zeta),$$

where $A(\zeta)$ is any function that is harmonic in $\overline{D}$. In particular, we may take $K(\zeta) = g(\zeta, z_0)$; then we obtain (5.1). ▷

2. Let us consider the problem of an equilibrium distribution of charge on a compact set.

DEFINITION. We say that a compact set E in $\mathbf{C}$ is *proper* if the connected component $D = D_E$ of $\overline{\mathbf{C}}\backslash E$, that contains the point at infinity, is proper. The *Green function* g_E of the compact set E is the Green function of the component of D_E that has a singularity at infinity.

PROPOSITION 5.2. *Let E be a proper compact set; then*

$$d\mu_E(\zeta) = \frac{1}{2\pi}\frac{\partial}{\partial n} g_E(\zeta)\, ds(\zeta), \qquad \zeta \in \Gamma = \partial D_E,$$

is a probability measure on E, and its logarithmic potential is equal to

$$V_E(z) = \begin{cases} \gamma - g_E(z) & \text{if } z \in D_E, \\ \gamma & \text{if } z \in \overline{\mathbf{C}}\backslash D_E, \end{cases}$$

where $g_E(z) = \ln|z| + \gamma + o(1)$, $z \to \infty$.

PROOF. The positivity of $d\mu_E$ follows from the properties of the Green function. From the Green formula, we obtain

$$\int_E d\mu_E(\zeta) = 1,$$

and also

$$\ln|t - z| - g_E(t) = \int_\Gamma \ln|\zeta - z| \frac{1}{2\pi}\frac{\partial}{\partial n} g(\zeta, t)\, ds(\zeta), \qquad t \in D_E, \tag{5.4}$$

where $z \in \overline{\mathbf{C}}\backslash\overline{D}_E$ is a specific point. If we take the limit as $t \to \infty$ in (5.4), we will have

$$-\gamma = \int_\Gamma \ln|\zeta - z|\, d\mu_E(\zeta). \tag{5.5}$$

The formula (5.5) is valid for all $z \in \overline{\mathbf{C}}\backslash D_E$.

We now fix $z \in D_E$. By the Green formula,

$$\begin{aligned} &\ln\frac{1}{|t - z|} + g_E(t) - g(t, z) \\ &\quad = \int_\Gamma \ln\frac{1}{|\zeta - z|}\frac{1}{2\pi}\frac{\partial}{\partial n} g(\zeta, t)\, ds(\zeta), \qquad t \in D_E. \end{aligned} \tag{5.6}$$

Letting $t \to \infty$ in (5.6), we have

$$\gamma - g_E(z) = \int_\Gamma \ln\frac{1}{|\zeta - z|}\, d\mu_E(\zeta). \quad \triangleright$$

DEFINITION. The number γ is the *Robin constant* of the compact set E, and $C_l(E) = e^{-\gamma}$ is its *logarithmic capacity.*

It follows from Proposition 5.2 that for a proper compact set this definition is equivalent to that given in §3.

3. Let E be any compact subset of $\mathbf{C}$, and D_E the connected component of $\overline{\mathbf{C}}\backslash E$ that contains the point at infinity. Following N. Wiener, we consider a *proper exhaustion* of D_E. There is a sequence of proper domains D_m, $m \in \mathbf{N}$, such that

$$\overline{D}_m \subset D_E, \qquad \overline{D}_m \subset D_{m+1}, \qquad \bigcup_m D_m = D_E.$$

Let $g_m(z)$ be the Green function of D_m with pole at infinity, and let γ_m be the Robin constant of the compact set $E_m = \overline{\mathbf{C}}\backslash D_m$. By the maximum principle for harmonic functions, for every point $z \in D_E$ the sequence $g_m(z)$, $m \geq m(z)$, increases; consequently, by Harnack's theorem we have the alternative:

a) as $m \to \infty$, $g_m(z)$ converges uniformly inside D_E to a harmonic function $g_E(z)$, and $\gamma_m \to \gamma < +\infty$;

b) $g_m(z) \to +\infty$ as $m \to \infty$, for $z \in D_E$, and $\gamma_m \to \gamma = +\infty$.

DEFINITION. In case a) the function $g_E(z)$ is called a *generalized Green function* for E. In case b), we say that no generalized Green function exists. In either case, γ is called the Robin constant for E, and $C_l(E) = e^{-\gamma}$ is its logarithmic capacity.

Therefore a generalized Green function exists when $C_l(E) > 0$. It is clear that g_E is independent of the choice of a proper exhaustion of D_E (by the maximum principle). A generalized Green function has the following properties:

(1) $g_E(z)$ is harmonic in $D'_E = D_E\backslash\{\infty\}$;
(2) $g_E(z) = \ln|z| + \gamma + o(1)$ as $z \to \infty$;
(3) the limits of the values of $g_E(z)$ on the boundary of D_E are nonnegative.

If E is a regular compact set, $g_E(z)$ is the ordinary Green function. In what follows, we always suppose that $C_l(E) > 0$.

LEMMA 5.1 (Bernshteĭn, Walsh). *Let E be a compact subset of $\mathbf{C}$, and $p_n(z)$ a polynomial of degree at most $n \in \mathbf{Z}_+$. If*

$$\|p_n\|_E \stackrel{\text{def}}{=} \max_{z\in E} |p_n(z)| \leq M,$$

then for all $z \in D_E$

$$|p_n(z)| \leq M e^{n g_E(z)}.$$

PROOF. The lemma is a direct consequence of the maximum principle. In fact, the function

$$\Phi_n(z) = \frac{1}{n} \ln \frac{|p_n(z)|}{M} - g_E(z) \tag{5.7}$$

is subharmonic in D_E (we assume that $p_n \not\equiv 0$), and its limiting values on the boundary of D_E are nonpositive; hence $\Phi_n(z) \leq 0$, $z \in D_E$. ▷

COROLLARY. *Let E be a compact subset of $\mathbf{C}$, and p_n a polynomial of degree $n \in \mathbf{Z}_+$ with leading coefficient unity. Then*

$$\|p_n\|_E^{1/n} \geq C_l(E).$$

PROOF. We set $M = \|p_n\|_E$. Then

$$\Phi_n(\infty) = -\frac{1}{n} \ln \|p_n\|_E - \gamma \leq 0. \quad \triangleright$$

4. Let us show that the Bernshteĭn-Walsh lemma cannot be strengthened.

DEFINITION. We say that a sequence $\{p_n\}_0^\infty$ of polynomials of the form $p_n(z) = z^n + \cdots$ is *extremal* for a compact set E if all the zeros of p_n belong to E and

$$\lim_{n\to\infty} \left(\frac{|p_n(z)|}{\|p_n\|_E}\right)^{1/n} = e^{g_E(z)}, \qquad z \in D_E' = D_E \setminus \{\infty\}. \tag{5.8}$$

PROPOSITION 5.3. *If all the zeros of the polynomials $p_n(z) = z^n + \cdots$ belong to a proper compact set E, then* (5.8) *is equivalent to each of the following conditions*:

$$\lim_{n\to\infty} \|p_n\|_E^{1/n} = C_l(E), \tag{5.9}$$

$$\lim_{n\to\infty} |p_n(z)|^{1/n} = C_l(E) e^{g_E(z)}, \qquad z \in D_E' \tag{5.10}$$

(*uniformly inside the domain*).

PROOF. Formula (5.8) means that $\Phi_n(z) \to 0$, $z \in D_E'$. Since the sequence $\{\Phi_n\}$ is uniformly bounded inside D_E, it converges to zero uniformly inside this domain; in particular, $\Phi_n(\infty) \to 0$, which is equivalent to (5.9).

Conversely, let $\Phi_n(\infty) \to 0$. If a subsequence Φ_{n_j} converges to Φ uniformly inside D_E, we have $\Phi(\infty) = 0$. Since we always have $\Phi_n(z) \leq 0$, $z \in D_E$, we have also $\Phi(z) \leq 0$ in D_E. Then, by the maximum principle for harmonic functions, $\Phi(z) \equiv 0$ in D_E. Thus the only limit point of the precompact sequence $\{\Phi_n\}$ is zero; it follows that the sequence converges to zero.

It is evident that (5.10) follows from (5.9) and (5.8). Let (5.10) be satisfied. Let us write $L_\varepsilon = \{z : g_E(z) = \varepsilon\}$, $\varepsilon > 0$. Then

$$\|p_n\|_E^{1/n} \leq \|p_n\|_{L_\varepsilon}^{1/n} \to C_l(E) e^\varepsilon, \qquad n \to \infty.$$

Since $\varepsilon > 0$ is arbitrary, we obtain (5.9). $\triangleright$

If there is an extremal sequence for the compact set E, this is a sequence of polynomials deviating least from zero.

PROBLEM D. *Find the polynomial $t_n(z)$ with the smallest norm $m_n = \|t_n(z)\|_E$ among the polynomials of degree $n \in \mathbf{Z}_+$, with leading coefficient 1, whose zeros belong to the compact set E.*

The problem has a solution, since $\|t\|_E$ is a continuous function of the set of roots of t. In general, t_n is not unique. Sometimes requiring the roots to belong to E is not a convenient hypothesis; consequently we consider the following problem.

PROBLEM E. *Find the polynomial* $t_n^*(z)$ *with the smallest norm* $m_n^* = \|t_n^*\|_E$ *among the polynomials* $n \in \mathbf{Z}_+$ *with leading coefficient* 1.

Problem E has a solution, since we may suppose that the zeros of the polynomials belong to the convex hull of E.

PROPOSITION 5.4. *For every compact set* E, *the limit*

$$\tau^*(E) = \lim_{n\to\infty} m_n^{*1/n}$$

exists and is equal to $C_l(E)$.

PROOF. Let E be a proper compact set; then $\Gamma = \partial D_E$ is the union of a finite number of piecewise analytic closed Jordan curves. We divide Γ in any way into n arcs γ_j so that

$$\int_{\gamma_j} d\mu_E = \frac{1}{n}.$$

There are, of course, only a bounded number of arcs, which might not be connected. Let z_j be any point of γ_j, and $p_n(z) = (z - z_1)\cdots(z - z_n)$. Then, by the definition of the Riemann integral, as $n \to \infty$ we have

$$-\frac{1}{n}\ln|p_n(z)| = \frac{1}{n}\sum_{j=1}^{n}\ln\frac{1}{|z - z_j|} \to \int_\Gamma \ln\frac{1}{|z-t|}\,d\mu_E(t) = \gamma - g_E(z),$$

uniformly inside D_E'. By Proposition 5.3, $\{p_n\}$ is an extremal sequence. In particular, $\tau^*(E) = C_l(E)$.

Let E be any compact set, $\{D_m\}$ a proper exhaustion of D_E, and $E_m = \overline{\mathbf{C}}\backslash D_m$. We have

$$\begin{aligned} C_l(E) &\le \varliminf_{n\to\infty}(m_n^*)^{1/n} \le \varlimsup_{n\to\infty}(m_n^*)^{1/n} \\ &\le \tau^*(E_m) = C_l(E_m) \to C_l(E), \qquad m\to\infty. \ \triangleright \end{aligned}$$

DEFINITION. The number $\tau^*(E)$ is called the *Chebyshev constant* of E.

Consequently the Chebyshev constant coincides with the capacity of the compact set.

We introduce the notation

$$V_n(\zeta_1, \dots, \zeta_n) = \prod_{1\le i\le j\le n}|\zeta_j - \zeta_i| = \left|\det\begin{pmatrix} 1 & \cdots & 1 \\ \zeta_1 & \cdots & \zeta_n \\ \cdots & \cdots & \cdots \\ \zeta_1^{n-1} & \cdots & \zeta_n^{n-1}\end{pmatrix}\right|$$

for the absolute value of a Vandermonde determinant. We set

$$V_n = \max V_n(\zeta_1, \dots, \zeta_n) : \zeta_1, \dots, \zeta_n \in E.$$

The maximum is evidently attained.

Proposition 5.5. *The limit*

$$d(E) = \lim_{n\to\infty} V_n^{2/n(n-1)}$$

exists.

Proof. Let $V_{n+1} = V_{n+1}(\tilde{\zeta}_1, \dots, \tilde{\zeta}_{n+1})$. For each $j = 1, \dots, n+1$, we have

$$V_{n+1} \le \prod_{i\neq j} |\tilde{\zeta}_i - \tilde{\zeta}_j| \cdot V_n .$$

If we multiply these inequalities, we obtain

$$V_{n+1}^{n+1} \le V_{n+1}^2 \cdot V_n^{n+1} .$$

Hence

$$V_{n+1}^{2/n(n+1)} \le V_n^{2/n(n-1)} .$$

The existence of the limit follows by monotonicity. ▷

Definition. The number $d(E)$ is called the *Fekete transfinite diameter.*

Proposition 5.6. *For every compact set E, the limit*

$$\tau(E) = \lim_{n\to\infty} m_n^{1/n} \tag{5.11}$$

exists, and $\tau(E) = \tau^*(E) = d(E)$.

Proof. Let $V_n = V_n(\tilde{\zeta}_1, \dots, \tilde{\zeta}_n)$; then

$$\frac{V_{n+1}}{V_n} \ge \max_{z\in E} \frac{V_{n+1}(z, \tilde{\zeta}_1, \dots, \tilde{\zeta}_n)}{V_n} \ge m_n^* .$$

Now let $V_{n+1} = V_{n+1}(\tilde{\zeta}_1, \dots, \tilde{\zeta}_{n+1})$. Using properties of the determinant, we obtain

$$V_{n+1} = \left|\det \begin{pmatrix} 1 & \cdots & 1 \\ \tilde{\zeta}_1 & \cdots & \tilde{\zeta}_{n+1} \\ \cdots & \cdots & \cdots \\ \tilde{\zeta}_1^{n-1} & \cdots & \tilde{\zeta}_{n+1}^{n-1} \\ t_n^*(\tilde{\zeta}_1) & \cdots & t_n^*(\tilde{\zeta}_{n+1}) \end{pmatrix}\right| \le m_n^*(n+1)V_n .$$

Thus $m_n^* \le V_{n+1}/V_n \le (n+1)m_n^*$, which implies the existence of the limit

$$\lim_{n\to\infty} \left(\frac{V_{n+1}}{V_n}\right)^{1/n} = d(E), \tag{5.12}$$

and the equation $d(E) = \tau^*(E)$. We can prove similarly that $m_n \le V_{n+1}/V_n \le (n+1)m_n$. From this and (5.12), there follow the existence of the limit (5.11) and the equation $d(E) = \tau(E)$. ▷

From Propositions 5.4 and 5.6, we have the following corollary.

COROLLARY. *For every compact set E, there exists an extremal sequence of polynomials; for example, the Chebyshev polynomials t_n with zeros in E.*

5. LEMMA 5.2. *Let $\Delta = [a, b]$ be an interval on the real line; E a compact subset of Δ; and $l = |E|$ the linear Lebesgue measure of E, $0 < l < |\Delta|$. If p_n is any polynomial of degree at most $n \in \mathbf{Z}_+$, such that $\|p_n\|_E \le 1$, we have*

$$\|p_n\|_\Delta \le \psi\left(\frac{2|\Delta|}{l} - 1\right)^n \stackrel{\text{def}}{=} \kappa^n,$$

where $\psi(x) = x + \sqrt{x^2 - 1} \ge 1$ (for $x \ge 1$).

PROOF. Let δ be an interval on the real line and let ψ_δ be the function that maps $\overline{\mathbf{C}}\backslash\delta$ on the exterior of the unit disk so that $\psi_\delta(\infty) = \infty$ and $\psi'_\delta(\infty) > 0$. Then $\psi_{[-1,1]} = \psi$ and $\psi_{[a,a+l]}(b) = \kappa$, an increasing function of the quotient $|\Delta|/l > 1$. Let $\{t_m\}$ be the Chebyshev polynomials of the compact set E, with zeros on E. From Lemma 5.1 and the corollaries of Propositions 5.4 and 5.6 we obtain

$$|p_n(b)|^{1/n} \le e^{g_E(b)} = \lim_{m\to\infty}\left(\frac{t_m(b)}{\|t_m\|_E}\right)^{1/m}. \tag{5.13}$$

Let $\tau_1, \dots, \tau_m$ be the zeros of t_m, arranged in nondecreasing order. We set

$$\tilde{\tau}_j = |E \cap [a, \tau_j]|, \qquad j = 1, \dots, m,$$

$\tilde{t}_m(x) = (x - \tilde{\tau}_1)\cdots(x - \tilde{\tau}_m)$. It is clear that

$$\frac{t_m(b)}{\|t_m\|_E} \le \frac{\tilde{t}_m(b)}{\|\tilde{t}_m\|_{[a,a+l]}} \le \kappa^m. \tag{5.14}$$

It follows from (5.13) and (5.14) that

$$|p_n(b)|^{1/n} \le \kappa. \tag{5.15}$$

Similarly

$$|p_n(a)|^{1/n} \le \kappa. \tag{5.16}$$

Let $\|p_n\|_\Delta = |p_n(x_0)|$. If $x_0 = a$ or $x_0 = b$, the lemma has already been established. Let $a < x_0 < b$, $\Delta_+ = [x_0, b]$, $\Delta_- = [a, x_0]$, and $E_\pm = \Delta_\pm \cap E$. There exists $j = \pm$ such that $|E_j|/|\Delta_j| \ge l/|\Delta|$. Then if we apply (5.15) or (5.16), we obtain

$$\|p_n\|_\Delta^{1/n} \le \psi\left(\frac{2|\Delta_j|}{|E_j|} - 1\right) \le \kappa. \quad \triangleright$$

We are going to make substantial use of the following result.

THEOREM 5.1. *Let μ be a positive measure on the interval Δ, such that*

$$\mu'(x) > 0, \quad \text{a.e. } x \in \Delta \tag{5.17}$$

(for almost all $x \in \Delta$, *with respect to Lebesgue measure). If* $\{p_n\}$ *is any sequence of polynomials with* $\deg p_n \le n$ *and* $p_n \not\equiv 0$, *and*

$$M_n = \int_\Delta |p_n(x)|\, d\mu(x),$$

we have

$$\lim_{n\to\infty} \left(\frac{M_n}{\|p_n\|_\Delta} \right)^{1/n} = 1. \tag{5.18}$$

PROOF. We set

$$e_n = \{x \in \Delta : |p_n(x)| > nM_n\}, \quad E_n = \Delta \backslash e_n, \quad n \in \mathbf{Z}_+.$$

By the Chebyshev inequality, $\mu(e_n) \le 1/n \to 0$. Consequently, by (5.17) and the properties of the Lebesgue integral we have $|e_n| \to 0$, i.e., $|E_n| \to |\Delta|$ as $n \to \infty$. By Lemma 5.2 (let $\mu(\Delta) = 1$)

$$M_n \le \|p_n\|_\Delta \le \|p_n\|_{E_n} \kappa_n^n \le nM_n \kappa_n^n, \tag{5.19}$$

where $\kappa_n = \psi(2|\Delta|/|E_n| - 1) \to 1$ when $n \to \infty$. Then (5.18) follows from (5.19). ▷

REMARKS

1. On Green functions and the Dirichlet problem, see, for example, [32, Chapter VI]; on transfinite diameter and the Chebyshev constant, ibid., Chapter VIII. Also see [103].
2. We have proved the Green formula under very severe restrictive conditions, which could be significantly weakened (see [32]). However, we make no use of the more general results.
3. The proof given here of Lemma 5.2 was communicated to us by A. A. Gonchar.

§6. Angelesco systems

The results in this section were proved by A. A. Gonchar and E. A. Rakhmanov [37].

1. Let $\Delta = (\Delta_1, \dots, \Delta_r)$ be a set of disjoint intervals on the real line, and $\sigma = (\sigma_1, \dots, \sigma_r)$, where σ_j is a measure on Δ_j; without loss of generality, $\sigma_j(\Delta_j) = 1$. We shall suppose that

$$\sigma_j'(x) > 0, \qquad \text{a.e. } x \in \Delta_j, \ j = 1, \dots, r. \tag{6.1}$$

We introduce the notation

$$\Delta_* = \Delta_1 \cup \dots \cup \Delta_r, \qquad G_j = \mathbf{C} \backslash \Delta_j, \qquad j = *, 1, \dots, r.$$

Fix a vector $p = (p_1, \dots, p_r)$ such that $p_j > 0$, $p_1 + \dots + p_r = 1$, and choose a sequence $\{n\}$ of r-dimensional multi-indices for which

$$\begin{gathered} n_1 + \dots + n_r = n \in \mathbf{Z}_+, \\ n_j/n \to p_j, \qquad n \to \infty, \ j = 1, \dots, r. \end{gathered} \tag{6.2}$$

We denote by Q_n the nth polyorthogonal polynomial corresponding to the set of measures σ (see Chapter 4, §3), so that Q_n has degree n and

$$\int_{\Delta_j} Q_n(x) x^k \, d\sigma_j(x) = 0, \qquad k = 0, \dots, n_j - 1, \quad j = 1, \dots, r. \tag{6.3}$$

We suppose that the leading coefficient of Q_n is 1.

We want to obtain the logarithmic asymptotics of the polynomials Q_n. The answer is expressed in terms of certain extremal measures. Let $C = (c_{ij})$, where

$$c_{ij} = \begin{cases} 2p_j^2 & \text{if } i = j, \\ p_i p_j & \text{if } i \neq j, \end{cases}$$

and J is the energy functional corresponding to C. As we know, there is a unique extremal measure $\overline{\mu}$ that minimizes the functional J on the set $\overset{\circ}{\mathfrak{M}}{}^+$ (see §4).

We set

$$V_j = V_{\overline{\mu}_j}, \quad j = 1, \dots, r, \qquad V_* = p_1 V_1 + \dots + p_r V_r.$$

THEOREM 6.1. *Uniformly on compact subsets of* G_*,

$$\lim_{n\to\infty} |Q_n(x)|^{1/n} = \exp\{-V_*(x)\}.$$

To prove Theorem 6.1, we need a number of auxiliary results. We turn to a discussion of these.

2. The polynomial Q_n has n_j simple zeros on the interval Δ_j, $j = 1, \dots, r$ (see Chapter 4, §3). Let $Q_{n,j}$ be the polynomial of degree n_j with leading coefficient 1, whose zeros coincide with the zeros of Q_n that belong to Δ_j, $j = 1, 2, \dots, r$. Then $Q_n = Q_{n,1}, \dots, Q_{n,r}$. We also set

$$\widetilde{Q}_{n,j} = Q_n / Q_{n,j}, \qquad j = 1, \dots, r.$$

Since $\widetilde{Q}_{n,j}$ is a polynomial with real coefficients, with no zeros on Δ_j, it has a fixed sign on this interval.

Using the preceding notation, we rewrite (6.3) in the form

$$\int_{\Delta_j} Q_{n,j}(x) x^k |\widetilde{Q}_{n,j}(x)|\, d\sigma_j(x) = 0, \qquad k = 0, \dots, n_j - 1, \ j = 1, \dots, r.$$

Therefore $Q_{n,j}$ is an ordinary n_jth orthogonal polynomial for the measure

$$d\tilde{\sigma}_{n,j}(x) = |\widetilde{Q}_{n,j}(x)|\, d\sigma_j(x)$$

on Δ_j. Consequently, $Q_{n,j}$ is a polynomial deviating least from zero in the space $L^2_{\tilde{\sigma}_{n,j}}$ among all polynomials of degree n_j with leading coefficient 1. Thus we have established the following lemma.

LEMMA 6.1. *The relations*

$$\int_{\Delta_j} Q_{n,j}(x)^2 |\widetilde{Q}_{n,j}(x)|\, d\sigma_j(x) = \min_Q \int_{\Delta_j} Q(x)^2 |\widetilde{Q}_{n,j}(x)|\, d\sigma_j(x)$$

are satisfied, where the minimum is taken over all polynomials Q *of degree* n_j *and leading coefficient* 1*, all of whose zeros belong to* Δ_j, $j = 1, \dots, r$.

3. We denote by $\mu_{n,j}$ the measure associated with the polynomial $Q_{n,j}$, i.e., the probability measure equidistributed over the zeros of this polynomial, $\mu_n = (\mu_{n,1}, \dots, \mu_{n,r})$.

LEMMA 6.2. *The sequence of measures μ_n converges weakly to the extremal measure $\overline{\mu}$.*

PROOF. Since $\mathring{\mathfrak{M}}^+$ is a weakly compact set, it is enough to prove that $\overline{\mu}$ is the unique limit point of the sequence $\{\mu_n\}$. Let μ be any limit point of this sequence, i.e., $\mu_n \to \mu$ as $n \to \infty$ for some subsequence $N \subset \mathbf{Z}_+$. Let us show that μ satisfies the system of relations

$$\min_{x\in\Delta_j}\left\{2p_j^2 V_{\mu_j}(x) + \sum_{i\neq j} p_i p_j V_{\mu_i}(x)\right\}$$
$$= \max_{\nu\in\mathring{\mathfrak{M}}^+(\Delta_j)} \min_{x\in\Delta_j}\left\{2p_j^2 V_\nu(x) + \sum_{i\neq j} p_i p_j V_{\mu_i}(x)\right\}, \qquad j = 1, \dots, r. \quad (6.4)$$

We introduce the functionals

$$w_j(\lambda) = \min_{x\in\Delta_j} W_j^\lambda(x), \qquad \lambda \in \mathfrak{M}^+, \ j = 1, \dots, r.$$

We fix $j = 1, \dots, r$ and a measure $\nu \in \mathring{\mathfrak{M}}^+(\Delta_j)$. We denote by $\mu^{(j)}$ the measure obtained from μ by replacing the jth component of μ_j by ν. According to (6.4), we need to show that

$$w_j(\mu^{(j)}) \le w_j(\mu). \quad (6.5)$$

We say that a probability measure λ is *n-atomic* if it is concentrated at n points of the complex plane (not necessarily distinct), and the measure of each point is equal to $1/n$. A polynomial Q of degree n with leading coefficient 1, whose zeros are points of the spectrum of λ, is *associated* with this measure. We note the relation

$$V_\lambda(x) = -\frac{1}{n}\ln|Q(x)|.$$

Let $\{\nu_n\}$ be any sequence of measures on Δ_j, converging weakly to ν, such that ν_n is an n_j-atomic measure; $\mu_n^{(j)}$ is obtained from μ_n by replacing $\mu_{n,j}$ by ν_n. Then $\mu_n^{(j)} \to \mu^{(j)}$ as $n \to \infty$, $n \in N$. Let $Q_n^{(j)}$ be the polynomial associated with ν_n. Using Lemma 6.1, we obtain

$$\begin{aligned}\max_{x\in\Delta_j}|(Q_n^{(j)}(x))^2\widetilde{Q}_{n,j}(x)| &\ge \int_{\Delta_j}(Q_n^{(j)}(x))^2|\widetilde{Q}_{n,j}(x)|\,d\sigma_j(x)\\ &\ge \int_{\Delta_j}(Q_{n,j}(x))^2|\widetilde{Q}_{n,j}(x)|\,d\sigma_j(x)\\ &= \kappa_n\cdot\max_{x\in\Delta_j}|(Q_{n,j}(x))^2\widetilde{Q}_{n,j}(x)|, \end{aligned} \quad (6.6)$$

where κ_n is the ratio of the corresponding numbers. Since (6.1) is satisfied, it follows from Theorem 5.1 that

$$\lim_{n\to\infty} \kappa_n^{1/n} = 1. \tag{6.7}$$

After taking logarithms, the relation (6.6) can be rewritten in the form

$$\begin{aligned} &\min_{x\in\Delta_j} \left\{ \frac{2n_j^2}{n^2} V_{\nu_n}(x) + \sum_{i\neq j} \frac{n_i n_j}{n^2} V_{\mu_{n,i}}(x) \right\} \\ &\qquad \le -\frac{n_j}{n^2} \ln \kappa_n + \min_{x\in\Delta_j} \left\{ \frac{2n_j^2}{n^2} V_{\mu_{n,j}}(x) + \sum_{i\neq j} \frac{n_i n_j}{n^2} V_{\mu_{n,i}}(x) \right\}. \end{aligned} \tag{6.8}$$

Taking limits in (6.8) as $n \to \infty$, $n \in N$, and using (6.2), (6.7), and Theorem 4.3, we obtain (6.5). According to Theorem 4.6, the only solution of these relations is the extremal measure $\overline{\mu}$. Consequently $\mu = \overline{\mu}$. ▷

Theorem 6.1 is a direct corollary of Lemma 6.2.

COROLLARY. *Uniformly on compact subsets of* G_j,

$$\lim_{n\to\infty} |Q_{n,j}(x)|^{1/n_j} = \exp\{-V_j(x)\}, \qquad j = 1, \dots, r.$$

COROLLARY. *Uniformly on compact subsets of* G_*,

$$\lim_{n\to\infty} |(Q_{n,j}(x))^2 \widetilde{Q}_{n,j}(x)|^{1/n} = \exp\left\{-\frac{1}{p_j} W_j(x)\right\}, \qquad j = 1, \dots, r.$$

COROLLARY. *For* $j = 1, \dots, r$

$$\lim_{n\to\infty} \left\{ \int_{\Delta_j} (Q_{n,j}(x))^2 |\widetilde{Q}_{n,j}(x)|\, d\sigma_j(x) \right\}^{1/n} = \exp\left\{-\frac{w_j}{p_j}\right\},$$

where, as before, $w_j = w_j(\overline{\mu})$.

4. Let us investigate the asymptotics of the function of the second kind,

$$R_{n,j}(z) = \int_{\Delta_j} \frac{Q_n(x)}{x - z}\, d\sigma_j(x), \qquad z \in G_j,\ j = 1, \dots, r.$$

THEOREM 6.2. *Uniformly on compact subsets of* G_j

$$\lim_{n\to\infty} |R_{n,j}(z)|^{1/n} = \exp\{p_j V_j(z) - w_j/p_j\}, \qquad j = 1, \dots, r. \tag{6.9}$$

PROOF. We have

$$\begin{aligned} R_{n,j}(z) &= \frac{1}{Q_{n,j}(z)} \int_{\Delta_j} \frac{Q_n(x)(Q_{n,j}(z) - Q_{n,j}(x))}{x - z}\, d\sigma_j(x) \\ &\quad + \frac{1}{Q_{n,j}(z)} \int_{\Delta_j} \frac{Q_n(x) Q_{n,j}(x)}{x - z}\, d\sigma_j(x) \\ &= \frac{1}{Q_{n,j}(z)} \int_{\Delta_j} \frac{(Q_{n,j}(x))^2 \widetilde{Q}_{n,j}(x)}{x - z}\, d\sigma_j(x). \end{aligned}$$

According to the first corollary of Theorem 6.1,

$$\lim_{n\to\infty} |Q_{n,j}(z)|^{-1/n} = \exp\{p_j V_j(z)\} \tag{6.10}$$

uniformly inside G_j. It remains to find the asymptotics of the numbers

$$\widetilde{R}_{n,j}(z) = \int_{\Delta_j} \frac{(Q_{n,j}(x))^2 \widetilde{Q}_{n,j}(x)}{x - z}\, d\sigma_j(x). \tag{6.11}$$

We estimate an upper bound of the absolute value

$$|\widetilde{R}_{n,j}(z)| \le \int_{\Delta_j} \frac{(Q_{n,j}(x))^2 |\widetilde{Q}_{n,j}(x)|}{|x - z|}\, d\sigma_j(x) \le \frac{M_{n,j}}{\rho_j(z)}, \tag{6.12}$$

where $M_{n,j}$ is the $L^1_{\sigma_j}$-norm of the polynomial $Q^2_{n,j}\widetilde{Q}_{n,j}$, and $\rho_j(z) = \min|x - z| : x \in \Delta_j$. Now we estimate a lower bound for (6.11). If z is real, we have

$$|\widetilde{R}_{n,j}(z)| = \int_{\Delta_j} \frac{(Q_{n,j}(x))^2 |\widetilde{Q}_{n,j}(x)|}{|x - z|}\, d\sigma_j(x) \ge \frac{M_{n,j}}{\rho^*_j(z)}, \tag{6.13}$$

where $\rho^*_j(z) = \max|x - z| : x \in \Delta_j$. If $\operatorname{Im} z \ne 0$, we have

$$|\widetilde{R}_{n,j}(z)| \ge \frac{|\operatorname{Im} z|}{\rho^*_j(z)^2} \cdot M_{n,j}. \tag{6.14}$$

From (6.12), (6.13), and (6.14), using the third corollary of Theorem 6.1, we find that

$$\lim_{n\to\infty} |\widetilde{R}_{n,j}(z)|^{1/n} = \exp\{-w_j/p_j\} \tag{6.15}$$

uniformly inside G_j. Then (6.9) follows from (6.10) and (6.15). ▷

Let $\pi_{n,1}, \dots, \pi_{n,r}$ be simultaneous Padé approximants of the Markov functions $\hat{\sigma}_1, \dots, \hat{\sigma}_r$ corresponding to the multi-index n. Then

$$\hat{\sigma}_j - \tau_{n,j} = R_{n,j}/Q_n, \qquad j = 1, \dots, r.$$

From Theorems 6.1 and 6.2, we have the following corollary.

COROLLARY. *Uniformly inside* G_*

$$\lim_{n\to\infty} |\hat{\sigma}_j(x) - \pi_{n,j}(x)|^{1/n} = \exp\left\{\frac{W_j(x) - w_j}{p_j}\right\}, \qquad j = 1, \dots, r.$$

We set

$$G^{\pm}_j = \{x \in G_* : \pm(w_j - W_j(x)) > 0\}.$$

Then we have the following corollary.

COROLLARY. *The approximants* $\pi_{n,j}$ *converge to* $\hat{\sigma}_j$, *uniformly inside* G_j^+, *and diverge to infinity on* G_j^-.

We note that G_j^- can be empty, whereas G_j^+ always contains a neighborhood of infinity. In [37], there is a detailed study, for two intervals, of how the set of divergence of G_j^- varies in terms of the geometry of the intervals, i.e., how it depends on the lengths of the intervals and the distance between them. We consider one result of this kind.

5. Let $r = 2$, $\Delta_j = [a_j, b_j]$, $j = 1, 2$; for definiteness, we take $a_2 \geq b_1$, $\Delta_j^* = S(\overline{\mu}_j)$, where $\overline{\mu} = (\overline{\mu}_1, \overline{\mu}_2)$ is the extremal measure. Let us show that Δ_j^* is an interval. In fact, W_j is continuous on Δ_j, equal to a constant w_j on Δ_j^*, and convex downward on $\Delta_j \backslash \Delta_j^*$, since we have, for the logarithmic potential of the measure ν with $S(\nu) \subset \mathbf{R}$,

$$V_\nu''(x) = \int_{S(\nu)} \frac{d\nu(t)}{(x-t)^2} > 0, \qquad x \in \mathbf{R} \backslash S(\nu). \tag{6.16}$$

Suppose that Δ_j^* is not an interval; then there are α, $\beta \in \Delta_j^*$, $\alpha < \beta$, for which $(\alpha, \beta) \cap \Delta_j^* = \varnothing$. From what was said above, $W_j(x) < w_j$ for $x \in (\alpha, \beta)$. This contradicts the equilibrium hypothesis.

Evidently Δ_j^* are intervals of the form

$$\begin{aligned} \Delta_1^* &= [a_1, b_1^*], \qquad a_1 < b_1^* \leq b_1, \\ \Delta_2^* &= [a_2^*, b_2], \qquad a_2 \leq a_2^* < b_2. \end{aligned}$$

In fact, in the opposite case a shift of an interval can decrease the distance between them and hence decrease the energy $J(\mu)$. (We could give a different proof that, as above, uses the equilibrium hypothesis.)

For two intervals, only one of the inclusions $\Delta_j^* \subset \Delta_j$, $j = 1, 2$, can be strict. In fact, let $b_1^* < b_1$ and $a_2^* > a_2$. It follows from the equilibrium hypothesis that $W_1'(b_1^* + 0) \geq 0$. According to (6.16), $W_1''(x) > 0$, $x \in (b_1^*, a_2^*)$. Consequently, $W_1'(x) > 0$, $x \in (b_1^*, a_2^*)$, so that $W_1'(a_2^* - 0) > 0$. In addition,

$$W_2 = \frac{p_2}{2p_1} W_1 + \frac{3}{2} p_2^2 V_2,$$

and therefore $W_2'(a_2^* - 0) > 0$. Therefore $W_2(x) < w_2$ in some left neighborhood of a_2^*. This contradicts the equilibrium hypothesis. It is clear from the preceding discussion that a domain of divergence arises, in particular, when there is "collision of charges," i.e., strict inclusion $\Delta_j^* \subset \Delta_j$. In this connection, it is interesting to analyze the limiting case when one of the intervals is unbounded (see [93]).

If $p_1 = p_2 = 1/2$ and $|\Delta_1| = |\Delta_2|$ (without loss of generality, $\Delta_2 = -\Delta_1$), we have, by the symmetry of the problem and the uniqueness of the extremal measure, $w_1 = w_2$, $\overline{\mu}_2(e) = \overline{\mu}_1(-e)$, where e is a Borel set on Δ_2 and

$V_2(x) = V_1(-x)$. In this case $\Delta_1^* = \Delta_1$ and $\Delta_2^* = \Delta_2$. Let us establish the convergence of the simultaneous Padé approximants in this symmetric case. The function $\overset{\circ}{W}_1(x) = w_1 - W_1(x)$ is continuous in $\mathbf{C}$, harmonic in $\mathbf{C}\backslash\Delta_*$, and tends to $+\infty$ as $x \to \infty$. If $x \in \Delta_1$, we have $\overset{\circ}{W}_1(x) = 0$ by the equilibrium hypothesis. In addition,

$$\overset{\circ}{W}_1 = \overset{\circ}{W}_2 + \tfrac{1}{4}(V_2 - V_1), \tag{6.17}$$

where $\overset{\circ}{W}_2(x) = w_2 - W_2(x)$. If $x \in \Delta_2$, we have $\overset{\circ}{W}_2(x) = 0$ by the equilibrium hypothesis; consequently, by (6.17),

$$\overset{\circ}{W}_1(x) = \frac{1}{4}\int_{\Delta_1} \ln\frac{|x-t|}{|x+t|}\, d\mu_1(t) \geq 0.$$

Then by the minimum principle for harmonic functions we have $\overset{\circ}{W}_1(x) > 0$, $x \in G_*$, i.e., $G_1^+ = G_*$. Similarly, $G_2^+ = G_*$. Hence we have established the following theorem.

THEOREM 6.3. *If $p_1 = p_2 = 1/2$, and the lengths of the intervals Δ_1 and Δ_2 are equal, then as $n \to \infty$ the approximants $\pi_{n,j}$ converge to $\hat{\sigma}_j$, uniformly inside the domain $\mathbf{C}\backslash(\Delta_1 \cup \Delta_2)$.*

§7. MT-systems

We shall investigate the behavior of linear forms with polynomial coefficients for a set of functions that form an MT-system. We consider only systems of two functions. This, however, does not mean a loss of generality; the case of an arbitrary MT-system can be analyzed in a similar way.

1. Let Δ and δ be disjoint intervals on the real line, and σ and τ measures on Δ and δ, respectively. Without loss of generality, we suppose that σ and τ are probability measures. We suppose that

$$\sigma'(x) > 0, \qquad \text{a.e. } x \in \Delta,$$
$$\tau'(t) > 0, \qquad \text{a.e. } t \in \delta.$$

Let

$$u(x) = \int_\delta \frac{d\tau(t)}{x-t}, \qquad x \in \overline{\mathbf{C}}\backslash\delta. \tag{7.1}$$

In addition,

$$f(z) = \int_\Delta \frac{d\sigma(x)}{z-x}, \qquad g(z) = \int_\Delta \frac{u(x)\,d\sigma(x)}{z-x}, \qquad z \in \overline{\mathbf{C}}\backslash\Delta. \tag{7.2}$$

For the set of functions (7.2), we pose Problem D (see Chapter 4, §3): for arbitrary nonnegative integers k and l, find polynomials $A_n(z)$ and $B_n(z)$ $(n = k+l)$ of respective degrees $k-1$ and l, such that for some polynomial $C(z)$ the linear form

$$R_n(z) = A_n(z)f(z) + B_n(z)g(z) + C_n(z) \tag{7.3}$$

has a zero at infinity at least of order $n+1$:

$$R_n(z) = \frac{D_n}{z^{n+1}} + \cdots . \tag{7.4}$$

We recall that a solution of the problem exists and that (7.4) is equivalent to the orthogonality relations

$$\int_\Delta L_n(x)x^m \, d\sigma(x) = 0, \qquad m = 0, \dots, n-1, \tag{7.5}$$

where

$$L_n(x) = A_n(x) + B_n(x)u(x).$$

It follows from the results of Chapter 4, §4, that for

$$k \geq l, \tag{7.6}$$

the solution of the problem is unique up to a normalizing factor, and $L_n(x)$ has n simple zeros inside Δ. We shall show later that (7.6) is satisfied. We shall adopt the normalizing condition that the leading coefficient of the polynomial B_n is unity. We denote by H_n a polynomial of degree n with leading coefficient 1, whose zeros are the zeros of L_n.

PROPOSITION 7.1. *The integral*

$$H = \int_\Delta |\mathscr{H}(x)|^2 \left| \frac{L_n(x)}{H_n(x)} \right| d\sigma(x), \tag{7.7}$$

where $\mathscr{H}$ *is any polynomial of degree* n *with leading coefficient* 1, *attains its minimum only on the polynomial* $\mathscr{H} = H_n$.

PROOF. The function $L_n(x)/H_n(x)$ is continuous on Δ, is real-valued, and has constant sign on this interval. The extremal polynomial for (7.7) is the unique polynomial that satisfies the orthogonality relations

$$\int_\Delta \mathscr{H}(x)x^m \frac{L_n(x)}{H_n(x)} \, d\sigma(x) = 0, \qquad m = 0, \dots, n-1. \tag{7.8}$$

If we set $\mathscr{H} = H_n$ in (7.8), we obtain (7.5). ▷

2. We consider the problem of restoring a form $L_n(x)$ from the knowledge of its zeros, i.e., from H_n. This is known as the *problem of multipoint Padé approximation.* It is also equivalent to certain orthogonality relations.

PROPOSITION 7.2. *The polynomial* B_n *is the unique polynomial of degree* n *with leading coefficient* 1 *that satisfies the orthogonality relations*

$$\int_\delta B_n(t)t^m \frac{d\tau(t)}{H_n(t)} = 0, \qquad m = 0, \dots, l-1. \tag{7.9}$$

All the zeros of B_n *are simple and lie inside* δ.

PROOF. The function $L_n(x)/H_n(x)$ is holomorphic in $\overline{\mathbf{C}}\backslash\delta$ and has a zero at infinity of order at least $l+1$. Consequently, for $m = 0, \dots, l-1$ the function $L_n(x)x^m/H_n(x)$ has residue zero at infinity. Let Γ be any closed Jordan contour that goes around δ counterclockwise and does not surround any points of Δ. If we apply the residue theorem to the domain outside Γ, we obtain

$$\frac{1}{2\pi i}\int_\Gamma \frac{L_n(x)}{H_n(x)}x^m\,dx = 0, \qquad m = 0, \dots, l-1.$$

On the other hand,

$$\frac{1}{2\pi i}\int_\Gamma \frac{L_n(x)}{H_n(x)}x^m\,dx = \frac{1}{2\pi i}\int_\Gamma \frac{A_n(x)}{H_n(x)}x^m\,dx + \frac{1}{2\pi i}\int_\Gamma \frac{B_n(x)u(x)}{H_n(x)}x^m\,dx, \tag{7.10}$$

where the first integral on the right-hand side of (7.10) is zero, by the Cauchy theorem applied to the domain inside Γ. If we insert (7.1) into the second integral and apply the Fubini theorem, we obtain

$$\begin{aligned} 0 &= \frac{1}{2\pi i}\int_\Gamma \frac{B_n(x)x^m}{H_n(x)}\int_\delta \frac{d\tau(t)}{x-t}\,dx \\ &= \int_\delta \frac{1}{2\pi i}\int_\Gamma \frac{B_n(x)x^m}{H_n(x)}\,\frac{dx}{x-t}\,d\tau(t) = \int_\delta \frac{B_n(t)t^m}{H_n(t)}\,d\tau(t). \quad \triangleright \end{aligned}$$

We have the following corollary.

PROPOSITION 7.3. *The integral*

$$I = \int_\delta |\mathscr{B}(t)|^2 \frac{d\tau(t)}{|H_n(t)|},$$

where $\mathscr{B}(t)$ is any polynomial of degree l with leading coefficient 1, attains its minimum value only on the polynomial $\mathscr{B} = B_n$.

3. We specify a sequence of multi-indices (k, l) such that $k \geq l$, $k/n \to p$, $l/n \to q$ $(n \to \infty)$, where $n = k + l$ and p and q are given numbers: $p + q = 1, p \geq q > 0$. We consider the corresponding sequences of forms $\{R_n\}$, $\{L_n\}$ and polynomials $\{H_n\}$, $\{B_n\}$. We denote by μ_n a measure associated with H_n, and by ν_n a measure associated with B_n.

PROPOSITION 7.4. *If $\mu_n \to \overline{\mu}$ and $\nu_n \to \overline{\nu}$, we have*

$$\Phi(x) \overset{\text{def}}{=} \lim_{n\to\infty} \frac{1}{n} \ln\left|\frac{L_n(x)}{H_n(t)}\right| = qV_{\overline{\nu}}(x) - w, \tag{7.11}$$

uniformly in x inside $\overline{\mathbf{C}}\backslash\delta$, where

$$w = \min_{t\in\delta}(2qV_{\overline{\nu}}(t) - V_{\overline{\mu}}(t)).$$

PROOF. We fix $x \in \overline{\mathbf{C}}\backslash\delta$. Let Γ be a contour that goes around δ counterclockwise and does not surround either x or Δ. Since $L_n(x)/H_n(x)$ is holomorphic in $\overline{\mathbf{C}}\backslash\delta$ and zero at infinity, by the Cauchy formula we have

$$\begin{aligned}\frac{L_n(x)}{H_n(x)} &= -\frac{1}{2\pi i}\int_\Gamma \frac{L_n(\xi)}{H_n(\xi)}\frac{d\xi}{\xi - x}\\ &= -\frac{1}{2\pi i}\int_\Gamma \frac{A_n(\xi)}{H_n(\xi)}\frac{d\xi}{\xi - x} - \frac{1}{2\pi i}\int_\Gamma \frac{B_n(\xi)}{H_n(\xi)}\frac{u(\xi)\,d\xi}{\xi - x}. \qquad (7.12)\end{aligned}$$

Since the first integral in (7.12) is zero by the Cauchy theorem,

$$\frac{L_n(x)}{H_n(x)} = -\int_\delta \frac{1}{2\pi i}\int_\Gamma \frac{B_n(\xi)}{H_n(\xi)}\frac{1}{\xi - x}\frac{1}{\xi - t}\,d\xi\,d\tau(t) = \int_\delta \frac{B_n(t)}{H_n(t)}\frac{d\tau(t)}{x - t}.$$

In addition, if we use the orthogonality relation (7.9), we obtain

$$\frac{L_n(x)}{H_n(x)} = \frac{1}{B_n(x)}\int_\delta \frac{B_n(t)^2}{x - t}\frac{d\tau(t)}{H_n(t)}.$$

If $\mu_n \to \overline{\mu}$ and $\nu_n \to \overline{\nu}$ as $n \to \infty$,

$$\left(-\frac{1}{n}\right)\ln|B_n(x)| \to qV_{\overline{\nu}}(x) \qquad (7.13)$$

uniformly inside $\overline{\mathbf{C}}\backslash\delta$. As in Theorem 6.2, we can show that

$$\frac{1}{n}\ln\left|\int_\delta \frac{B_n(t)^2}{x - t}\frac{d\tau(t)}{H_n(t)}\right| \to -w, \qquad n \to \infty. \qquad (7.14)$$

Then (7.11) follows from (7.13) and (7.14). ▷

4. Using Propositions 7.1, 7.3, and 7.4, we consider the following system of extremal problems:

Find measures $\mu \in \mathring{\mathfrak{M}}^+(\Delta)$ *and* $\nu \in \mathring{\mathfrak{M}}^+(\delta)$ *for which*

$$\begin{aligned} w &\stackrel{\text{def}}{=} \min_{t\in\delta}(2qV_\nu(t) - V_\mu(t)) = \max_{\tilde\nu\in\mathring{\mathfrak{M}}^+(\delta)}\min_{t\in\delta}(2qV_{\tilde\nu}(t) - V_\mu(t)),\\ W &\stackrel{\text{def}}{=} \min_{x\in\Delta}(2V_\mu(x) - qV_\nu(x)) = \max_{\tilde\mu\in\mathring{\mathfrak{M}}^+(\Delta)}\min_{x\in\Delta}(2V_{\tilde\mu}(x) - qV_\nu(x)). \end{aligned} \qquad (7.15)$$

It follows from Theorem 4.6 that the only solution of the problems (7.15) is an extremal pair of measures (μ, ν) for the energy function J with the matrix

$$C = \begin{pmatrix} 2 & -q \\ -q & 2q^2 \end{pmatrix}.$$

THEOREM 7.1. *We have the weak convergence of measures*

$$\mu_n \to \mu, \quad \nu_n \to \nu \qquad (n \to \infty).$$

PROOF. Let $(\overline{\mu}, \overline{\nu})$ be any limit point (in the weak topology) of the sequence $\{(\mu_n, \nu_n)\}$. We show that $(\overline{\mu}, \overline{\nu})$ is a solution of (7.15); then it follows from the uniqueness of the solution that $\overline{\mu} = \mu$ and $\overline{\nu} = \nu$; then the theorem will be proved.

Let $\tilde{\nu}$ be any measure in $\mathring{\mathfrak{M}}^+(\delta)$, $\{\tilde{\nu}_n\}$ any sequence of l-atomic measures that converges to $\tilde{\nu}$, and $\mathscr{B}_n$ the polynomial associated with $\tilde{\nu}_n$. Using Proposition 7.3, we obtain

$$\max_{t\in\delta}\left|\frac{\mathscr{B}_n(t)^2}{H_n(t)}\right| \geq \int_\delta \frac{|\mathscr{B}_n(t)|^2}{|H_n(t)|}\,d\tau(t) \geq \int_\delta \frac{|B_n(t)|^2}{|H_n(t)|}\,d\tau(t) = \kappa_n\cdot\max_{t\in\delta}\left|\frac{B_n(t)^2}{H_n(t)}\right|, \tag{7.16}$$

where κ_n is the corresponding ratio. As in Theorem 5.1, it is shown that

$$\lim_{n\to\infty}\kappa_n^{1/n} = 1. \tag{7.17}$$

Taking logarithms in (7.16), we obtain

$$\min_{t\in\delta}\left(\frac{2l}{n}V_{\tilde{\nu}_n}(t) - V_{\mu_n}(t)\right) \leq -\frac{1}{n}\ln\kappa_n + \min_{t\in\delta}\left(\frac{2l}{n}V_{\nu_n}(t) - V_{\mu_n}(t)\right). \tag{7.18}$$

If we take the limit in (7.18) as $n\to\infty$, and use (7.17) and Theorem 4.3, we will have

$$\min_{t\in\delta}(2qV_{\tilde{\nu}}(t) - V_{\overline{\mu}}(t)) \leq \min_{t\in\delta}(2qV_{\overline{\nu}}(t) - V_{\overline{\mu}}(t)).$$

Thus $(\overline{\mu}, \overline{\nu})$ satisfies the first relation in (7.15). The second relation is proved similarly by using Propositions 7.1 and 7.4. ▹

It follows from (7.3) and (7.5) that

$$R_n(z) = \int_\Delta \frac{L_n(x)}{z-x}\,d\sigma(x) = \frac{1}{H_n(z)}\int_\Delta \frac{(H_n(x))^2}{z-x}\frac{L_n(x)}{H_n(x)}\,d\sigma(x). \tag{7.19}$$

From Theorem 7.1 (using (7.19)) we obtain the following results.

COROLLARY. *Uniformly inside* $\overline{\mathbf{C}}\backslash\delta$

$$\lim_{n\to\infty}|B_n(x)|^{1/n} = \exp\{-qV_\nu(x)\}.$$

COROLLARY. *We have the relation*

$$\lim_{n\to\infty}\left(\int_\delta |B_n(t)|^2\frac{d\tau(t)}{|H_n(t)|}\right)^{1/n} = e^{-w}.$$

COROLLARY. *Uniformly inside* $\overline{\mathbf{C}}\backslash\Delta$

$$\lim_{n\to\infty}|H_n(x)|^{1/n} = \exp\{-V_\mu(x)\}.$$

COROLLARY. *The following relation is valid*:

$$\lim_{n\to\infty}\left(\int_\Delta |H_n(x)|^2\left|\frac{L_n(x)}{H_n(x)}\right|\,d\sigma(x)\right)^{1/n} = e^{-w-W}.$$

COROLLARY. *Uniformly inside* $\overline{\mathbf{C}}\backslash(\Delta\cup\delta)$

$$\lim_{n\to\infty}|L_n(x)|^{1/n} = \exp\{qV_\nu(x) - V_\mu(x) - w\}.$$

COROLLARY. *Uniformly inside* $\overline{\mathbf{C}}\setminus\Delta$

$$\lim_{n\to\infty} |R_n(z)|^{1/n} = \exp\{V_\mu(z) - w - W\}.$$

For MT-systems of arbitrary order $r+1$, the discussion can be carried out on the same lines, and leads to consideration of an extremal problem with the triangular matrix

$$C = \begin{pmatrix} 2q_0^2 & -q_0q_1 & & & 0 \\ -q_0q_1 & 2q_1^2 & -q_1q_2 & & \\ & \cdots & \cdots & & \\ & -q_{r-2}q_{r-1} & 2q_{r-1}^2 & -q_{r-1}q_r & \\ 0 & & & & \\ & & -q_{r-1}q_r & 2q_r^2 & \end{pmatrix},$$

where $1 = q_0 > q_1 > \cdots > q_r > 0$.

REMARK

We present an additional result that can be obtained by applying potential theory.

Let $\Delta = [0, 1]$, and σ a finite positive Borel measure on Δ such that $\sigma'(t) > 0$ for a.e. (in Lebesgue measure) $t \in \Delta$.

Let $\widetilde{Q}_n(x)$ be a polynomial of degree n with leading coefficient 1, satisfying the orthogonality relations

$$\int_\Delta Q_n(x)x^k\,d\sigma(x) = 0, \qquad k = 0, \ldots, n-1,$$

where $Q_n(x) = \widetilde{Q}_n(x^r)$, and r is a fixed positive integer. The polynomial $\widetilde{Q}_n(x)$ exists and is unique. The polynomial $Q_n(x)$ is invariant under the transformation

$$x \to \zeta^m x, \tag{$*$}$$

where $\zeta = \exp(2\pi i/r)$, the rth roots of unity.

Let Δ_m denote the interval obtained from Δ by the transformation $(*)$, and $\Delta_* = \Delta_0 \cup \cdots \cup \Delta_{r-1}$. We consider the following sequence of extremal problems:

Find a set of probability measures $\mu_0, \ldots, \mu_{r-1}$ supported on the intervals $\Delta_0, \ldots, \Delta_{r-1}$ such that for each $m = 0, \ldots, r-1$ the relations

$$w_m = \min_{x\in\Delta_m}\left(2V_{\mu_m} + \sum_{\substack{l=0\\ l\neq m}}^{r-1} V_{\mu_l}\right)(x) = \max_\nu \min_{x\in\Delta_m}\left(2V_\nu + \sum_{\substack{l=0\\ l\neq m}}^{r-1} V_{\mu_l}\right)(x) \tag{$**$}$$

are satisfied, where the maximum is taken over all probability measures ν on Δ_m.

As in §6, it can be shown that the solution of $(**)$ exists and is unique. Moreover, by the symmetry of the problem we have

$$d\mu_m(x) = d\mu(x\bar{\zeta}^m), \qquad \mu = \mu_0,$$
$$w_0 = \cdots = w_{r-1} \stackrel{\text{def}}{=} w.$$

We set $V_m = V_{\mu_m}$, $V_* = V_0 + \cdots + V_{r-1}$, and $W_m = V_* + V_m$. The potentials W_m satisfy the following equilibrium conditions: $W_m(x) = w$, $x \in \Delta_m$. These conditions completely characterize the extremal measures.

THEOREM A. *Uniformly on compact subsets of* $\mathbf{C}\backslash\Delta_*$, *we have the asymptotic formula*

$$\lim_{n\to\infty} |Q_n(x)|^{1/n} = \exp(-V_*(x)).$$

We define the approximants $\pi_n(x)$ of the Markov function

$$f(x) = \int_0^1 \frac{d\sigma(t)}{x-t}, \qquad x \in \overline{\mathbf{C}}\backslash\Delta,$$

by the formula $\pi_n = P_n/Q_n$, where

$$P_n(x) = \int_0^1 \frac{Q_n(x) - Q_n(t)}{x-t}\, d\sigma(t)$$

is the polynomial of the second kind.

THEOREM B. *Uniformly on compact subsets of* $\overline{\mathbf{C}}\backslash\Delta_*$, *we have the asymptotic formula*

$$\lim_{n\to\infty} |f(x) - \pi_n(x)|^{1/n} = \exp(W_0(x) - w).$$

THEOREM C. *For all* $x \in \overline{\mathbf{C}}\backslash\Delta_*$, *the inequality*

$$W_0(x) < w$$

is satisfied. Consequently, as $n \to \infty$ *the approximants* π_n *converge to the function* f, *uniformly on compact subsets of the domain* $\overline{\mathbf{C}}\backslash\Delta_*$.

In contrast to the case of several arbitrary disjoint intervals on the real line, in the symmetric problem $(**)$ we can give an explicit form of the extremal measure μ and its potential.

THEOREM D. *Let*

$$\begin{aligned} h(\xi) &= (\sin\xi)^{1+1/r}((1+r)\sin(r\xi/(1+r)))^{-1} \\ &\quad \times ((1+r)\sin(\xi/(1+r)))^{-1/r}, \qquad 0 < \xi < \pi. \end{aligned}$$

Then the following results hold:

1) *The extremal measure* μ *is absolutely continuous with respect to Lebesgue measure, and*

$$d\mu(t) = (-1/\pi)\, dh^{-1}(t), \qquad 0 < t < 1,$$

where h^{-1} *is the inverse of* h.

2) *The potentials* V_0 *and* V_* *satisfy the formulas*

$$V_0(x) = -\frac{1}{\pi}\int_0^\pi \ln|x - h(\xi)|\, d\xi,$$

$$V_*(x) = -\frac{1}{\pi}\int_0^\pi \ln|x^r - h(\xi)^r|\, d\xi.$$

3) $w = (1+1/r)\ln((1+r)^{1+r}r^{-r})$.

In particular, when $r = 1$ Theorem D defines the Chebyshev measure on $[0, 1]$:

$$d\mu(t) = (1/\pi)t^{-1/2}(1-t)^{-1/2}\, dt.$$

We can give another characterization of μ.

THEOREM E. *The extremal measure* μ *is the unique solution of the following Hausdorff moment problem*:

$$\int_0^1 x^n\, d\mu(x) = \tau^{-n}\binom{n+n/r}{n}, \qquad n \in \mathbf{Z}_+,$$

where $\tau = (1+r)^{1+1/r}r^{-1}$.

We set $\rho = (1+r)r^{-r/(1+r)}$. Let

$$\mathscr{J}_0(z) = \sum_{n=0}^\infty \frac{(z/\rho)^{(1+1/r)n}}{n!\,\Gamma(n/r+1)}$$

be the generalized Bessel function (it is expressed in terms of generalized hypergeometric functions) and

$$\mathscr{U}(\lambda) = \int_0^\infty \mathscr{I}_0(x)e^{-\lambda x}\,dx$$

its Laplace transform. We have the following theorem.

THEOREM F. *The Markov function*

$$\mathscr{W}(x) = \int_0^1 \frac{d\mu(t)}{x-t}, \qquad x \in \overline{\mathbf{C}}\setminus[0, 1],$$

of the extremal measure μ *has the form*

$$\mathscr{W}(x) = x^{-1/(1+r)}\mathscr{U}(x^{r/(1+r)}).$$

References

1. A. I. Aptekarev, *Asymptotics of Hadamard determinants and the convergence of rows of Padé approximants for sums of exponentials*, Mat. Sb. **113 (155)** (1980), 520–537; English transl. in Math. USSR Sb. **41** (1982), 427–442.

2. ——, *Convergence of rational approximations to a set of exponential functions*, Vestnik Moskov. Univ. Ser. I Mat. Mekh. **1981**, no. 1, 68–74; English transl. in Moscow Univ. Math. Bull. **36** (1981).

3. ——, *On Padé approximants for the set* $\{{}_1F_1(1, c; \lambda_i z\}_{i=1}^k$, Vestnik Moskov. Univ. Ser. I Mat. Mekh. **1981**, no. 2, 58–62; English transl. in Moscow Univ. Math Bull. **36** (1981).

4. ——, *Asymptotic properties of polynomials orthogonal on a system of contours, and periodic motions of Toda lattices*, Mat. Sb. **125 (167)** (1984), no. 2, 231–258; English transl. in Math. USSR Sb. **53** (1986), 233–260.

5. ——, *Analytic properties of functions represented by* 2*-dimensional continued fractions with constant coefficients, and their simultaneous Padé-Hermite approximants* (*type* P), Preprint, M. V. Keldysh Inst. Prikl. Mat. Akad. Nauk SSSR, no. 79, 12 pp. (Russian)

6. A. I. Aptekarev and E. M. Nikishin, *The scattering problem for a discrete Sturm-Liouville operator*, Mat. Sb. **121 (163)** (1983), no. 3, 327–358; English transl. in Math. USSR Sb. **49** (1984), 325–356.

7. F. V. Atkinson, *Discrete and continuous boundary problems*, Academic Press, New York-London, 1964.

8. N. I. Akhiezer, *The classical moment problem and some related topics in analysis*, Fizmatgiz, Moscow, 1961; English transl., Hafner, New York, 1965.

9. ——, *Lectures on the theory of approximation*, OGIZ, Moscow, 1947; 2nd rev. ed., "Nauka", Moscow, 1965; English transl., Ungar, New York, 1956.

10. N. I. Akhiezer and I. M. Glazman, *Theory of linear operators in Hilbert space*, GITTL, Moscow, 1950; 2nd rev. ed., "Nauka", Moscow, 1966; English transl. vols. I, II, Ungar, New York, 1961.

11. N. I. Akhiezer and M. G. Kreĭn, *Some questions in the theory of moments*, Khar′kov, 1938; English transl., Amer. Math. Soc., Providence, R.I., 1962.

12. M. B. Balk, *The Padé approximation process for certain analytic functions*, Studies on Contemporary Problems of the Theory of Functions of a Complex Variable (A. I. Markushevich, ed.), Fizmatgiz, Moscow, 1960, pp. 234–257.

13. N. K. Bari, *Trigonometric series*, Fizmatgiz, Moscow, 1961.

14. A. Erdélyi, W. Magnus, F. Oberhettinger, and F. G. Tricomi, *Higher transcendental functions*, vol. 2, McGraw-Hill, New York-London-Toronto, 1953.

15. Yu. M. Berezanskiĭ, *Expansions in eigenfunctions of selfadjoint operators*, "Naukova Dumka", Kiev, 1965; English transl., Amer. Math. Soc., Providence, R. I., 1968.

16. Z. I. Borevich and I. R. Shafarevich, *Number theory*, "Nauka", Moscow, 1964; 2nd rev. ed., 1972; English transl., Academic Press, New York, 1966.

17. N. G. de Bruijn, *Asymptotic methods in analysis*, 2nd ed., North-Holland, Amsterdam; Noordoff, Groningen, 1961.

18. M. Brelot, *Eléments de la théorie classique du potentiel*, Centre de Documentation Universitaire, Paris, 1959.

19. V. V. Vavilov, *On the convergence of the Padé approximants of meromorphic functions*, Mat. Sb. **101 (143)** (1976), no. 1, 44–56; English transl. in Math. USSR Sb. **30** (1976), 17–38.

20. G. N. Watson, *A treatise on the theory of Bessel functions*, Cambridge Univ. Press, Cambridge; Macmillan, New York, 1944.

21. I. M. Vinogradov, *Foundations of the theory of numbers*, 8th. ed., "Nauka", Moscow, 1972; English transl., Pergamon Press, New York, 1955.

22. V. S. Vladimirov, *Generalized functions in mathematical physics*, "Nauka", Moscow, 1979; English transl., "Mir", Moscow, 1979.

23. ——, *Equations of mathematical physics*, 4th ed., "Nauka", Moscow, 1981; English transl., Dekker, New York, 1971.

24. A. I. Galochkin, *Lower bounds of linear forms in the values of certain hypergeometric functions*, Mat. Zametki **8** (1970), 19–28; English transl. in Math. Notes **8** (1970).

25. ——, *Estimates from below of polynomials in the values of analytic functions of a certain class*, Mat. Sb. **95 (137)** (1974), 396–417; English transl. in Math. USSR Sb. **24** (1974), 385–408.

26. ——, *Lower bounds of linear forms in the values of certain G-functions*, Mat. Zametki **18** (1975), 541–552; English transl. in Math. Notes **18** (1975).

27. ——, *Improvement of estimates of certain linear forms*, Mat. Zametki **20** (1976), 35–45; English transl. in Math. Notes **20** (1976).

28. ——, *Diophantine approximations of the values of certain entire functions with algebraic coefficients*. I, Vestnik Moscov. Univ. Ser. I Mat. Mekh. **1978**, no. 6, 25–32; English transl. in Moscow Univ. Math. Bull. **33** (1978).

29. J. B. Garnett, *Bounded analytic functions*, Academic Press, New York-London, 1981.

30. E. Hecke, *Vorlesungen über die Theorie der algebraischen Zahlen*, Akademische Verlagsgesellschaft, Leipzig, 1954.

31. Ya. L. Geronimus, *Polynomials orthogonal on a circle and on an interval*, Fizmatgiz, Moscow, 1958; English transl., *Orthogonal polynomials: Estimates, asymptotic formulas, and series of polynomials orthogonal on the unit circle and on an interval*, Consultants Bureau, New York, 1961.

32. G. M. Goluzin, *Geometric theory of functions of a complex variable*, 2nd ed., "Nauka", Moscow, 1966; English transl., Amer. Math. Soc., Providence, R.I., 1969.

33. A. A. Gonchar, *On convergence of Padé approximants for some classes of meromorphic functions*, Mat. Sb. **97 (139)** (1975), no. 4, 607–629; English transl. in Math. USSR Sb. **26** (1975), 555–575.

34. ——, *On the convergence of generalized Padé approximants of meromorphic functions*, Mat. Sb. **98 (140)** (1975), no. 4, 564–577; English transl. in Math. USSR Sb. **27** (1975), 503–514.

35. ——, *On the speed of rational approximation of some analytic functions*, Mat. Sb. **105 (147)** (1978), no. 2, 147–163; English transl. in Math. USSR Sb. **34** (1978), 131–146.

36. A. A. Gonchar and Guillermo López Lagomasino, *On Markov's theorem for multipoint Padé approximants*, Mat. Sb. **105 (147)** (1978), no. 4, 512–524; English transl. in Math. USSR Sb. **34** (1978), 449–460.

37. A. A. Gonchar and E. A. Rakhmanov, *On the convergence of simultaneous Padé approximants for systems of functions of Markov type*, Trudy Mat. Inst. Steklov. **157** (1981), 31–48; English transl. in Proc. Steklov Inst. Math. **1983**, no. 3.

38. ——, *Equilibrium measure and the distribution of zeros of extremal polynomials*, Mat. Sb. **125 (167)** (1984), no. 1, 117–127; English transl. in Math. USSR Sb. **53** (1986), 119–130.

39. ——, *The equilibrium problem for vector potentials*, Uspekhi Mat. Nauk **40** (1985), no. 4, 155–156. (Russian)

40. U. Grenander and G. Szegő, *Toeplitz forms and their applications*, Univ. of California Press, Berkeley and Los Angeles, 1958.

41. W. B. Jones and W. J. Thron, *Continued fractions. Analytic theory and applications*, Addison-Wesley, Reading, Mass., 1980.

42. V. K. Dzyadyk, *Introduction to the theory of uniform approximation of functions by polynomials*, "Nauka", Moscow, 1977. (Russian)

43. A. Zygmund, *Trigonometric series*, Cambridge Univ. Press, Cambridge, 1959.

44. V. A. Kalyagin, *On a class of polynomials defined by two orthogonality relations*, Mat. Sb. **110 (152)**(1979), no. 4, 609–627; English transl. in Math. USSR. Sb. **38** (1981), 562–580.

45. ——, *A note on the structure of the Padé table*, Vestnik Moskov. Univ. Ser. I Mat. Mekh. **1980**, no. 5, 38–41; English transl. in Moscow Univ. Math. Bull. **35** (1980).

46. L. Carleson, *Selected problems on exceptional sets*, Van Nostrand, Princeton, Toronto, and London, 1967.

47. S. Karlin and W. J. Studden, *Tchebycheff systems: with applications in analysis and statistics*, Interscience, New York, London, and Sydney, 1976.

48. J. W. S. Cassels, *An introduction to Diophantine approximation*, Cambridge Univ. Press, Cambridge, 1957; augment. Russian transl., IL, Moscow, 1961.

49. F. Klein, *Elementarmathematik vom höheren Standpunkte aus*, Springer-Verlag, Berlin, 1924; reprint, 1968; English transl., Macmillan, New York, 1932; reprint, Dover, New York, 1945, 1953.

50. N. I. Koblitz, *p-adic numbers, p-adic analysis, and zeta-functions*, Springer-Verlag, New York and Heidelberg, 1977.

51. A. N. Kolmogorov and S. V. Fomin, *Elements of the theory of functions and functional analysis*, 3rd ed., "Nauka", Moscow, 1972; English transl., Prentice-Hall, Englewood Cliffs, N.J., 1970.

52. M. A. Krasnosel′skiĭ and M. G. Kreĭn, *Fundamental theorems on the extension of Hermitian operators and some of their applications to the theory of orthogonal polynomials and the moment problem*, Uspekhi Mat. Nauk **2** (1947), no. 3, 60–106. (Russian)

53. M. G. Kreĭn, *Infinite J-matrices and a matrix moment problem*, Dokl. Akad. Nauk SSSR **69** (1949), 125–128. (Russian)

54. ——, *Fundamental propositions of the representation theory of Hermitian operators with deficiency index (m, m)*, Ukrain. Mat. Zh. **1** (1949), no. 2, 3–66; English transl., Amer. Math. Soc. Transl. (2) **97** (1971), 75–143.

55. M. G. Kreĭn and A. A. Nudel′man, *The Markov moment problem and extremal problems*, "Nauka", Moscow, 1973; English transl., Amer. Math. Soc., Providence, R.I., 1977.

56. A. G. Kurosh, *A course of higher algebra*, 10th ed., "Nauka", Moscow, 1975; English transl., "Mir", Moscow, 1975.

57. P. Koosis, *Introduction to H^p spaces*, Cambridge Univ. Press, Cambridge and New York, 1980.

58. M. A. Lavrent′ev and B. V. Shabat, *Methods of the theory of functions of a complex variable*, 4th ed., "Nauka", Moscow, 1973; German transl. (of 3rd ed.), VEB Deutscher Verlag Wiss., Berlin, 1967.

59. N. S. Landkof, *Foundations of modern potential theory*, "Nauka", Moscow, 1966; English transl., Springer-Verlag, Berlin and New York, 1972.

60. S. Lang, *Introduction to Diophantine approximations*, Addison-Wesley, Reading, Mass., 1966.

61. A. A. Markov, *Selected papers on the theory of continued fractions and the theory of functions deviating least from zero*, OGIZ, Moscow-Leningrad, 1948. (Russian)

62. A. I. Markushevich, *Theory of analytic functions*, 2nd ed., "Nauka", Moscow, 1967–1968; English transl. (of 1st ed.), Prentice-Hall, Englewood Cliffs, N.J., 1965, 1967.

63. J. K. Moser, *Various aspects of integrable Hamiltonian systems*, Dynamical Systems, Progr. Math., vol. 8, Birkhäuser, Boston, 1980, pp. 233–289.

64. I. P. Natanson, *Theory of functions of a real variable*, GITTL, Moscow-Leningrad, 1950; 2nd ed., 1957; English transl., Ungar, New York, 1955, 1961.

65. A. F. Nikiforov and V. B. Uvarov, *Special functions of mathematical physics*, "Nauka", Moscow, 1978; English transl., Birkhäuser, Basel and Boston, 1988.

66. E. M. Nikishin, *A system of Markov functions*, Vestnik Moskov. Univ. Ser. I Mat. Mekh. **1979**, no. 4, 60–63; English transl. in Moscow Univ. Math. Bull. **34** (1979).

67. ——, *On irrationality of the values of the functions $F(x, s)$*, Mat. Sb. **109(151)** (1979), no. 3, 410–417; English transl. in Math. USSR Sb. **37** (1980), 381–388.

68. ——, *On logarithms of natural numbers*, Izv. Akad. Nauk SSSR Ser. Mat. **43** (1979), no. 6, 1319–1327; correction, ibid., **44** (1980), no. 4, 972; English transl. in Math. USSR Izv. **15** (1980), no. 3, 523–530.

69. ——, *On simultaneous Padé approximants*, Mat. Sb. **113 (155)** (1980), no. 4(12), 499-519; English transl. in Math. USSR Sb. **41** (1982), 409–426.

70. ——, *Arithmetic properties of the Markov function for the Jacobi weight*, Anal. Math. **8** (1982), no. 1, 39–46.

71. ——, *The discrete Sturm-Liouville operator and some problems of function theory*, Trudy Sem. Petrovsk. **10** (1984), 3–77; English transl. in J. Soviet Math. **35** (1987), no. 5.

72. ——, *Asymptotics of linear forms for simultaneous Padé approximants*, Izv. Vyssh. Uchebn. Zaved. Mat. **1986**, no. 2, 33–41; English transl. in Soviet Math. (Iz. VUZ) **30** (1986).

73. A. V. Ozerskiĭ, *Chain algorithms*, Izdat. Latv. Univ., Riga, 1981. (Russian)

74. D. V. Pannikov, *Poles of Padé approximants to* ${}_1F_1(1\,;\,c\,;\,z)$, Vestnik Moskov. Univ. Ser. I Mat. Mekh. **1982**, no. 1, 11–14; English transl. in Moscow Univ. Math. Bull. **37** (1982).

75. V. I. Parusnikov, *The Jacobi-Perron algorithm and simultaneous approximation of functions*, Mat. Sb. **114 (156)** (1981), no. 2, 322–333; English transl. in Math. USSR Sb. **42** (1982), 287–296.

76. ——, *Limit periodic multidimensional continued fractions*, Preprint, M. V. Keldysh Inst. Prikl. Mat. Akad. Nauk SSSR, no. 62, 1983. (Russian)

77. ——, *Weakly complete systems of functions and multidimensional continued fractions*, Vestnik Moskov. Univ. Ser. I Mat. Mekh. **1984**, no. 2, 13–17; English transl. in Moscow Univ. Math. Bull. **39** (1984).

78. L. R. Pineĭro, *Simultaneous Padé approximants for a class of Markov functions*, Author's abstract, Candidate's Dissertation, Moscow State University, 1985. (Russian)

79. I. I. Privalov, *Superharmonic functions*, Gostekhizdat, Moscow, 1940. (Russian)

80. ——, *Boundary properties of analytic functions*, 2nd ed., GITTL, Moscow-Leningrad, 1950; German transl., VEB Deutscher Verlag Wiss., Berlin, 1956.

81. ——, *Introduction to the theory of functions of a complex variable*, 12th ed., "Nauka", Moscow, 1977; German transl., Teubner, Leipzig, 1959.

82. E. A. Rakhmanov, *Convergence of diagonal Padé approximants*, Math. Sb. **104 (146)** (1977), 271–291; English transl. in Math. USSR Sb. **33** (1977), 243–260.

83. ——, *On the asymptotics of the ratio of orthogonal polynomials*. I, II, Mat. Sb. **103 (145)** (1977), no. 2, 237–252; **118 (160)** (1982), 104–117; English transl. in Math. USSR Sb. **32** (1977), 199–214; **46** (1983), 105–118.

84. ——, *On asymptotic properties of polynomials orthogonal on the real axis*, Mat. Sb. **119 (161)** (1982), 163–203; English transl. in Math. USSR Sb. **47** (1984), 155–194.

85. F. Riesz and B. Szökefalvi-Nagy, *Leçons d'analyse fonctionnelle*, Gauthier-Villars, Paris; Akadémiai Kiadó, Budapest, 1955; English transl., Ungar, New York, 1955.

86. A. G. Sveshnikov and A. N. Tikhonov, *Theory of functions of a complex variable*, "Nauka", Moscow, 1979; English transl., "Mir", Moscow, 1982.

87. G. Szegő, *Orthogonal polynomials*, 2nd ed., Amer. Math. Soc., Providence, R.I., 1959.

88. Yu. V. Sidorov, M. V. Fedoryuk, and M. N. Shabunin, *Lectures on the theory of functions of a complex variable*, "Nauka", Moscow, 1976; 2nd ed., 1982; English transl., "Mir", Moscow, 1985.

89. V. I. Smirnov and N. A. Lebedev, *Functions of a complex variable: Constructive theory*, "Nauka", Moscow, 1964; English transl., M.I.T. Press, Cambridge, Mass., 1968.

90. V. N. Sorokin, *Asymptotics of linear functional forms of a class of Markov functions*, Moscow State Univ., 1982. Deposited with VINITI 16/Aug/82, no. 4526. (Russian)

91. ——, *Asymptotics of linear functional forms of two logarithms*, Uspekhi Mat. Nauk **38** (1983), no. 1 (229), 193–194. (Russian)

92. ——, *Simultaneous approximation of several linear forms*, Vestnik Moskov. Univ. Ser. I Mat. Mekh. **1983**, no. 1, 44–47, 101; English transl. in Moscow Univ. Math. Bull. **38** (1983).

93. ——, *Simultaneous Padé approximants for finite and infinite intervals*, Izv. Vyssh. Uchebn. Zaved. Mat. **1984**, no. 8, 45–52; English transl. in Soviet Math. (Iz. VUZ) **28** (1984).

94. ——, *On the irrationality of the values of hypergeometric functions*, Mat. Sb. **127 (1969)** (1985), 245–258; English transl. in Math. USSR Sb. **55** (1986), 243–258.

95. ——, *Generalization of classical orthogonal polynomials and convergence of simultaneous Padé approximants*, Trudy Sem. Petrovsk. **11** (1986), 125–165 ; English transl. in J. Soviet Math. **45** (1989), no. 6.

96. ——, *Generalization of Laguerre polynomials and the convergence of simultaneous Padé approximants*, Uspekhi Mat. Nauk **41** (1986), no. 1, 209–210. (Russian)

97. ——, *Asymptotics of linear forms and polynomial coefficients for some functions of Stieltjes type*, Sibirsk. Mat. Zh. **27** (1986), no. 1, 157–169; English transl. in Siberian Math. J. **27** (1986).

98. T. J. Stieltjes, *Recherches sur les fractions continues*, Ann. Fac. Sci. Univ. Toulouse **8** (1894), J1–J122; **9** (1895), A1–A47; reprinted in *Oeuvres complètes*, vol. 2, Noordhoff, Groningen, 1918, pp. 402–566.

99. S. Stoïlow, *Teoria functiieor de o variabilă complexă*, Ed. Acad. Repub. Pop. Romîne, Bucharest, vol. 1, 1954, vol. 2, 1957; Russian transl., IL, Moscow, 1962.

100. P. K. Suetin, *Classical orthogonal polynomials*, "Nauka", Moscow, 1976. (Russian)

101. S. P. Suetin, *On the convergence of rational approximations to polynomial expansions in domains of meromorphy of a given function*, Mat. Sb. **105 (147)** (1978), no. 3, 413–430; English transl. in Math. USSR Sb. **34** (1978), 367–381.

102. ——, *On Montessus de Ballore's theorem for nonlinear Padé approximants of orthogonal expansions and Faber series*, Dokl. Akad. Nauk SSSR **253** (1980), no. 6, 1322–1325; English transl. in Soviet Math. Dokl. **22** (1980), 274–277.

103. J. L. Walsh, *Interpolation and approximation by rational functions in the complex domain*, Amer. Math. Soc., Providence, R.I., 1966.

104. J. Wermer, *Potential theory*, Springer-Verlag, Berlin and New York, 1974.

105. N. I. Fel′dman, *Estimates from below for certain linear forms*, Vestnik Moskov. Univ. Ser. I Mat. Mekh. **1967**, no. 2, 63–72; English transl. in Moscow Univ. Math. Bull. **22** (1967).

106. ——, *An effective refinement of the exponent in Liouville's theorem*, Izv. Akad. Nauk SSSR Ser. Mat. **35** (1971), 973–990; English transl. in Math. USSR Izv. **5** (1971), 985–1002.

107. ——, *A certain linear form*, Acta Arith. **21** (1972), 347–355. (Russian)

108. ——, *Approximations of algebraic numbers*, Izdat. Moskov. Univ., Moscow, 1981. (Russian)

109. ——, *Hilbert's seventh problem*, Izdat. Moskov. Univ., Moscow, 1982. (Russian)

110. P. R. Halmos, *Measure theory*, Van Nostrand, New York, 1950.

111. ——, *A Hilbert space problem book*, Van Nostrand, Princeton, N.J., 1967.

112. A. Ya. Khinchin [A. Ja. Hinčin], *Continued fractions*, 4th ed., "Nauka", Moscow, 1987; English transls. (of 3rd ed.), Noordhoff, Groningen, 1963; Univ. of Chicago Press, Chicago, Ill., 1964.

113. A. N. Khovanskiĭ [Hovanskiĭ], *The application of continued fractions and their generalizations to problems in approximation theory*, GTTI, Moscow, 1956; English transl., Noordhoff, Groningen, 1963.

114. P. L. Chebyshev, *Complete collected works*, vol. 3, Izdat. Akad. Nauk SSSR, Moscow, 1948. (Russian)

115. A. B. Shidlovskiĭ, *A criterion for the algebraic independence of a class of entire functions*, Izv. Akad. Nauk SSSR. Ser. Mat. **23** (1959), 35–66. (Russian)

116. ——, *Diophantine approximations and transcendental numbers*, Izdat. Moskov. Univ., Moscow, 1982.

117. W. M. Schmidt, *Diophantine approximation*, Springer-Verlag, Berlin, 1980.

118. G. K. Engelis, *On polynomials defined by the Rodrigues formula*, Uchen. Zap. Latv. Gos. Univ. **20** (1958), 137–143. (Russian)

119. N. Abramesco, *Sulle serie di polinomi di una variabile complessa. Le serie di Darboux*, Ann. Mat. Pura Appl. (3) **31** (1922), 207–249.

120. M. A. Angelesco, *Sur deux extensions des fractions continues algébriques*, C. R. Acad. Sci. Paris **168** (1919), 262–263.

121. P. Appell, *Sur une suite des polynômes ayant toutes leurs racines réelles*, Arch. Math. Phys. (3) **1** (1901), 69–71.

122. A. Baker, *Rational approximations to $\sqrt[3]{2}$ and other algebraic numbers*, Quart. J. Math. Oxford Ser. (2) **15** (1964), 376–383.

123. G. A. Baker, *Essentials of Padé approximants*, Academic Press, New York, 1975.

124. M. G. de Bruin, *Generalized C-fractions and a multidimensional Padé table*, Thesis, Amsterdam, 1974.

125. ——, *Three new examples of generalized Padé tables, which are partly normal*, Department of Math., University of Amsterdam, Report 76–11, pp. 1–13.

126. ——, *Convergence in the Padé table for* ${}_1F_1(1\,;\,c\,;\,x)$, Nederl. Akad. Wetensch. Proc. Ser. A **79**=Indag. Math. **38** (1976), no. 5, 408–418.

127. ——, *Convergence along steplines in a general Padé table*, Padé and Rational Approximation, Academic Press, New York, 1977, pp. 15–22.

128. ——, *Convergence of generalized C-fractions*, J. Approx. Theory **24** (1978), no. 3, 177–207.

129. ——, *The interruption phenomenon for generalized continued fractions*, Bull. Austral. Math. Soc. **19** (1978), no. 2, 245–272.

130. N. Ghosh, *On a generalisation of Legendre polynomials*, Bull. Calcutta Math. Soc. **21** (1929), 147–154.

131. J. Coates, *On the algebraic approximation of functions*. I, II, III, IV, Nederl. Akad. Wetensch. Proc. Ser. A **69**=Indag. Math. **28** (1966), 421–434, 435–448, 449–461; **70**=**29** (1967), 205–212.

132. H. Davenport and K. F. Roth, *Rational approximations to algebraic numbers*, Mathematika **2** (1955), 160–167.

133. L. Euler, *De relatione inter ternas pluresve quantitates instituenda*, Leonardi Euleri Commentationes Arithmeticae Collectae, vol. 2, Petropoli, 1849, p. 99.

134. L. Fejér, *Asymptotikus értékek meghatározásárol*, Mat. és Term. Értesitö **27** (1909), 1–33.

135. G. Frobenius, *Ueber Relationen zwischen den Naherungsbrüchen von Potenzreihen*, J. Reine Angew. Math. **90** (1881), 1–17.

136. J. Hadamard, *Essai sur l'étude des fonctions données par leur développement de Taylor*, J. Math. Pures Appl. **4** (1982), no. 8, 101–186.

137. C. Hermite, *Sur la fonction exponentielle*, C.R. Acad. Sci. Paris **77** (1873), 18–24, 74–79, 226–233, 285–293.

138. ——, *Sur la généralisation des fractions continues algébriques*, Oeuvres, vol. IV, Gauthier-Villars, Paris, 1917, 357–377.

139. P. Humbert, *Sur les équations de Didon*, Nouvelles Annales de Math. (4) **19** (1919), 443–451.

140. C. G. J. Jacobi, *Ueber die Auflösung der Gleichung* $\alpha_1 x_1 + \alpha_2 x_2 + \cdots + \alpha_n x_n = \mathrm{f.\ u.}$ J. Reine Angew. Math. **69** (1868), 21–28.

141. ——, *Allgemeine Theorie der Kettenbruchähnlichen Algorithmen, in welchen jede Zahl aus drei vorgegebenden gebildet wird*, J. Reine Angew. Math. **69** (1868), 29–64.

142. H. Jager, *A multidimensional generalization of the Padé table*. I–VI, Nederl. Akad. Wetensch. Proc. Ser. A. **67**=Indag. Math. **26** (1964), 193–225, 227–249.

143. S. Lang, *Report on Diophantine approximations*, Bull. Soc. Math. France **93** (1965), 177–192.

144. S. Lang and H. Trotter, *Continued fractions for some algebraic numbers*, J. Reine Angew. Math. **225** (1972), 112–134.

145. F. Lindemann, *Ueber die Zahl* π, Math. Ann. **20** (1882), 213–225.

146. J. H. Loxton and A. J. Poorten, *A note on simultaneous polynomial approximations of exponential functions*, Bull. Austral. Math. Soc. **11** (1974), 333–338.

147. A. Magnus, *Expansion of power series into P-fractions*, Math. Z. **80** (1969), no. 3, 209–216.

148. K. Mahler, *Perfect systems*, Compositio Math. **19** (1968), no. 2, 95–166.

149. ——, *Lectures on transcendental numbers*, Lecture Notes in Math., vol. 546, Springer-Verlag, Berlin-New York, 1976.

150. R. de Montessus, *Sur les fractions continues algébriques*, Bull. Soc. Math. France **30** (1902), 28–36.

151. J. K. Moser, *Three integrable Hamiltonian systems connected with isospectral deformations*, Adv. in Math. **16** (1975), 197–220.

152. J. Nuttall, *The convergence of Padé approximants of meromorphic functions*, J. Math. Anal. Appl. **41** (1973), 775–780.

153. H. Padé, *Sur la représentation approchée d'une fonction par des fractions rationelles*, Ann. Sci. Ecole Norm. Sup. (3) **9** (1892), Supplément, 3–93.

154. ——, *Mémoire sur les développements en fractions continues de la fonction exponentielle, pouvant servir d'introduction à la théorie des fractions continues algébriques*, Ann. Sci. Ecole Norm. Sup. (3) **16** (1899), 395–426.

155. E. B. Saff and R. S. Varga, eds., *Padé and rational approximation. Theory and applications*, Academic Press, New York and London, 1977.

156. O. Perron, *Grundlagen für eine Theorie des Jacobischen Kettenbrüchalgorithmus*, Math. Ann. **64** (1907), no. 1, 1–76.

157. ——, *Die Lehre von den Kettenbrüchen*, Teubner, Leipzig and Berlin, 1929.

158. C. Pommerenke, *Padé approximants and convergence in capacity*, J. Math. Anal. Appl. **41** (1973), 775–780.

159. A. Pringsheim, *Über Konvergenz und funktionentheoretischen Charakter gewisser limitär-periodischer Kettenbrüche*, Sitzungsber. Bayer Akad. Wiss. München, Math.-Phys. Kl. **1910**, 6. Abh.

160. K. F. Roth, *Rational approximations to algebraic numbers*, Mathematika **58**, (1955), no. 2, 1–20, 168.

161. H. van Rossum, *Systems of orthogonal and quasi-orthogonal polynomials connected with the Padé table*. I, II, III, Nederl. Akad. Wetensch. Proc. Ser. A **61**=Indag. Math. **17** (1955), 517–525, 526–534, 675–682.

162. E. B. Saff, *An extension of Montessus de Ballore's theorem on the convergence of interpolating rational functions*, J. Approx. Theory **6** (1972), 63–68.

163. E. B. Saff and R. S. Varga, *On the zeros and poles of Padé approximants to* e^z, Numer. Math. **25** (1975), no. 1, 1–14.

164. ——, *Zero-free parabolic regions for sequences of polynomials*, SIAM J. Math. Anal. **7** (1976), 344–357.

165. ——, *On the zeros and poles of Padé approximants to* e^z. II, Padé and Rational Approximation, Academic Press, New York, 1977, pp. 195–213.

166. C. L. Siegel, *Transcendental numbers*, Princeton Univ. Press, Princeton, N.J., 1949.

167. G. A. Baker and J. L. Gammel, eds., *The Padé approximant in theoretical physics*, Academic Press, New York, 1970.

168. M. Tsuji, *On a power series which has only algebraic singularities on its convergence circle*, Japan J. Math. **3** (1926), 69–85.

169. ——, *Potential theory in modern function theory*, Maruzen, Tokyo, 1959.

170. E. B. Van Vleck, *On the convergence of algebraic continued fractions, whose coefficients have limiting values*, Trans. Amer. Math. Soc. **5** (1904), no. 3, 253–262.

171. H. S. Wall, *Analytic theory of continued fractions*, Van Nostrand, New York, 1948.

172. H. Widom, *Extremal polynomials associated with a system of curves in the complex plane*, Adv. Math. **3** (1969), 127–232.

173. P. Wynn, *Upon a convergence result in the theory of the Padé table*, Trans. Amer. Math. Soc. **165** (1972), 239–249.

Subject Index